T0310398

MACHINING OF STAINLESS STEELS AND SUPER ALLOYS

MACHINING OF STAINLESS STEELS AND SUPER ALLOYS

TRADITIONAL AND NONTRADITIONAL TECHNIQUES

Helmi A. Youssef
Alexandria University, Egypt

WILEY

This edition first published 2016

Registered Office
John Wiley & Sons, Ltd, The Atrium, Southern Gate, Chichester, West Sussex, PO19 8SQ, United Kingdom

For details of our global editorial offices, for customer services and for information about how to apply for permission to reuse the copyright material in this book please see our website at www.wiley.com.

Library of Congress Cataloging-in-Publication Data

Youssef, Helmi A.
Machining of stainless steels and super alloys : traditional and nontraditional techniques / Helmi A. Youssef, Alexandria University, Alexandria, Egypt.
pages cm
Includes bibliographical references and index.
ISBN 978-1-118-91956-9 (cloth)
1. Metal-cutting. 2. Stainless steel. 3. Chromium-cobalt-nickel-molybdenum alloys. I. Title.
TJ1185.Y67 2015
671.3′5–dc23

2015020392

A catalogue record for this book is available from the British Library.

Cover Image: tombaky/iStockphoto

Set in 10/12pt Times by SPi Global, Pondicherry, India
Printed and bound in Singapore by Markono Print Media Pte Ltd

1 2016

Knowledge is a treasure, but
Practice is the key of it
Fuller

Dedicated
To my little angles

Youssef J., Nour, Anorine, Fayrouz, and Youssra
With lot of love

Contents

Preface

The term machining covers a large collection of manufacturing methods for shaping parts by removing a portion from workpiece material, so as to obtain a finished product. Machining is used to convert forgings, castings, welded, and sintered products into desired shapes, sized, and finished to fulfill design requirements. Compared to other manufacturing techniques, machining is characterized by its versatility and capability of achieving the highest accuracy and surface quality in the most economical way.

The development of new tool materials opened a new era for the machining industry in which a parallel development in machine tools took place. In the last century, nontraditional machining techniques offered alternative methods for machining parts of complex shapes in extra-hard and tougher exotic materials that were difficult to machine by traditional methods.

In highly developed industrial countries, the yearly cost associated with metal removal has been estimated at about 10% of the gross national production. For this reason, rational approach and minor improvements in productivity of material removal processes are of major importance in high volume production.

During recent decades, engineering materials have been greatly developed. Materials such as hardened steels, stainless steels, super alloys, carbides, ceramics, and fiber-reinforced composite materials are frequently used in modern industry. The cutting speed and material removal rate when machining such materials using traditional methods such as turning, milling, grinding, and so on, tend to be falling. Sometimes, it is difficult to machine hard materials to certain shapes using these traditional methods. By adopting a unified program, and utilizing the results of basic and applied research, it will become possible in the very near future to process many of these engineering materials that were formerly considered difficult-to-cut using traditional or nontraditional methods.

This book mainly covers the machining technologies applicable for stainless steels and super alloys, which represent two of the main difficult-to-cut materials frequently used in modern industrial applications. The treatment of its topics starts from traditional machining processes, and extends to the most recent technologies of nontraditional and hybrid machining techniques. The book may therefore be regarded as a reference text for both senior undergraduate and postgraduate students enrolled in manufacturing, materials, and mechanical engineering programs. The book covers elective courses of manufacturing technology taught in engineering institutions all over the world. Additionally, the book can be used by students in other disciplines concerned with design and manufacturing such as automotive, aerospace engineering, and gas turbine plants. Besides being a reference, this book offers broadly based fundamental information for selecting the appropriate machining parameters and tooling, developing successful machining strategies in both traditional and nontraditional machining processes. It focuses on scientific and practical developments affecting the present and future of machining processes for stainless steels and super-alloys.

The instructor may tailor the course to adapt specific needs of students. Moreover, I hope that engineers and senior technologists working in the fields of machining technology, too, will find many useful ideas and thoughts in this book, if not in their specialty then in the ever-broadening fields of alternative and competitive processes and materials.

The book has two main aims: first to deal with the characteristics of stainless steels and super alloys that have been proved to be useful as engineering materials for many applications; second, it explains how and why these materials differ in machining performance and machining methods used, which represent important topics addressed in this book.

The book is written in nine chapters describing the machining possibilities of stainless steels and super alloys. These materials may be machined either traditionally, or nontraditionally, considering economical, and ecological measures. Traditional machining operations such as turning, milling, drilling, reaming, broaching, and grinding are generally used, whenever possible. Sometimes, it is no longer possible to find suitable tool materials that are sufficiently hard to cut these materials at economical speeds; then the use of nontraditional machining represents the solution. Nontraditional machining processes, such as jet machining, abrasive flow machining, electrochemical machining, laser beam machining, and so on, are capable of machining these alloys precisely, irrespective of their hardness and strength.

Outline of the Book

Writing such a book is a difficult task, since it is dealing with two complex materials (stainless steels and super alloys); each is composed of many alloys which are required to be machined using two different machining techniques (traditional and nontraditional machining processes). Moreover, being a book of such a small size, its material should be carefully selected to clearly express and highlight the different topics dealt with. Although its is intended to be as thorough and insightful as possible, the book provides mathematical modeling and equations only as needed to enhance the basic understanding of the material at hand. I have been selective in citing literature that seems to supports points of view closest to the fact in each problem considered. Ideas and experimental results have been considered from a sufficient number of sources that I have tried to clearly acknowledge through the book. Individuals desiring additional information on specific items of the book are directed to various references at the end of each chapter.

The table below gives at a glance how the chapters of the book have been organized to realize its objectives. Moreover, the topics covered by individual chapters of the book are briefly described below.

Book chapters	Topics dealt with •			
	TM	NTM	SSs	SAs
CH1	•	•	•	•
CH2	—	—	•	—
CH3	—	—	—	•
CH4	•	—	•	•
CH5	•	—	•	—
CH6	•	—	—	•
CH7	—	•	—	—
CH8	—	•	•	•
CH9	•	•	•	•

Chapter titles: see table of contents, TM: traditional machining, NTM: nontraditional machining, SSs: stainless steels, and SAs: super alloys.

Chapter 1 introduces stainless steels and super alloys as an important category of Difficult-to-Cut Materials. Historical background of stainless steels and super alloys, as well as their applications, are presented. The machining processes are then classified into traditional and nontraditional machining processes. Finally, the machining variables of a machining process are defined.

Chapters 2 and ***3*** consider the types and classifications of SSs and SAs according to the AISI- and the UNS-designation systems. In ***Chapter 2***, the effect of alloying on the properties of SSs has been presented, then the SSs are grouped into their different categories. In ***Chapter 3***, SAs are similarly classified into Ni-, Fe-, and Co-base alloys. It is self-evident that to perform any machining process correctly, the material properties must be known. The mechanical and thermal characteristics of SSs and SAs have been briefly presented, and their chemical compositions have been provided. Finally, their industrial applications have been highlighted.

Chapters 4–6 deal with the traditional methods of machining stainless steels and super alloys. Three issues related to traditional machining of stainless steels and super alloys are to be dealt with in ***Chapter 4***. These are machinability, cutting tools, and cutting fluids. After the basic machinability of TM has been considered, methods of enhancement of machinability of stainless steels and super alloys are presented. Then, tool materials such as high speed steels (HSSs), satellites, UCON, coated and carbides, ceramics, and cermets that are used for economical machining of difficult-to-cut material such as stainless steels and super alloys are presented. More recently developed tool materials such as polycrystalline cubic boron nitride (PCBN), and polycrystalline synthetic diamond (PCD) are also provided. Many tables listing the composition and properties of tool material are given. Finally, suitable cutting fluids for machining both stainless steels and super alloys have been suggested. ***Chapter 5*** covers traditional machining processes of stainless steels. The machinability of free-machining and enhanced stainless alloys as compared to the nonfree-machining alloys are investigated. A proposed 10-level machinability chart is developed to rank important grades of stainless steels. Traditional machining processes of stainless steels are then presented. The relevant machining parameters are tabulated. ***Chapter 6*** covers traditional machining processes of super alloys. The traditional machining of these important alloys are presented similarly to their treatment in ***Chapter 5***. Relevant machining parameters are tabulated and tooling and cutting fluids for traditional techniques of super alloys are considered.

Chapters 7–8 deal with nontraditional machining methods. ***Chapter 7*** provides a survey of nontraditional machining processes, in which the basics of nontraditional machining operations, such as jet machining, ultrasonic machining, electrochemical machining, electric discharge machining, laser beam machining, and so on, are considered. Furthermore, the capabilities, advantages and limitations of each process are presented. ***Chapter 8*** describes briefly, as based on available data in literature, specialized company information, and handbooks, some of the nontraditional machining processes which have been applied successfully to stainless steels and super alloys.

Chapter 9 deals with current and recent developments regarding both traditional and nontraditional machining of stainless steels and super alloys. This chapter relies mainly on notable academic publications and recent conference proceedings, to fill the gap of the advanced knowledge between industry and the latest and frequently cited research documents. The main objective of these researches was the enhancement of the productivity of traditional and nontraditional machining of stainless steels and super alloys by increasing machinability through many strategies.

At the end of the book, review questions are provided to make the students aware of the importance of relevant topics.

Prerequisite Knowledge

The student should have acquired the following as prerequisite courses:

- Manufacturing technology
- Fundamentals of machining processes
- Material science
- Heat treatment of metals and alloys

Features of the Book

The book provides the following features:

1. It provides a description for machining technologies of stainless steels, and super alloys in today's industrial applications.
2. It presents stainless steels, and super alloys, described and classified in a tabulated form (84) and illustrations (114).
3. The bibliography at the end of each chapter contains books and periodicals that serve as thoroughly updated references to the reader.
4. It provides rational selection of cutting tools along with recommended cutting fluids for traditional machining of stainless steels, and super-alloys.
5. It presents important guidelines for machining stainless steels and super-alloys.
6. It covers the most recent machining technologies that are not covered in other books.
7. It provides up-to-date and recent developments regarding both traditional and nontraditional machining of stainless steels and super alloys as based on notable academic publications and recent conference proceeding.

8. It covers the basics as well as the most recent advances in manufacturing technology.
9. It offers for the first time one easy-to-use reference for traditional and nontraditional machining stainless steels and super alloys.

Finally, I have done my best to eliminate all possible errors. I do not expect complete success in this regard and would appreciate being informed of errors which still persist. Individual questions or comments may be directed to me personally at youssef_helmi@yahoo.com.

Helmi A. Youssef
Alexandria, Egypt
November 2015

About the Author

Professor Helmi A.A. Youssef, born in August 1938 in Alexandria, Egypt, acquired his B.Sc. Honors degree in Production Engineering from Alexandria University in 1960. He then completed his scientific building in the Carolo-Welhelmina, TH Braunschweig in Germany during 1961–1967. In June 1964 he acquired his Dipl.-Ing degree; then in December 1967, completed his Dr.-Ing degree in the domain of Non-traditional Machining. In 1968, he returned to Alexandria University, Production Engineering Department as an assistant professor. In 1973 he was promoted to associate, and in 1978 to full professor. In 1995–1998, Professor Youssef was chairman of the Production Engineering Department, Alexandria University. Since 1989, he has been a member of the scientific committee for promotion of professors in Egyptian universities. Between 1975 and 1995, Professor Youssef was a visiting professor in El-Fateh University in Tripoli, the Technical University in Baghdad, King Saud University (KSU) in Riyadh, and Beirut Arab University (BAU) in Beirut. He established laboratories and supervised many PhD and MSc theses. Professor Youssef has organized and participated in many international conferences. He has published many scientific papers in specialized journals. He authored many books in his fields of specialization, two of which are single authored. One book, co-authored on Machining Technology, was published in 2008 by CRC, USA. Another co-authored book, *Manufacturing Technology*, was also published by CRC in 2011. Currently, he is emeritus professor in PED, Alexandria University. His work at the university involves developing courses and conducting research in the areas of metal cutting and nontraditional machining.

About the Author

Acknowledgments

First of all, I wish to acknowledge with thanks the efforts of my dear colleagues, who have effectively contributed to the development of this book. It is a pleasure to express my deep gratitude to Professor A. Visser, University of Bremen, Germany, and Professor H. Attia, McGill University, Canada, for supplying valuable materials. Also, heartfelt thanks are due to Professor H. El-Hofy, Dean of Innovative Design Engineering School, Egypt-Japan University, and Professor M. Ahmed, Alexandria University, for the mutual discussions and continuing interest to the topics of the book. I would like to appreciate the efforts of Dr. Khaled youssef for his continual assistance in tackling software problems. The assistance of Mr. Saied Teileb of Lord Alexandria Razor Company for his valuable, and nice Auto-CAD drawings is highly appreciated. Thanks and apologies to others whose contributions have been overlooked.

Heartfelt thanks are due to my wife, Nabielah Youssef, for her help, understanding, and patience through many long months that it took to write this book.

Finally, I sincerely appreciate the support, the dedication, and continued help of the editorial board and production staff of Wiley & Sons, for their efforts in ensuring that this book is produced in the best form. Also, I am grateful to the authors of all sources referenced in this book. I appreciate the help of many associations, organizations and publishers that supplied me with the necessary permissions, and courtesies to reproduce many illustrations, photographs, and tables. These are:

- ASM, International Materials Park, OH.
- Eugen G. Leuze Verlag KG, Bad Saulgau, Würtenberg, Germany
- Krupp, Widia, GmbH, Essen, Germany
- Kennametal
- Chemical Corporation
- VDI-Verlag, Düsseldorf, Germany
- CIRP-Annals, Paris, France
- Elsevier Ltd, Oxford, UK

- Springer Verlag
- El-Fath Press, Alexandria, Egypt.
- CRC Press, Taylor & Francis, Boca Raton
- VDI Verlag, Düsseldorf, Germany, 1990
- US-Patent
- Alexandria Engineering Journal, Egypt

Nomenclature

Symbol	Definition	Unit
Ac	Uncut chip cross-sectional area	mm^2
C	Capacitance	μF
C	Taylor constant	—
c	Acoustic speed in horn material	m/s
c	Electrolyte specific heat	cal/kg·°C
c_l	Specific heat of WP material	N m/kg·°C
C_d	Coefficient of thermal diffusivity = $k_t/\rho.\ c_l$	m^2/s
D	Workpiece diameter	mm
df	Electron beam focusing diameter	mm
dg	Abrasive grain diameter	μm
du	Chemical undercut	mm
E	Young's modulus	MPa
Ed	Energy of individual discharge	J
EF	Etch factor	—
F	Force	N
F	Faraday's constant = 96 487	A·s/mol
f	Feed rate	mm/rev
f r	Natural frequency (resonant frequency)	Hz
Fa	Axial force	N
Fc	Main cutting force	N
fe	Frequency of exciting vibration	Hz
Ff	Feed force	N
Fr	Radial force	N
h	Undeformed chip thickness	mm
H	Thermal energy	cal
hc	Chip thickness	mm
hg	Frontal gap thickness in EDM	μm

Symbol	Definition	Unit
I	Machining current	A
I	Plasma current	A
ib	Electron beam current	A
i_b	Electron beam current	A
ic	Charging current	A
id	Discharging current	A
Ip	Pre-magnetizing current	A
ks	Specific cutting energy	N/mm^2
k_t	Thermal conductivity	N/s·°C
Li	Electron beam energy	A·s
m	Mass of anodic dissolution of work material	g
m_e	Electrolyte mass flow in time t	kg
n	Taylor exponent	—
N	Atomic weight of work material	g/mol
n, N	Spindle rotational speed	rpm
N/n	EC-equivalent of work material	g/mol
P	Laser power	W
P_e	Power of electron beam	N m/s
pt	Stagnant pressure	MPa
q	Electrolyte flow rate	m^3/s
R	Resultant cutting force	N
R	Radius	mm
R	Resistance	Ω
rn	Tool nose radius	mm
RT	Machinability rating as based on the tool life	—
RV	Machinability rating as based on the cutting speed	—
T	Tool life	min
t	Depth of cut	mm
t	Time	s
t	Depth of cut in CH-Milling	μm
t	Plate thickness	m
T	Drilling torque	N·m
t_1	plate thickness	m
T_b	Boiling temperature of electrolyte	°C
tc	Charging time	μs
td	Discharging time	μs
Te	Chemical etch depth	mm
te	Etching time	min
Ti	Inlet temperature of electrolyte	°C
ti	Pulse duration	μs
To	Initial bulk temperature	K
Ts	Surface temperature	K
u	Undercut in CH-Milling	μm
v	Cutting speed	m/min
Vb	Electron beam accelerating voltage	kV

Symbol	Definition	Unit
VB	Flank wear	mm
vf	Feed rate in ECM	mm/min
vf	Traverse speed in EBM	mm/min
vfp	Limiting sinking speed	m/min
vg	Peripheral speed of grinding wheel	m/s
vj	Jet velocity	m/s
v_t	Traverse cutting speed	m/min
Z	Material removal rate	$mm^3 \cdot s^{-1}$
z	Kienzle exponent	—

Greek symbols	Definition	Unit
σb	Bending strength	N/mm^2
η	Overall efficiency of the machine tool, and current efficiency	—
ρ	Density	kg/m^3
σf	Flow stress	MPa
θ_m	Melting point of WP material	°C
ρ_s	Electrical resistivity of the gap	Ω·m
ρ_e	Electrolyte density	kg/m^3
χ	Angle of approach	degree
ξ	Oscillation amplitude	μm
λ	Wave length	m

Glossary

Abbreviation	Description
ac	Alternating current
AFM	Abrasive flow machining
AISI	American Iron and Steel Institute
AJM	Abrasive jet machining
ANSI	American National Standards Institute
ASME	American Society of Mechanical Engineers
ASTM	American Society for Testing and Materials
AWJM	Abrasive water jet machining
BHN	Brinell hardness number
BUE	Built-up edge
CBN	Cubic boron nitride
CH-Milling	Chemical milling
CNC	Computerized numerical control
CNT	Carbon nano tube
dc	Direct current electrochemical machining
DCECM	Direct current
DIN	Deutsche Institut für Normung
DOC	Depth of cut
DS	Directionally solidified
DTC	Difficult-to-cut
DTCMs	Difficult-to-cut materials
EBM	Electron beam machining
ECG	Electrochemical grinding
ECM	Electrochemical machining
EDM	Electric discharge machining
ED-Milling	Electric discharge milling
ESM	Electrolytic stream machining
EWR	Electrode wear rate
GW	Grinding wheel

Abbreviation	Description
HAZ	Heat affected zone
HB	Hardness Brinell
HF	High frequency
HMP	Hybrid machining process
HRB	Hardness Rockwell B
HRC	Hardness Rockwell C
HSM	High speed machining
HSS	High speed steel
IBR	Integrally bladed rotor
ISO	International Organization for Standardization
JM	Jet machining
Laser	Light amplification by stimulated emission of radiation
LAT	Laser assisted turning
LBM	Laser beam machining
MAM	Magnetic abrasive machining
MQL	Minimum quantity lubrication
MRR	Material removal rate
MSD	Multiple shot drilling
NC	Numerical control
Nd:Yag	Neodymium-doped yttrium aluminum garnet
NTD	Nozzle-tip distance
NTMPs	Nontraditional machining processes
OCV	Open circuit voltage
ODS	Oxide-dispersed strengthened
PAM	Plasma arc machining
PAT	Plasma assisted turning
PBM	Plasma beam machining
PCD	Polycrystalline diamond
PCM	Photochemical machining
PD	Percussion drilling
RC	Resistance capacitance
RCL	Recast layer
SAs	Super alloys
SE	Spray etching
SEDM	Sinking electric discharge machining
SEM	Scanning electron microscope
SI	Surface integrity
SOD	Stand-off distance
SSs	Stainless steels
STEM	Shaped tube electrolytic machining
TAM	Thermal assisted machining
TMPs	Traditional machining processes
USALBM	Ultrasonic assisted laser beam machining
USM	Ultrasonic machining
WEDM	Wire electric discharge machining
WJM	Water jet machining
WP	Workpiece

1

Introduction

1.1 Stainless Steels and Super Alloys as Difficult-to-Cut Materials

In recent decades, engineering materials have greatly developed. At the same time, the cutting speed and the material removal rate (MRR) in machining such materials using traditional methods such as turning, milling, drilling, grinding, and so on, has been going down. In many cases, it has been challenging to machine these materials such as stainless steels, refractory metals and alloys, Ti-alloys, super alloys, carbides, ceramics, composites, and even diamond, using traditional methods. It is no longer possible to find tool materials that are sufficiently hard to cut such materials.

To meet these challenges, new processes with advanced methodology and tooling have needed to be developed. These are the nontraditional processes, which are capable of machining a wide spectrum of these difficult-to-cut materials irrespective of their hardness. The increasing use of ceramics, high strength polymers, and composites will also necessitate the use of nontraditional methods of machining. In addition, grinding will be applied to a greater extent than in the past, with greater attention to creep feed grinding (CFG), and the use of polycrystalline diamond (PCD) and polycrystalline cubic boron nitride (PCBN) [1].

The question now is why both stainless steels and super alloys, as difficult-to-cut materials, have both been selected in regard to their machinability in this book? The reasons are as follows:

1. There are diverse and important industrial applications necessitating the use of special materials and alloys, characterized by high strength, high temperature strength, and high corrosion, and oxidation resistance.
2. There is difficulty associated with machining both materials, especially as they comprise dozens of grades of different machining characteristics.

Machining of Stainless Steels and Super Alloys: Traditional and Nontraditional Techniques,
First Edition. Helmi A. Youssef.

3. Both materials are characterized by low thermal conductivity, high coefficient of thermal expansion, high ductility, and high work-hardening rate, making their machining a tedious task. Their low thermal conductivity leads to an increase in tool temperature, and consequently reduces the tool life. The high work-hardening rate and low thermal conductivity affect chip formation leading to segmented chips. Also, the high coefficient of thermal expansion of these alloys leads to serious difficulties in maintaining machining tolerances.
4. The high ductility favors development of built-up edge (BUE) on the tool, thus destroying surface finish, and promoting vibration and chatter.

The basic issue in achieving optimum machining of stainless steels and super alloys is to select adequate cutting speed and correct tool for each work material. In order to realize this objective, a good understanding of the effect of cutting speed on mechanical and thermal properties of the work material and cutting tool should be considered. Some other techniques which are used to enhance the machinability of stainless steels and super alloys will be presented in relevant chapters of this book.

1.1.1 Historical Background of Stainless Steels and Super Alloys

1.1.1.1 Stainless Steels

Stainless steels (SSs) were introduced at the beginning of the twentieth century as a result of pioneering work in England, Germany, and France. However, the development had started several decades before. In 1821, the French Berthier found that iron-chromium alloy was resistant to some acids. Others studied the effects of Cr in an iron matrix, but using a low percentage of Cr. In 1875, another Frenchman, Brustline recognized the importance of carbon levels in addition to Cr.

In 1904, Leon Guilet published a research on martensitic and ferritic SS-alloys with composition that today would be known as 410, 442, 446, and 440 C. In 1906, he also published a detailed study of an Fe-Ni-Cr austenitic alloy; that was equivalent to the 300 series of SS. In Germany, in 1908, Monnartz and Borchers found evidence of the relationship between a minimum level of Cr (10.5%) on corrosion resistance as well as the importance of low carbon content and the role of Mo in increasing corrosion resistance to chlorides.

Harry Brearley of Sheffield was generally accredited as the initiator of the industrial era of SS. He was trying to develop a new material for barrels for heavy guns that would be resistant to abrasive wear. He noted that materials with high Cr-contents did not take an etch. This discovery had led to the patent of a steel with 9–16% Cr and less than 0.7% C; the first stainless steel had been born. Most of his work was on stainless 430, patented in 1919. The first product was the table cutlery that is still used today.

Parallel with the work in England and Germany, F.M. Becket was working in Niagara Falls, to find a cheap and scaling-resistant material for furnaces that run up to 1000 °C. He found that at least 20% Cr was necessary to achieve resistance to oxidation or scaling. That was the first development of heat-resistant steels.

The world's first free-machining stainless, invented by Frank Pahlmer (1928) [2], was a straight-grade with sulfur (0.15%S). It was the forerunner of today's martensitic 416 stainless. Sulfur and phosphorous were both added to make the austenitic 303 stainless which is the first free-machining Cr-Ni grade in the early 1930s. Selenium (Se) additions instead of sulfur have been favored in the States.

1.1.1.2 Super Alloys

Stainless steels served as a starting point for the satisfaction of high temperature engineering requirements. Moreover, they were soon found to be limited in their strength capabilities. The metallurgists responded to the increased needs by making what might be termed *super alloys (SAs)* of stainless varieties. Of course, it was long before the *hyphen* was dropped and the improved iron-base materials became known as one type of super alloy.

The term super alloy was first coined shortly after the second World War to describe a group of alloys developed for use in turbo-superchargers and aircraft turbine engines that required high performance at elevated temperatures. For more than six decades now, super alloys have provided the most reliable and cost-effective means of achieving high operating temperatures and stress conditions in aircraft, and also land gas turbines. As we move towards the third decade of the twenty-first century, super alloys seem to be extending their useful temperature range along with their excellent inherent characteristics. The development of these materials continues to this day with optimization of chemical composition and production methods. This will lead to the development of a new class of material tailored to meet the need for better mechanical properties at elevated temperatures.

Although patents for Al- and Ti-additions to Nichrome type alloys were issued in the 1920s, the super alloy industry emerged with the adoption of Co-base super alloys (Haynes, Stellite 31) to satisfy the increasing demands of higher temperature strength of aircraft engines. Some Ni-Cr-alloys (Inconels and Nimonics), based more or less on toaster wire and developed in the first decade of the twentieth century, were also available for engineering applications. So the race was on to make superior metal alloys available for the insatiable thirst of the designer for higher temperature strength capability. The race still continues.

1.1.2 Industrial Applications of Stainless Steels and Super Alloys

1.1.2.1 Stainless Steels

The average person has no idea what stainless steel is, but it is all around us. Most of us use stainless steel table ware and wear a wristwatch with a stainless steel case. There are stainless steel racks in refrigerators and ovens and there are stainless steel toasters, tea kettles, and even kitchen sinks. Cars have stainless steel exhaust systems that last for ten years instead of the three years that would be expected if they had been ordinary steel. The industrial applications of specific types of stainless steel alloys will be presented afterwards in the relevant locations of the book.

Stainless steels are defined as steel alloys characterized primarily by their corrosion resistance, high strength and ductility and high chromium content. They are called stainless because in the presence of O_2 (air), they develop a colorless thin, hard, adherent film of chromium oxide, and remain lustrous. This film builds up again in the event the surface is scratched (i.e., self-healing). For passivation to occur, a minimum chromium content of about 10.5% by mass should be present. Several other alloying elements such as carbon, nickel, manganese, silicon, titanium, molybdenum, aluminum, sulfur, phosphorous, nitrogen, and so on, can be added to the Cr-Fe matrix to form well over 150 different compositions of SSs, of which about 15 are most commonly used.

These alloys are milled into coils, sheets, plates, bars, wire, and tubing to be used in cookware, cutlery, surgical instruments, major appliances, industrial equipment (e.g., in sugar refineries), and as an automotive and aerospace structural alloy and construction material in large buildings. Storage tanks and tankers used to transport orange juice and other foods are often made of SS, due to its corrosion resistance and antibacterial properties. This also influences its use in commercial kitchens and food processing plants, as it can be steam-cleaned, sterilized, and does not need painting or application of other surface finishes.

1.1.2.2 Super Alloys

Super alloys or heat resistant super alloys (HRSA) constitute a category that straddles the ferrous and nonferrous metals. Some of them are based on iron, whereas others are based on nickel and cobalt. In fact, many super alloys contain substantial amounts of three or more metals, rather than consisting of one base metal plus alloying elements. Although the tonnage of these alloys is not significant compared with most of the other metals, they are nevertheless commercially important because they are very expensive; and they are technologically important because of what they can do.

Super alloys are a group of high-performance alloys designed to meet very demanding requirements for strength and resistance to surface degradation (corrosion and oxidation) at high service temperatures. Conventional room temperature strength is usually not the important criterion for these metals, and most of them possess room temperature strength properties that are good but not outstanding. Their high temperature performance is what distinguishes them; tensile strength, hot hardness, creep resistance, and corrosion resistance at very elevated temperatures are the mechanical properties of interest. Operating temperatures are often in the vicinity of 1100 °C, without a damaging reduction in strength and hardness [3].

The main fields of application of super alloys generally embrace aircraft gas turbines, space vehicles, steam turbine power plants, reciprocating engines, heat-treatment equipment, metal processing forming and casting dies, nuclear power systems, medical applications, chemical and petrochemical industries, pollution control equipment, coal gasification, and liquefaction systems. Super alloys (also some types of SSs, classified as super alloys) are frequently used in advanced aero-engine components such as turbine blades, turbine vanes, turbine vane rings, turbine nozzles, engine, and turbine casings.

1.2 Traditional and Nontraditional Machining Processes

1.2.1 Importance of Machining in Manufacturing Technology

While technological advancements continue to take place throughout the developed manufacturing industry, machining still remains the most important process used to shape metals and alloys. Compared to other manufacturing processes, machining is characterized by its versatility and capability of achieving the highest accuracy and surface integrity in the most economical way [4]. Most materials and alloys, hard or soft, cast or wrought, ductile or brittle, are machined. Most engineering products, in terms of size from watch parts to aircraft wing spares (over 30 m long), or ship propellers, are produced by machining. Such versatility of machining processes can be attributed to many factors, including the following:

- Machining does not require elaborate tooling.
- It can be employed to most engineering materials.
- Tool wear is kept within limits, and tools are not costly.
- The large number of machining parameters can be suitably controlled to overcome techno-economical difficulties.

The development of new tool materials opened a new era for the machining industry in which a parallel development in machine tools took place. In the last century, nontraditional machining techniques offered alternative methods for machining parts of complex shapes in extra-hard and tougher exotic materials that were difficult-to-machine by traditional methods.

In highly developed industrial countries, the yearly cost associated with metal removal has been estimated at about 10% of the gross national production. Metal cutting machine tools form about 70% of the operating production machines, and are characterized by their high accuracy and productivity. For these reasons, rational approach and minor improvements in productivity of material removal processes are of major importance in high volume production. It is also known that about 10% of the materials produced by machining industry goes into waste. This is not an exclusive feature of the machining process as it is also present in all other methods related to manufacturing. Therefore, machining should not be identified, in most cases, as a method yielding a high loss of material [5].

Machining is generally used as a final finishing operation for parts produced by casting and forming before they are ready for use. However, there are a number of reasons that makes machining processes an obligatory solution as compared with other manufacturing techniques. These are:

- If closer dimensional control and tight tolerances may be required than are available by casting and forming.
- If special surface quality may be required for proper functioning of a part.
- If the part has external and internal geometric features that cannot be produced by other manufacturing operations.
- If it is more economical to machine the part rather than to produce it by other manufacturing operations.

On the other hand, machining has the following limitations:

- It generally necessitates a longer time to remove material than to shape the part by forming and casting.
- Unless carried out properly, machining can have adverse effects on the properties of the surface quality of the product.
- Machining is generally energy-, capital-, and labor-intensive.
- Machining necessitates highly qualified operators and specialized personnel. A wrong decision causes high production cost and less machining quality.
- Machining necessitates sophisticated measuring tools.

1.2.2 Classification of Machining Processes

Engineering materials have been recently developed whose hardness and strength are considerably increased, such that the cutting speed and the MRR tend to fall when machining such materials using traditional methods like turning, milling, grinding, and so on. In many cases, it is impossible to machine hard materials to certain shapes using these traditional methods. Sometimes it is necessary to machine alloy steel components of high strength in a hardened condition. It is no longer possible to find tool materials that are sufficiently hard to cut at economical speeds, such as hardened steels, austenitic steels, Nimonics, carbides, ceramics, and fiber-reinforced composite materials. The traditional methods are unsuitable to machine such materials economically, and there is no possibility that they can be further developed to do so because most of these materials are harder than the materials available for use as cutting tools.

By utilizing the results of relevant applied research, it is now possible to process many of the engineering materials that were formerly considered to be nonmachinable using traditional methods. The newly developed machining processes are often called nonconventional or nontraditional machining processes (NTMPs). These are nontraditional in the sense that traditional cutting tools are not employed; instead, energy in its direct form is utilized. These processes cover recent research and development in techniques that focus on achieving high accuracies and good surface finishes of parts machined without burrs or residual stresses, especially with hard-to-cut materials that cannot be machined by traditional means.

Therefore, the machining processes are classified into traditional machining processes (TMPs) and NTMPs [1] (Figure 1.1).

1. TMPs, in which chips are formed by the interaction of a cutting tool with the material being machined. These processes employ traditional tools of a basic wedge form to penetrate into the workpiece. These tools must be harder than the material to be machined. TMPs comprises two categories:
    a. Cutting: (chipping processes) that use tools of definite geometry such as turning, planing, drilling, milling, broaching, and so on.
    b. Abrasion processes that use tools of nondefinite geometry such as grinding, honing, lapping, and so on.
2. NTMPs, in which the machining energy is utilized in its direct form. These processes are less familiar, and are desired to meet the increasingly difficult demands for which TMPs cannot be used.

NTMPs also comprise two categories (Figure 1.2):

- Abrasion processes, where the mechanical energy is used for machining of the work materials; these processes include ultrasonic machining (USM), water jet machining (WJM), abrasive jet machining (AJM), abrasive water jet machining (AWJM), and so on.
- Erosion processes using chemical and electrochemical energy such as chemical machining (CHM), electrochemical machining (ECM), and so on, or using thermal energy such as electric discharge machining (EDM), laser beam machining (LBM), electron beam machining (EBM), plasma beam machining (PBM), and so on (Figure 1.1). EDM has firmly established its use in the production of forming tools, dies, molds and effectively machining of advanced materials such as SAs, and SSs.

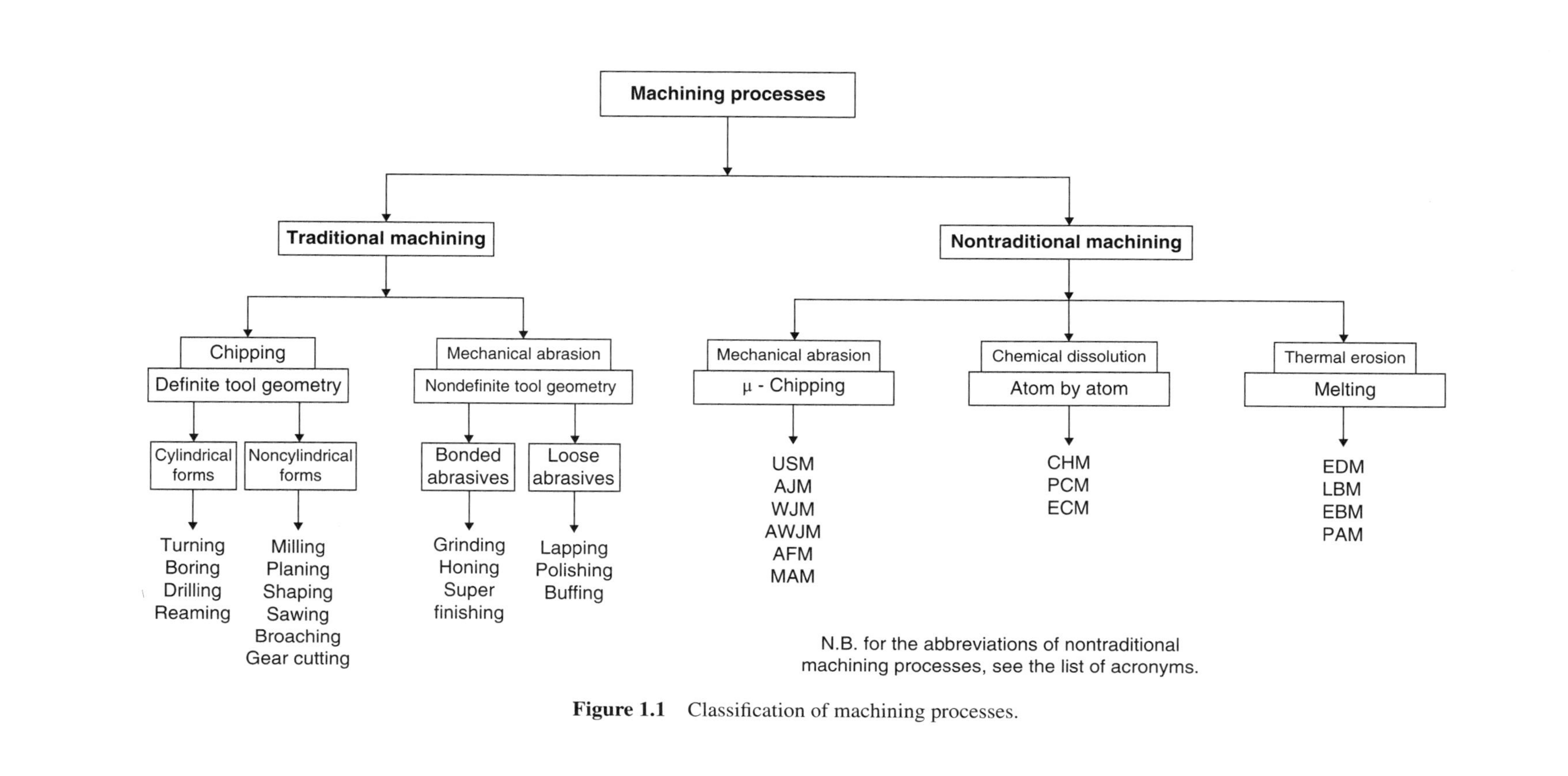

Figure 1.1 Classification of machining processes.

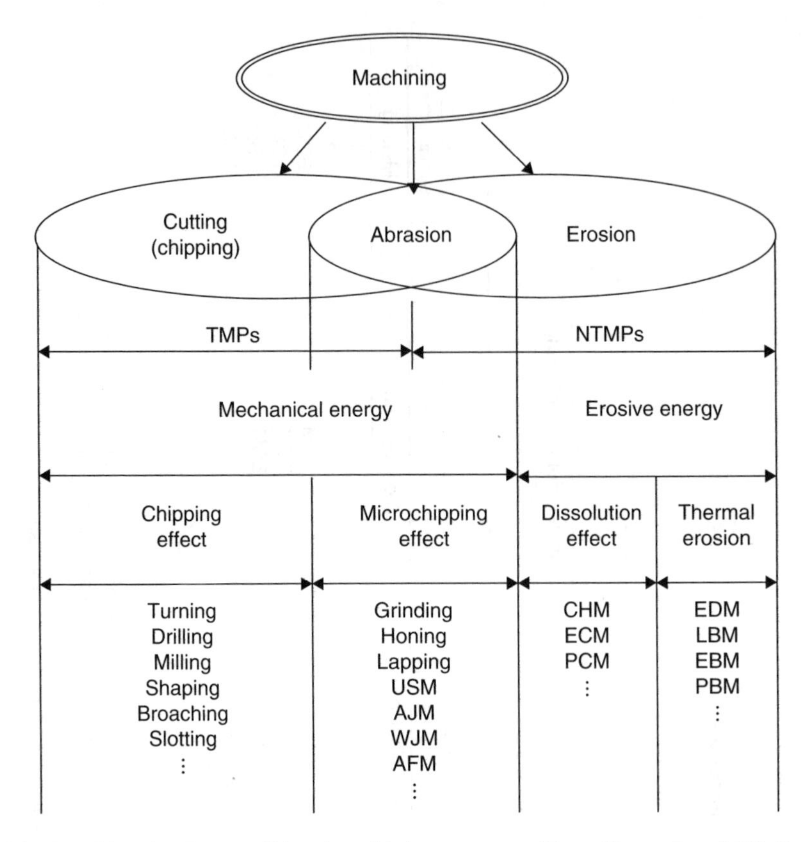

Figure 1.2 Traditional and nontraditional machining processes (From: Youssef *et al.* [1]. Reproduced with permission).

NTMPs are mostly restricted to small-scale removal of materials. They have specifically the following characteristics as compared to traditional processes:

- They are capable of machining a wide spectrum of metallic and nonmetallic materials irrespective of their hardness or strength.
- Complex and intricate shapes, in hard and extra-hard materials, can be readily produced with high accuracy and surface quality and commonly without burrs.
- The hardness of cutting tools is of no relevance, especially in many NTMPs, where there is no physical contact between the work and the tool.
- Simple kinematic movements are needed in the NTM equipment, which simplifies the machine design.
- Micro and miniature holes and cavities can be readily produced by NTM.

However, it should be emphasized that:

1. NTMPs cannot replace TMPs. They can be used only when they are economically justified or it is impossible to use TMPs.

2. A particular NTMP found suitable under given conditions may not be equally efficient under other conditions. A careful selection of the NTMP for a given machining job is therefore essential. The following aspects must be considered in that selection:
   a. Properties of the work material and the form geometry to be machined
   b. Process parameters
   c. Process capabilities
   d. Economic and environmental considerations.

1.2.3 Variables of Machining Processes

Any machining process has two types of interrelated variables. These are input (independent) and output (dependent) variables (Figure 1.3) [4]:

1.2.3.1 Input (Independent) Variables

- Workpiece material, like composition and metallurgical features
- Starting geometry of the workpiece, including preceding processes
- Selection of process, which may be TMP or NTMP
- Tool material and tool geometry
- Cutting parameters
- Work-holding devices ranging from general purpose vise to specially designed jigs and fixtures
- Cutting fluids.

1.2.3.2 Output (Dependent) Variables

- Cutting force and power. Cutting force influences deflection and chattering; both affect part size and accuracy. The power influences heat generation and consequently tool wear.
- Geometry of finished product, thus obtaining a machined surface of desired shape, tolerance, and mechanical properties.

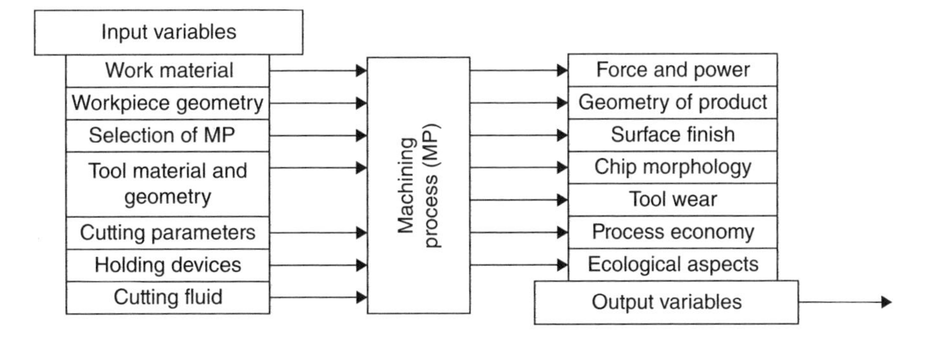

Figure 1.3 Input and output variables of a machining process.

- Surface finish: it may be necessary to specify multiple cuts to achieve a desired surface finish.
- Tool failure due to the increased power consumption.
- Economy of the machining process is governed by cutting speed, and other variables, as well as cost and economical factors. Machining economy represents an important aspect.
- Ecological aspects and health hazards must be considered and eliminated by undertaking necessary measures.

References

[1] Youssef, H.A., El-Hofy, H., Ahmed, M.H. (2011) *Manufacturing Technology – Materials, Processes, and Equipment*. CRC Press, Boca Raton, FL.

[2] Palmer, F.R. (1934) Ferrous alloys. US Patent 1961,777.

[3] Groover, M.P. (2010) *Fundamentals of Modern Manufacturing, Materials, Processes, and Systems*, John Wiley & Sons, Inc., 4th edn.

[4] Youssef, H.A., El-Hofy, H. (2012) *Principles of Traditional and Nontraditional Machining*. El-Fath Publishing Press, Alexandria.

[5] Youssef, H.A., El-Hofy, H. (2008) *Machining Technology – Machine Tools and Operations*, CRC Press, Boca Raton, FL.

2

Types and Classifications of Stainless Steels

2.1 Role of Alloying Elements in Stainless Steels

The addition of the following elements has the following effects on stainless steels (SSs):

1. *Carbon, C* Carbon is always present in SS. In all categories, except martensitic, the level of carbon is kept quite low. In martensitic grades, the level is deliberately increased to obtain high strength and hardness. Heat treating by heating to a high temperature, quenching, and then tempering develops the martensitic structure.

   Carbon does affect the corrosion resistance. If carbon is allowed to combine with Cr to form chromium carbides, it may have a detrimental effect on the ability of the passive film to form. If in localized areas, Cr is reduced below 10.5%, the passive film will not form.
2. *Chromium, Cr* Chromium is a highly reactive element and accounts for the passive nature of all stainless steels. The resistance to corrosion is the direct result of the presence of Cr. As stated, once the SS contains at least 10.5% Cr, the passive film of Cr_2O_3 is instantaneously formed, which prevents further diffusion of O_2 into the surface. The higher the Cr-level, the greater is the protection (Figure 2.1).
3. *Nickel, Ni* It is an essential alloying element in stainless steels. Its presence results in the formation of an austenitic structure that gives these grades their strength, ductility, and toughness, even at cryogenic temperatures. It also makes the steel nonmagnetic. While the role of Ni has no direct influence on the development of the passive surface layer, it results in significant improvement in resistance to acid attack, particularly with sulfuric acid.
4. *Molybdenum, Mo* The addition of molybdenum to the Cr-Fe-Ni matrix adds resistance to localized pitting attack and better resistance to crevice corrosion, particularly in ferritic grades. It helps the detrimental effects of chlorides (A1S1-316 with 2% Mo is preferred

Machining of Stainless Steels and Super Alloys: Traditional and Nontraditional Techniques, First Edition. Helmi A. Youssef.

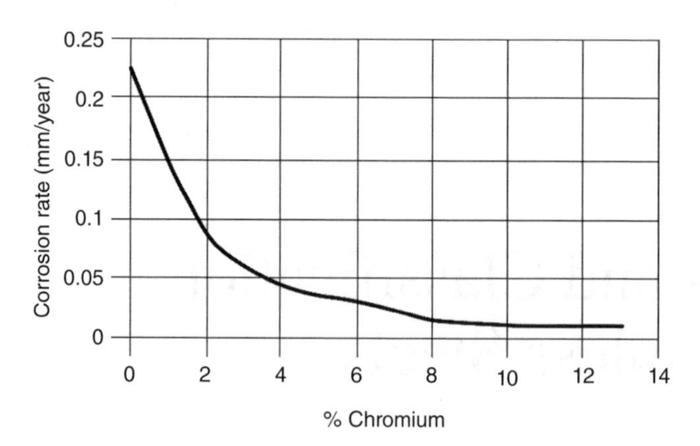

Figure 2.1 Effect of chromium content on corrosion rate according to evidence found by Monnartz and Borchers, Germany, 1908.

over, 304 in coastal and de-icing salt situations). The higher the Mo-content (some SSs contain up to 6% Mo), the better is the resistance to higher chloride levels.

5. *Manganese, Mn* Generally, manganese is added to stainless steels to assist in deoxidation during melting, and to prevent the formation of iron sulfide inclusions, which can cause hot cracking problems. It is also an austenite stabilizer and when added in higher levels (4–15% Mn), it replaces some of the Ni in the AISI-200 series SS-grades of reduced cost.
6. *Silicon, Si and Copper, Cu* Small amounts of silicon and copper are usually added to austenitic stainless steels containing Mo to improve corrosion resistance to sulfuric acid. Silicon also improves oxidation resistance and is a ferrite stabilizer. In austenitic SSs, high Si-content improves resistance to oxidation and also prevents carburizing at elevated temperatures (AISI-309 and AISI-310 are examples). In castings, higher Si-content increases the fluidity of the molten metal thus improving the castability. High Si-welding wire (metal inert gas MIG) promotes washing and wetting behavior. Copper enhances machinability of austenitic types by reducing work hardening tendency. Cu works like Ni as a powerful, "austenite" phase stabilizer, reducing the formation of strain induced martensite during cold working. Cold forging grades are Cu-bearing types.
7. *Nitrogen, N* In austenitic and duplex stainless steels, nitrogen increases the resistance to localized pitting attack and intergranular corrosion. Nitrogen when added, it provides strengthening to SSs.
8. *Niobium, Nb* Its addition prevents intergranular corrosion, particularly in the HAZ (heat affected zone) after welding. Nb helps preventing the formation of chromium carbides that can rob the microstructure of the required amount of Cr for passivation. In ferritic stainless steels, the addition of Nb is an effective way to improve the thermal fatigue resistance.
9. *Titanium, Ti* Titanium is the main element used to stabilize stainless steel before the use of Argon-Oxygen Decarburization (AOD) vessels. When SS is melted in air, it is difficult to reduce the carbon levels. The most common grade (AISI-302) before AODs was allowed to a maximum carbon level of 0.15%. At this high level, Ti reacts with the carbon to form titanium carbides, and prevents the formation of chromium carbides, which could affect the formation of the passive film of Cr_2O_3.

10. *Sulfur, S or Selenium, Se* S or Se is generally kept to low levels as it can form sulfide inclusions. It is mainly used to improve machinability (where these inclusions act as chip breakers). The addition of S or Se, however, reduces the resistance to pitting corrosion.

2.2 Types of Stainless Steels

Stainless steel does not constitute a single well-defined material, but instead comprises, depending on the alloying elements additions, several families of alloys, each generally having its own characteristic, microstructure, alloying elements, and properties, that are suited to a wide range of applications.

Several classification schemes have been devised to categorize stainless steels. The American Iron and Steel Institute (AISI) and the Unified National Standard (UNS) grouped these alloys by chemistry and assigned a three-digit number (AISI), or a five-digit number (UNS) that identifies stainless steels. Accordingly, stainless steels have been subdivided into five families, three basic families (*ferritic*, *martensitic,* and *Austenitic*), and two derived families (*duplex and precipitation hardened*) (Figure 2.2). All are correlated with their specific microstructures (Tables 2.1, 2.2, 2.3, 2.6, and 2.7).

2.2.1 Basic Alloys of Stainless Steels (Ferritic, Martensitic, Austenitic)

These comprise the following three categories.

2.2.1.1 Ferritic Stainless Steels of AISI-Designations

Series 400 [405-409-429-430-434-436-439-442-444-446] (Table 2.1).

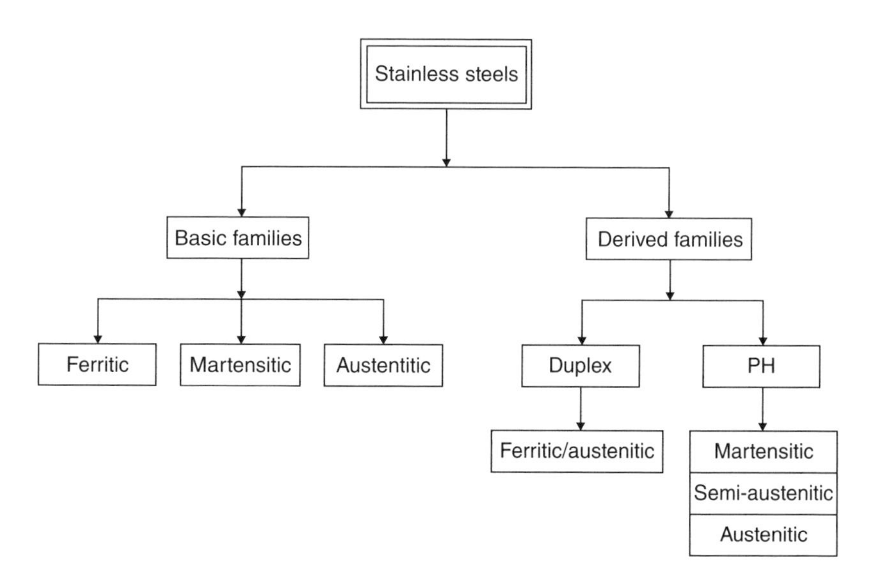

Figure 2.2 Basic (standard) and derived families of stainless steel alloys.

Table 2.1 Selected free- and nonfree-machining ***ferritic*** stainless steels according to AISI-and UNS-designation

AISI(UNS)-designation	Composition (wt%)										Remarks
	C	Mn	Si	P	S	Cr	Ni	Mo	N	Others	
Nonfree-machining ferritic alloys											
405 (S40500)	0.08	1	1	0.04	0.03	11.5–14.5	—	—	—	0.1–0.3 A1	
409 (S40900)	0.08	1	1	0.045	0.045	10.5–11.75	0.5	—	—	(6 C–0.75) Ti	
								—	—	—	
430 (S43000)	0.12	1	1	0.04	0.03	16–18	—	0.75–1.25	—	—	Improve corrosion resistance due to Mo
434 (S43400)	0.12	1	1	0.04	0.03	16–18	—		—	—	Improve corrosion resistance due to Cr
442 (S44200)	0.2	1	1	0.04	0.03	18–23	0.5	—	0.025	[0.2 + 4 (C + N)] − 0.8 = Ti + Nb	Improve corrosion resistance due to Mo
444 (S44400)	0.025	1	1	0.04	0.03	17.5–19.5	1	1.75–2.5	0.25	—	Improve corrosion resistance due to Cr
446 (S44600)	0.2	1.5	1	0.03	0.03	23–27	—	—			
Free-machining ferritic alloys											
[XM-34] (S18200)	0.08	1.25–2.5	1	0.04	0.15 min	17.5–19.5	—	1.5–2.5	—		
(S18235)	0.025	0.5	1	0.03	0.15–0.35	17.5–18.5	1	2–2.5	0.025	0.3–1 Ti	Ti for improving corrosion resistance
430 F (S43020)	0.12	1.25	1	0.06	0.15 min	16–18	—	0.6 min	—		Mo (optional)
430 FSe (S43023)	0.12	1.25	1	0.06	0.06	16–18	—	—	—	0.15 min Se	

[ ] Designation used by ASTM.

Table 2.2 Selected free- and nonfree-machining ***martensitic*** stainless steels according to AISI- and UNS-designation

AISI(UNS)-designation	Composition (wt%)										Remarks
	C	Mn	Si	P	S	Cr	Ni	Mo	N	Others	
Nonfree-machining martensitic alloys											
403 (S40300)	0.15	1	0.5	0.04	0.03	11.5–13	—	—	—	—	
410 (S41000)	0.15	1	1	0.04	0.03	11.5–13	—	—	—	—	
414 (S41400)	0.15	1	1	0.04	0.03	11.5–13.5	1.5–2.5	—	—	—	
420 (S42000)	0.15 min	1	1	0.04	0.03	12–14	—	—	—	—	STAVAX
(S42010)	0.15–0.3	1	1	0.04	0.03	13.5–15	0.25–1	0.4–1	—	—	TrimRite
431(S43100)	0.2	1	1	0.04	0.03	15–17	1.25–2.5	—	—	—	
440 A (S44002)	0.6–0.75	1	1	0.04	0.03	16–18	—	0.75	—	—	
440 B (S44003)	0.75–0.95	1	1	0.04	0.03	16–18	—	0.75	—	—	
440 C (S44004)	0.95–1.2	1	1	0.04	0.03	16–18	—	0.75	—	—	
Free-machining martensitic alloys											
416 (S41600)	0.15	1.25	1	0.06	0.15 min	12–14	—	0.6	—	—	
[XM-6] (41610)	0.15	1.5–2.5	1	0.06	0.15 min	12–14	—	0.6	—	—	Enhanced m/c-ing version (Mn)
416 Se (S41623)	0.15	1.25	1	0.06	0.06	12–14	—	—	—	0.15 min Se	
420 F (S42020)	0.15 min	1.25	1	0.06	0.15 min	12–14	—	0.6	—	—	
420 FSe (S42023)	0.3–0.4	1.25	1	0.06	0.06	12–14	—	0.6	—	0.5 min Se, 0.6 Zr or Cu	
440 F (S44020)	0.95–1.2	1.25	1	0.04	0.1–0.35	16–18	0.75	0.4–0.6	0.08	—	
440 FSe (S44023)	0.95–1.2	1.25	1	0.04	0.03	16–18	0.75	0.6	0.08	0.15 min Se	

[ ] Designation used by ASTM.

Table 2.3 Selected free- and nonfree-machining ***austenitic*** stainless steels according to AISI-and UNS-designation

AISI(UNS)-designation	Composition (wt%)										Remarks
	C	Mn	Si	P	S	Cr	Ni	Mo	N	Others	
Nonfree-machining austenitic alloys											
201 (S20100)	0.15	5.5–7.5	1	0.06	0.03	16–18	3.5–5.5	—	0.25	—	
[XM-19] (S20910)	0.06	4–6	1	0.04	0.03	20.5–23.5	11.5–13.5	1.5–3	0.2–0.4	0.1–0.3 Nb, 0.1–0.3 V	Nitronic 50
[XM-11] (S21904)	0.04	8–10	1	0.06	0.03	19–21.5	5.5–7.5	—	0.15–0.4	—	
[XM-28] (S24100)	0.15	11–14	1	0.06	0.03	16.5–19.5	0.5–2.5	—	0.2–0.45	—	Nitronic 32 [18-2-Mn]
18-18 Plus (S28200)	0.15	17–19	1	0.045	0.03	17–19	—	0.5–1.5	0.4–0.6	0.5–1.5 Cu	—
310 (S31000)	0.15	2	1	0.045	0.03	16–18	6–8	—	—	—	High work-hardening rate
302 (S30200)	0.15	2	1	0.045	0.03	17–19	8–10	—	—	—	
304 (S30400)	0.08	2	1	0.045	0.03	18–20	8–10	—	—	—	Most commonly used SS in the world
304 L (S30403)	0.03	2	1	0.045	0.03	18–20	8–12	—	—	—	Weldable version
[XM-21] (S30452)	0.08	2	1	0.045	0.03	18–20	8–10.5	—	0.16–0.3	—	
305 (S30500)	0.12	2	1	0.045	0.03	17–19	10–13	—	—	—	
309 (S30900)	0.12	2	1	0.045	0.03	22–24	12–15	—	—	—	
309S (S30908)	0.08	2	1	0.045	0.03	22–24	12–15	—	—	—	
310 (S31000)	0.25	2	1.5	0.045	0.03	24–26	19–22	—	—	—	
310S (S31008)	0.08	2	1	0.045	0.03	24–26	19–22	2–3	—	—	
316 (S31600)	0.08	2	1	0.045	0.03	16–18	10–14	2–3	—	—	Second commonly used austenitic SS
316 L (S31603)	0.03	2	1	0.045	0.03	16–18	10–14	2–4	—	—	Weldable version
317 (S31700)	0.08	2	1	0.045	0.03	18–20	11–15	—	—	5× C = Ti	
321 (S32100)	0.08	2	1	0.045	0.03	17–19	9–12	—	—	10× C = Nb	

347 (S34700)	0.08	2	1	0.045	0.03	17–19	9–13	2–3	—	(8 C-1) = Nb, 3–4 Cu	
[20Cb-3] (N08020)	0.07	2	1	0.045	0.035	19–21	32–38		—		
Free-machining austenitic alloys											
[XM-1] (S20300)	0.08	5–6	1	0.04	0.1–0.35	16–18	5–6.5	0.5	—	1.75–2.25 Cu	
303 (S30300)	0.15	2	1	0.2	0.15 min	17–19	8–10	0.6	—	Mo (optional)	
[XM-5] (S30310)	0.15	2.5–4.5	1	0.2	0.25 min	17–19	7–16	0.75	—	—	303 Plus X
303 Se (S30323)	0.15	2	1	0.2	0.06	17–19	8–10	—	—	0.15 min Se	
303 Cu (S30330)	0.15	2	1	0.15	0.1	17–19	6–10	—	—	2.5–4 Cu, 0.1 Se	—
[xm-2] (S30345)	0.15	2	1	0.05	0.11–0.16	17–19	8–10	0.4–0.6	—	0.6–1 Al	—
[XM-3] (S30360)	0.15	2	1	0.04	0.12–0.3	17–19	8–10	0.75	—	0.12–0.3 Pb	—
316 F (S31620)	0.08	2	1	0.2	0.1 min	17–19	12–14	1.7–2.5	—	—	—

[ ] Designation used by ASTM.

These have carbon levels below 0.12% (442 and 446 are at 0.2% C), and high Cr-content (10.5–27%), and relatively small amounts of other alloying elements. Higher levels of Cr in alloys 442 and 446 promote their corrosion and oxidation resistance. Molybdenum is added to 434 and 444 to improve corrosion resistance particularly in chloride containing solution. Ferritic alloys can be degraded by the presence of σ-chromium, an intermetallic phase which can precipitate upon welding.

These grades are magnetic and cannot be hardened by heat treatment; however, they may be hardened by cold working, but not to the same extent as austenitic alloys (Figure 2.3). Ferritic alloys have reduced corrosion resistance compared to austenitic. They are generally not chosen for toughness. In the annealed condition, they have a TYS (tensile yield strength) of 275–350 MPa.

The last four alloys in Table 2.1 are free-machining ferritic alloys since they contain free-machining additives.

Since ferritic alloys are the cheapest type of stainless steels, they should be given first consideration when SS-alloy is required. Ferritic stainless steels are generally used for non-structural applications such as kitchen and restaurant equipment, automobile trims, heaters, dish washers, and annealing baskets.

2.2.1.2 Martensitic Stainless Steels also of AISI-Designation

Series 400 [403-410-414-416-420-422-431-440] (Table 2.2).

Martensitic alloys have a relatively high carbon-level (0.15–1.2% C) as compared to ferritic and austenitic grades, and Cr-level from 11.5 to 18%. Mo (<1%) can be added to improve mechanical properties and corrosion resistance (UNS-42010). Nickel (<2.5%) can be added for the same reason (414 and 431).

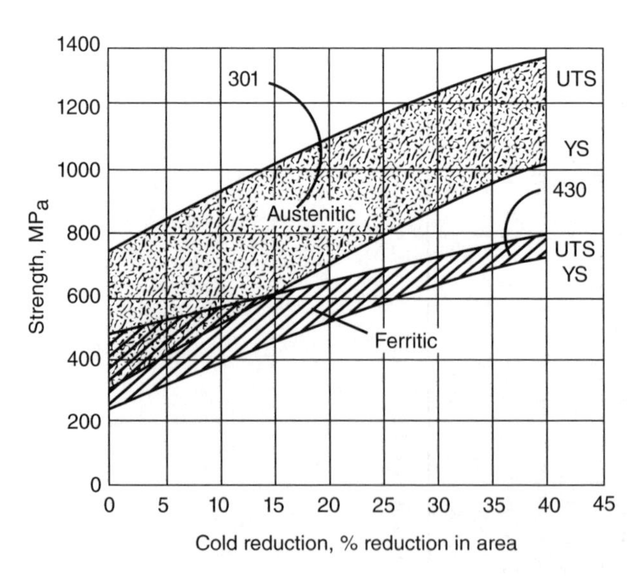

Figure 2.3 Comparison between cold-working rates of ferritic 430 and austenitic 301 stainless steel. (Adapted from ASM [1].)

Martensitic alloys are also magnetic. They are not as corrosion resistant as the other two basic classes. In the annealed condition, the TYS is about 275 MPa, and thus these alloys can be moderately hardened by cold working similar to ferritic alloys. However, when hardened and tempered, their TYS increases up to 1900 MPa, depending primarily on carbon content. In the annealed condition, they are machinable. The last seven alloys in Table 2.2 are free-machining martensitic alloys, since they contain considerable amounts of free-machining additives of S or Se (minimum 0.15% each).

Martensitic alloys cost about 1.5 times as much as ferritic, part of this being due to additional heat treatment, which generally consists of austenitization, quenching, stress relief, and tempering. Martensitic alloys are used for applications such as cutlery, surgical tools, instruments, valves, rivets, screws, hand tools, vegetable choppers, razor blades, riffle barrels, mining machinery, bolts, nuts, and aircraft fittings.

2.2.1.3 Austenitic Stainless Steels of AISI-Designation

Of Series 300 [Fe-Cr-Ni] and Series 200 [Fe-Cr-Mn] (Table 2.3).

This category makes up over 70% of the total stainless steel production. Austenitic stainless steels contain maximum 0.15% C, a minimum of 16% Cr, and sufficient Ni and/or Mn to retain an austenitic structure at all temperatures from the cryogenic region to the melting point of the alloy. These alloys cannot be hardened by heat treatment and are nonmagnetic. They all exhibit excellent corrosion resistance but are susceptible to stress corrosion cracking (Mo is added for resistance to chlorides). Austenitic stainless steels are the most ductile of all SSs; hence they can be easily formed, which is a characteristic of fcc. However, with increasing cold work, their formability and ductility are reduced and strengthen significantly (Figure 2.3). The response of the popular 304-alloy (also known as 18–8 of composition 18% Cr and 8% Ni) to a small amount of cold work (15%) is illustrated in Table 2.4.

Austenitic alloys are often used in the water quench condition, where the water-quench serves to retain the alloy in solid solution. No phase transformation occurs during quenching, since austenite is the stable phase for all temperatures involved.

Corrosion resistance of austenitic alloys varies from good to excellent, depending on the alloy contents. Higher Cr, Ni, Mo, or Cu can be added to improve corrosion and oxidation resistance [316, 317, 309, 310, and UNS-N08020]. The latter has Ni-levels high enough to classify it as Ni-base super-alloy, which is known by its very high creep strength. To prevent intergranular corrosion after high temperature exposure, Ti or Nb is added to stabilize carbon

Table 2.4 Yield and tensile strength of AISI 304 stainless steel at water – quench and cold rolled conditions

AISI 304	Water quench	15% cold rolled
YTS (MPa)	260	805
UTS (MPa)	620	965
Elongation.% (50 mm)	68	11

YTS: yield tensile strength and UTS: ultimate tensile strength.
Source: Adapted from DeGarmo's Materials and Processes in Manufacturing [2].

in alloys such as 321 and 347. Carbon levels are reduced to low values to produce weldable alloys such as 304 L and 309 S.

Austenitic stainless steels are classified into two groups (Table 2.3). These are:

1. Standard group of AISI-designation (Series 300), where Ni is the austenite stabilizer. With sufficient amounts of both Cr and Ni; it is possible to produce a stainless steel in which austenite is stable all over the temperature range such as 301, 302, and so on. Nitrogen can also be used to provide strengthening in Cr-Ni standard group such as [XM-21] (Table 2.3).
2. The Mn-group of AISI-designation (Series 200), where a substantial quantity of Mn, usually with higher levels of N, and in many cases N is added, such as [XM-11], [XM-19], [XM-28], and UNS28200 (Table 2.3).

The last eight alloys in Table 2.3 are free-machining versions, since they contain free-machining additives such as S and Se.

As Austenitic alloys are costly they should not be specified where less expensive ferritic or martensitic alloys would be adequate. Austenitic alloys of standard series 300 may cost twice as much as the ferrite variety due to their expensive alloying elements (Ni and Cr). Mn and N (series 200) are substituted for some of the Ni to produce a lower cost, but of somewhat lower-quality alloy.

Austenitic alloys are used in a wide variety of applications, such as kitchenware, fittings, welded constructions, lightweight transportation equipment, furnace and heat exchanger parts, and equipment for several chemical environments. The austenitic grade 304 is the most commonly used SS in the world. The excellent forming and welding characteristics make it the standard steel for many applications in industry, architecture, and transportation.

The grade 316 is the second most commonly used austenitic SS. Like 304, it has excellent forming characteristics, but the added Mo gives 316 an improved corrosion resistance, so it is usually regarded as "marine steel grade." Its low carbon version 316 L is also immune from grain boundary carbide precipitation after welding.

Table 2.5 finally summarizes typical compositions of the basic alloys of SSs. The microstructure that a stainless steel attains depends primarily on its composition, in which the main alloy components Cr and Ni are most important. In reality, the variation can be wide, due to

Table 2.5 Typical compositions (wt%) of the basic ferritic, martensitic, and austenitic SSs

Element	Ferritic (bcc)	Martensitic (bcc)	Austenitic (fcc)
Carbon	0.08–0.2	00.15–1.2	0.03–0.25
Chromium	11–27	11.5–18	16–26
Nickel	0–1	0–2.5	3.5–22
Manganese	1–1.5	1–1.5	2 (series 200, 5.5–10)
Silicon	1	1	1–1.5
Molybdenum	0–2.5	0–1	Some cases: 1.5–3
Phosphorus and sulfur[a]	0.075	0.075	0.075
Titanium	0–1	—	0–0.4

[a] In free-machining alloys, the minimum S (or Se) content is 0.15%.

Adapted from Degarmo *et al.* [2].

the influence of other alloying elements that strive to stabilize either austenite or ferrite. The effect of the alloying elements on the structure of SSs is summarized in the Schäffler-Delong diagram (Figure 2.4). So, it becomes possible to calculate the total ferrite an austenite stabilizing effect of the alloying elements, giving the so-called Cr- and Ni-equivalents in the Schäffler-Delong diagram as follows:

$$\text{Cr-equivalent} = \%\text{Cr} + 1.5\%\text{Si} + \%\text{Mo}$$
$$\text{Ni-equivalent} = \%\text{Ni} + 30(\%\text{C} + \%\text{N}) + 0.5(\%\text{Mn} + \%\text{Cu} + \%\text{Co})$$

Different diagrams have been published with slightly different equivalents, phase limits, or general layout. Figure 2.5 shows one of the more simplified diagrams for identifying the locations of some different types of SSs.

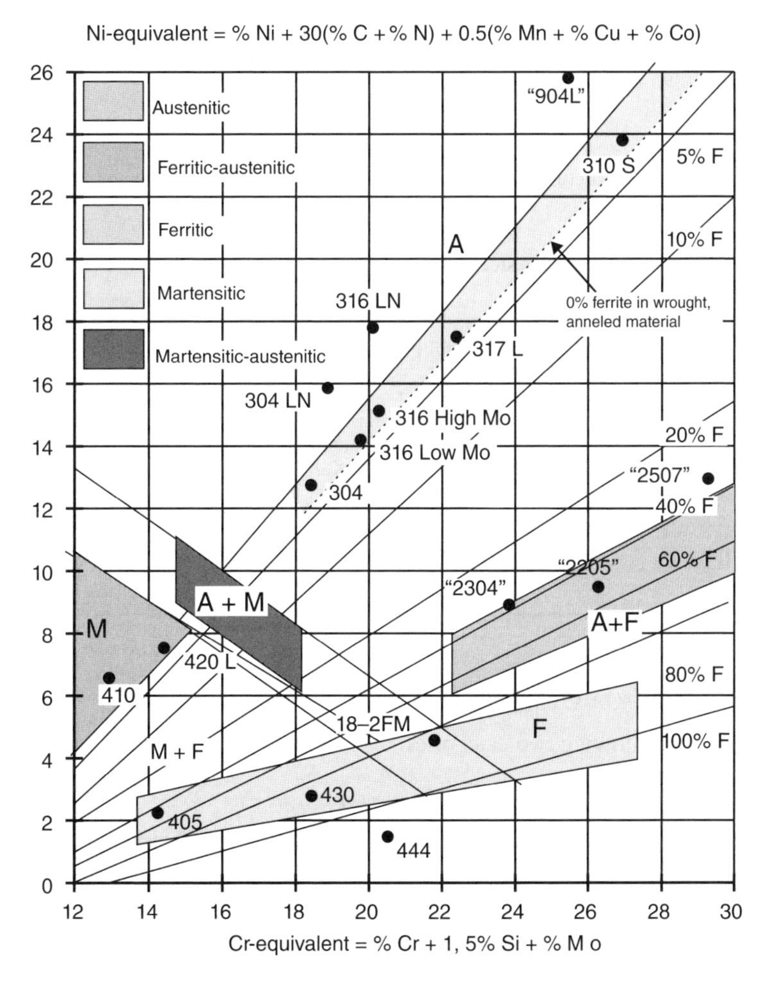

Figure 2.4 Schäffler-Delong diagram.

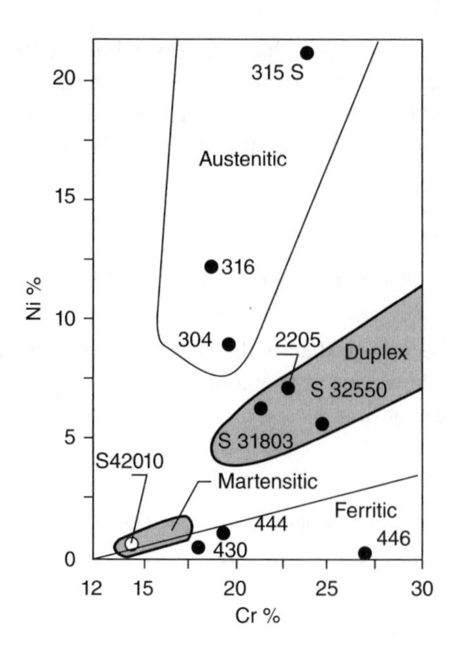

Figure 2.5 Microstructure of stainless steel alloys as depending on Cr- and Ni-percent.

Table 2.6 Selected ***duplex*** stainless steels, according to AISI-and UNS-designation. Duplex alloys not containing free-machining versions

AISI(UNS)-designation	Composition (wt%)										Remarks
	C	Mn	Si	P	S	Cr	Ni	Mo	N	Others	
(S31803)	0.03	2	1	0.03	0.02	21–23	4.5–6.5	2.5–3.5	0.08–0.2	—	2205
(S32550)	0.04	1.5	1	0.04	0.03	24–27	4.5–6.5	2–4	0.1–0.25	1.5–2.5 Cr	Ferralium
329(S32900)	0.2	1	0.75	0.04	0.03	23–28	2.5–5	1–2	—	—	—
(S32950)	0.03	2	0.6	0.035	0.01	26–29	3.5–5.2	1–2.5	0.15–0.35	—	7-Mo Plus

2.2.2 *Derived Alloys of Stainless Steels (Duplex, PH-Alloys)*

These comprise the following two categories:

Duplex alloys: Of designations shown in Table 2.6.

These contain 21–29% Cr and 2.5–6.5% Ni (Figure 2.5). They are water-quenched from 960 to 1020 °C to produce ferritic/austenitic microstructure; hence they are magnetic. Duplex alloys have a TYS of about 550 MPa in the annealed condition, which is about twice that of the standard water-quenched austenitic alloy. They generally have good ductility and toughness, along with higher resistance to both corrosion and stress-corrosion cracking than austenitic alloys. The desired ferritic/austenitic balance (1:1) of the most common duplex alloy 329 is achieved by adding Cr (23–28%), Mo (1–2%), and sufficient

Ni (2.5–5%). Nitrogen is added for strengthening (UNS-31803, UNS-32950) (Table 2.6). Duplex alloys do not embrace any free-machining categories.

Typical applications of duplex alloys are in water treatment plants, and heat exchanger components. It is used for chemical, food, medical, and papermaking industries, and in processes that include acids or chlorine, and equipment related to off-shore oil and gas industry.

Precipitation hardenable alloys: Of designations shown in Table 2.7.

These alloys contain Cr and Ni, along with Cu, Al, Ti, Nb, or Mo. Molybdenum is added to improve the mechanical properties and corrosion resistance. Cu is also added to enhance corrosion resistance (UNS-45000). Precipitation hardenable (PH) alloys are characterized by their ability to be age-hardened to various strength levels. PH-alloys can attain a TYS of up to 1700. Cold-working prior aging leads to even higher yield strengths. These alloys have generally good ductility, toughness, and strength at elevated temperature, with acceptable to good corrosion resistance.

Depending on their composition, these alloys (Table 2.7) may be subdivided into the following:

1. Martensitic (magnetic)
2. Semi-austenitic (magnetic)
3. *Austenitic (nonmagnetic).*

A better combination of strength and corrosion resistance is achieved by austenitic and semi-austenitic types than martensitic PH-alloys. The most common PH-alloy UNS-17400 contains Cr and Ni as do all alloys, with Cu for age-hardening and Nb for stabilizing the carbon. Similar to duplex, PH-alloys do not include a free-machining category; however, the alloy UNS-17400 has enhanced machining characteristics (it is machined at high speed without chattering). Except in the semi-austenitic PH-alloy such as UNS-35500, carbon is normally restricted to provide the desired phase transformation. The additional alloys and extra processing make the PH-alloys some of the most expensive stainless steels; therefore, they should be only used, when absolutely required.

The main applications of these alloys are in aircraft, aerospace, and structural components. The ultra strength alloy 17–7 PH (UNS 17700) is frequently used in spring, pressure vessels and where high strength coupled with good corrosion resistance is required.

2.3 Concluding Comments and Comparative Characteristics

In conclusion, the following comments and comparative characteristics of different SS-alloys can be outlined:

1. Chromium sets stainless steels apart from other steels. The unique self-healing passive surface layer on the steel is a result of chromium oxide. Commercially available grades contain around minimum 11% Cr, and may be either ferritic or martensitic depending on carbon range control. More Cr enhances corrosion and oxidation resistance.
2. Cr-levels over 20% provide improved aqueous corrosion resistance for the duplex and higher alloyed austenitics. In addition, Nickel widens the environmental scope that stainless steel can handle.

Table 2.7 Selected precipitation hardened stainless steels, according to AISI-und UNS-designation. PH-alloys not containing free-machining versions

AISI(UNS)-designation	Composition (wt%)									Remarks
	C	Mn	Si	P	S	Cr	Ni	Mo	N	Others	
Martensitic PH-alloys											
[XM-13] (S13800)	0.05	0.2	0.1	0.01	0.008	12.25–13.25	7.5–8.5	2–2.5	0.01	0.9–1.35 Al	PH13-8Mo
[XM-12] (S15500)	0.07	1	1	0.04	0.03	14–15.5	3.5–5.5	—	—	0.15–0.45 Nb, 2.5–4.5 Cu	15–5 PH
(630) (S17400)	0.07	1	1	0.04	0.03	15.5–17.5	3–5	—	—	0.15–0.45 Nb, 3–5 Cu (enhanced machining alloy)	17–4 PH
[XM-25] (S45000)	0.05	1	1	0.03	0.03	14–16	5–7	0.5–1	—	8C=Nb, 1.25–1.75 Cu	Custom 450
[XM-16] (S45500)	0.5	0.5	0.5	0.04	0.03	11–12	7.5–9.5	0.5	—	0.1–0.5 Nb, 1.5–2.5 Cu, 0.8–1.4 Ti	Custom 455
Semiaustenitic PH-alloys: Better combination of strength and corrosion resistance than martensitic PH-alloys											
(631) (S17700)	0.09	1	1	0.04	0.04	16–18	6.5–7.75	—	—	0.75–1.5 Al	17–7 PH
(633) (S35000)	0.07–0.11	0.5–1.25	0.5	0.04	0.03	16–17	4–5	2.5–3.25	0.07–0.13	Best machining, if supplied in over temperature condition	Pyromet 350
(634) (S35500)	0.1–0.15	0.5–1.25	0.5	0.04	0.03	15–16	4–5	2.5–3.25	0.07–0.13	Best machining, if supplied in over temperature condition	Pyromet 355
Austenitic PH-alloys: Better combination of strength and corrosion resistance than martensitic PH-alloys											
(660) (S66286)	0.08	2	1	0.04	0.03	13.5–16	24–27	1–1.5	—	0.35 Al, 0.001–0.01 B, 1.9–2.35 Ti, 0.1–0.5 V	Poor machinability

[ ] Designation used by ASTM and ( ) designation used by AISI.

3. The addition of 2% Ni to the martensitic type AISI-421 improves corrosion resistance marginally. Additions of 4.5–6.5% Ni enhance greatly the corrosion resistance of duplex types, The corrosion resistance is not only related to Ni-level. The duplex with 5% Ni has better corrosion resistance than that of austenitic AISI-304 with its 8% Ni.
4. More specific alloy additions are also made to enhance corrosion resistance. These include Mo and N for pitting and crevice corrosion resistance. The AISI-316 is the main Mo bearing austenitic alloy. Many of the currently available duplex grades contain additions of both elements Mo and N. Copper is also used to enhance corrosion resistance in hazardous environments such as intermediate concentration ranges of sulfuric acid. Grades containing Cu include the austenitic 18–18 Plus (S28200).
5. Only the martensitic stainless steels are hardened by heat treatment, like other alloy steels. PH-SSs are strengthened by heat treatment, but they use a different mechanism to the martensitic types. The ferritic, austenitic, and duplex types cannot be strengthened or hardened by heat treatment, but respond to varying degrees to cold working as a strengthening mechanism.
6. Ferritic types have useful mechanical properties at ambient temperatures, but have limited ductility, compared to austenitics. They are not suitable for cryogenic applications and lose strength at elevated temperatures over about 600 °C, although they have been used for application such as automotive exhaust systems very successfully. Austenitic types, with their characteristic fcc-lattice, have quite distinct properties. Mechanically, they are more ductile and impact tough at cryogenic temperatures. They are nonmagnetic. They are also characterized by their lower thermal conductivity, and higher thermal expansion rates than other types of stainless steels.
7. Duplex types, which have a mixed structure of austenite and ferrite, share some of the properties of those types but fundamentally are mechanically stronger than either ferritic or austenitic types.
8. Depending on their type and heat treatment condition, wrought stainless steels are formable and machinable. Stainless steels can also be cast or forged into shape. Most of the available types can be joined using appropriate thermal methods, including soldering, brazing, and welding.
9. Austenitics are suitable for a wide range of application involving flat product forming (pressing, drawing, stretch forming, spinning, and so on). Although ferritic and duplex types are also useful for these forming methods, the excellent ductility and work-hardening characteristic of austenitic alloys is controlled through the Ni-level. Austenitic AISI-301 has a low Ni-content (7% Ni), and so work hardens when cold worked. Ni-levels of around 8.5% Ni make austenitics ideally suited to deep drawing operations, for example, in the manufacture of stainless steel-sinks.
10. Martensitics are not readily formable, but are used extensively for blanking in the manufacture of cutting blades (Sandvik 13C26).
11. Most stainless steels can be traditionally machined. Techniques involve control of feed and speed to undercut work hardening layers with good lubrication and cooling systems. Where high production volume systems are employed, free-machining stainless steels may be needed. Sulfur additions being the tradition approach in grades like AISI-303 and others.

Table 2.8 illustrates the main advantages and disadvantages of various types of stainless steels.

Table 2.8 Advantages and disadvantages of various types of SSs

	Type	Advantages	Disadvantages
Basic SS-alloys	Ferritic	Low cost, moderate corrosion resistance, and good formability	Limited corrosion resistance, formability, and of elevated temperature strength compared to austenitics
	Martensitic	Hardened by heat treatment	Limited corrosion resistance and weldability compared to austenitics. Limited formability compared to ferritics
	Austenitics	Widely available, good and acceptable corrosion resistant, good cryogenic toughness, excellent formability, and weldability	Work hardening can limit formability, and machinability. Limited resistance to stress corrosion cracking
Derived SS-alloys	Duplex	Good stress corrosion cracking resistance, good mechanical strength in annealed condition	Temperature range is more restricted than austenitics
	PH-SS Martensitic 17/4 PH	Hardened by heat treatment, but with better corrosion resistance than basic martensitics	Limited availability, corrosion resistant, restricted formability, and weldability compared to austenitics

NB: Free-machining alloys are only available in basic SS-Alloys.
Adapted from: engineer on line. Ws/stainless steel selection.htm.[3].

References

[1] Kosa, T., Ney, R.P. (1989) *ASM Handbook: Machining*, Vol. **16**, Materials Park, OH.

[2] Degarmo, P., Black, J.T., Kohser, R.A. (2012) *DeGarmo's Materials and Processes in Manufacturing,* 11th edn, John Wiley & Sons, Inc.

[3] http://engineeronline.ws/stainless.htm (accessed December 3, 2013).

3

Types and Classifications of Super Alloys

3.1 General Features and Classifications

Super alloys are a relatively new class of materials which exhibit high mechanical strength, ductility, creep strength at high operating temperatures, high fatigue strength, and typically superior resistance to corrosion and oxidation even at elevated temperatures. These features make super alloys ideal for applications in aircraft, submarines, nuclear reactors, dies for hot working of metals, and petrochemical equipment. It will be noted that not all applications require elevated temperature strength capacity. Their high strength, coupled with corrosion resistance, has made certain super alloy standard materials applicable for biomedical joint implants, and cryogenic applications [1].

These alloys are usually classified into three main groups which are Fe-based, Ni-based, and Co-based alloys (Figure 3.1). The physical, mechanical, and machining behavior of each group varies considerably due to the chemical compositions of the alloy and the metallurgical processing it receives during manufacturing. Super alloys are also classified into wrought, cast, and powder metallurgy (PM) super alloys. The wrought may be solid-solution-strengthened or precipitation hardened alloys. Representative listing of super alloys and compositions, emphasizing alloys developed and used in USA, are given in Tables 3.1–3.5, and are also visualized in Figures 3.2 and 3.3. These are generally identified by trade names or by special numbering systems. Appropriate compositions of super alloys can be forged, rolled into sheets/bars, or otherwise produced in a variety of shapes. Bar and plate stock materials are the easiest form of raw materials to deal with. They are bought straight from material producers, and are intended for the production of shafts, sleeves, and similar components. The more highly alloyed compositions are normally processed as castings (Table 3.3). It should be noted that refractory metals have higher melting

Machining of Stainless Steels and Super Alloys: Traditional and Nontraditional Techniques,
First Edition. Helmi A. Youssef.

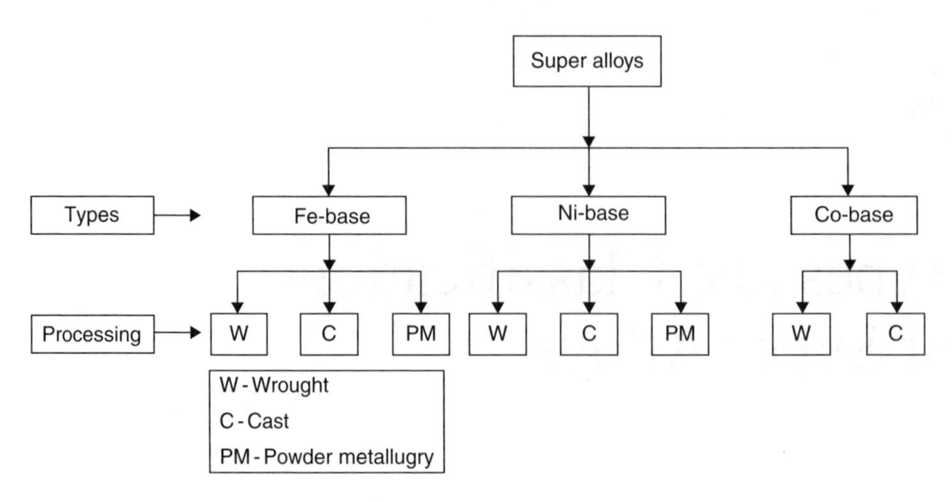

Figure 3.1 Classification of super alloys.

points than super alloys; however, they do not have the same desirable characteristics as super alloys and hence they are not widely used.

Super alloys generally contain Ni, Cr, Co, Mo, and Fe as major alloying elements. Others are Al, W, Ti, and so on. The role of these alloying elements is to enhance the characteristics of super alloys in the following manner [2, 3]:

- Ni stabilizes alloy structure and properties at high temperatures.
- Co, Mo, and W increase strength at elevated temperature.
- Cr, Al, Si enhance resistance to oxidation and provide high temperature corrosion.
- C increases creep strength.

It should be noted, however, that super alloy properties are directly related not only to the alloy chemistry, but also to melting procedures, forging and working processes, casting techniques, and above all to heat treatment following forging or casting.

The coating technology of super alloys is an integral part of super alloy development and application. Lack of a coating means less ability to use super alloys for extended times at elevated temperatures. Some engineering applications have already exceeded the temperature limits of these alloys, and still others await the development of materials that would make them feasible. It is estimated that the exhaust temperatures of future jet engines will exceed 1500 °C. In such applications, the super alloys are protected by coatings that isolate them from gases in their operating environment [4].

3.2 Types of Super Alloys

Super alloys are typically characterized by an austenitic face-centered crystalline structure. They are classified into three grades which are Fe-base, Ni-base, and Co-base alloys.

Table 3.1 Nominal compositions of commonly used solid-solution strengthened wrought super alloys covered in this book and their UNS (Unitied National Standard)-designation (Figure 3.2)

Super alloy (trade name)	UNS no.	Composition (wt%)										
		Cr	Ni	Co	Mo	W	Nb	Ti	Al	Fe	C	Other
Fe-base alloys												
M-155 (Multimet)	R30155	21.0	20.0	20.0	3.0	2.5	1.0	—	—	32.2	0.15	0.15 N, 0.2 La, 0.02 Zr
Haynes 556	—	22.0	21.0	20.0	3.0	2.5	0.1	—	0.3	29.0	0.10	0.5 Ta, 0.02 La, 0.002 Zr
19-9 DL	K63198	19.0	9.0	—	1.25	1.25	0.4	0.3	—	66.8	0.30	1.10 Mn, 0.60 Si
Incoloy 800	N08800	21.0	32.5	—	—	—	—	0.38	0.38	45.7	0.05	—
Incoloy 801	N08801	20.5	32.0	—	—	—	—	1.13	—	46.3	0.05	—
Incoloy 802	—	21.0	32.5	—	—	—	—	0.75	0.58	44.8	0.35	—
Ni-base alloys												
Haynes 214	—	16.0	76.5	—	—	—	—	—	4.5	3.0	0.03	—
Inconel 600	N06600	15.0	76.0	—	—	—	—	—	—	8.0	0.08	0.25 Cu
Inconel 625	N06625	21.5	61.0	—	9.0	—	3.6	0.2	0.2	2.5	0.05	—
Hastelloy X	N06002	22.0	49.0	<1.5	9.0	0.6	—	—	2.0	18.8	0.15	—
Nimonic 75	—	19.5	75.0	—	—	—	—	0.4	0.15	2.5	0.12	<0.25 Cu
Co-base alloys												
AiResist 213	—	19.0	<0.5	65.0	—	4.5	—	—	3.5	<0.5	0.17	6.5 Ta, 0.15 Ze, 0.02 Zr
Haynes 25 (L605)	R30605	20.0	10.0	50.0	—	15.0	—	—	—	3.0	0.10	1.5 Mn
Haynes 188	R30188	22.0	22.0	37.0	—	14.5	—	—	—	<3.0	0.10	0.9 La
S-816	R30816	20.0	20.0	42.0	4.0	4.0	4.0	—	—	4.0	0.38	—
MP-159	—	19.0	25.5	35.7	7.5	—	0.6	—	0.2	9.0	—	0.3 Ti

Table 3.2 Nominal compositions of commonly used precipitation hardened wrought super alloys covered in this book and their UNS-designation (Figure 3.2)

Super alloy (trade name)	UNS no.	Composition (wt%)										
		Cr	Ni	Co	Mo	W	Nb	Ti	Al	Fe	C	Other
Fe-base alloys												
A-286 (pyromet)	K63198	15.0	26.0	—	1.25	—	—	2.0	0.2	55.2	0.04	0.005 B, 0.3 V
Discaloy	K66220	14.0	26.0	—	3.0	—	—	1.7	0.25	55.0	0.06	—
Incoloy 903	—	<0.1	38.0	15.0	0.1	—	3.0	1.4	0.7	41.0	0.04	—
Incoloy 907	—	—	38.4	13.0	—	—	4.7	1.5	0.03	42.0	0.01	0.15 Si
Incoloy 909	—	—	38.0	13.0	—	—	4.7	1.5	0.03	42.0	0.01	0.40 Si
V-57	—	14.8	27.0	—	1.25	—	—	3.0	0.25	48.6	<0.08	0.01 B, (0.5 max)
Ni-base alloys												
Astroloy	—	15.0	56.5	15.0	5.25	—	—	3.5	4.4	<0.3	0.06	0.03 B, 0.06 Zr
Inconel 100	—	10.0	60.0	15.0	3.0	—	—	4.7	5.5	<0.6	0.15	1.0 V, 0.06 Zr, 0.015 B
IN-100	—	10.0	60.0	15.0	3.0	—	—	4.7	5.5	<0.6	0.15	0.06 Zr, 1.0 V
Inconel 718	N07718	19.0	52.5	—	3.0	—	5.1	0.9	0.5	18.5	<0.08	<0.15 Cu
M 252	—	19.0	56.5	10.0	10.0	—	—	2.6	1.0	<0.75	0.15	0.005 B
Nimonic 80A	N07080	19.5	73.0	1.0	—	—	—	2.2	1.4	1.5	0.05	<0.1 Cu
Nimonic 90	N07090	19.5	55.5	18.0	—	—	—	2.4	1.4	1.5	0.06	—
René 41	N07041	19.0	55.0	11.0	10.0	—	—	3.1	1.5	<0.3	0.09	0.01 B
René 95	—	14.0	61.0	8.0	3.5	3.5	3.5	2.5	3.5	<0.3	0.16	0.01 B, 0.05 Zr
Udimet 500	N07500	19.0	48.0	19.0	4.0	—	—	3.0	3.0	<0.4	0.8	0.005 B
Udimet 700	—	15.0	53.0	18.5	5.0	—	—	3.4	4.3	<1.0	0.07	0.03 B
Waspaloy	N07001	19.5	57.0	13.5	4.3	—	—	3.0	1.4	<2.0	0.07	0.006 B, 0.09 Zr

Co-base alloys are not included in this group.

Table 3.3 Nominal compositions of commonly used Fe-based cast super alloys covered in this book and their UNS-designation (Figure 3.3)

ACI-code	UNS no.	Compositions (wt%)				
		C	Cr	Ni	Fe	N
HC	J92605	0.2–0.5	26–30	4(max)	Remainder	—
HD	J93005	0.2–0.5	26–30	4–7	Remainder	—
HE	J93403	0.2–0.5	26–30	8–11	Remainder	—
HH	J93503	0.2–0.5	24–28	14–18	Remainder	0.2 max
HN	J94213	0.2–0.5	19–23	23–27	Remainder	—
HT	J94605	0.35–0.75	13–17	33–37	Remainder	—

ACI (Alloy Casting Institute) complementing ASTM designation.
All types of Fe-base cast super alloys contain: 2% Si, 0.5% Mo, 2% Mn except HC (1% Mn), and HD (1.5% Mn), P + S not exceed 0.04% for all types.

The physical metallurgy of these alloys is extremely complex, perhaps more challenging than that of any other alloying system. In addition, as demonstrated in Tables 3.1 and 3.2, the compositions of these alloys are complex as well.

3.2.1 *Fe-Base Alloys*

These alloys are sometimes designated as Fe-Ni-base alloys. They have been developed from austenitic steels. This group is typically the easiest to be machined, partly because it has the poorest hot strength properties of all groups. It consists of an iron-base with larger amounts of Ni and Cr than typical types of stainless steels. These alloys generally contain 29–67% Fe, up to 22% Cr, and 9–44% Ni. This grade includes: Incoloy, Ascoloy, and crucible A-286 (Tables 3.1 and 3.2, and Figure 3.2). Some Fe-base alloys have very low thermal coefficient of expansion (Incoloy 909), which make them especially suited for shafts, rings, and casings. At lower temperatures, and depending on strength needs for an application, Fe-Ni-base alloys find more use than Co-base and Ni-base alloys.

Table 3.5 presents the composition of the so-called mechanically alloyed super alloys (produced through PM-techniques). Incoloy MA 956 is an Fe-base alloy of no Ni-content at all.

Figure 3.4 visualizes the effect of operating temperature on the yield strength for all types of Fe-base super alloys. The features and specific applications of some important Fe-base super alloys are listed as follows:

1. ***A-286*** is a precipitation hardened alloy. It is designed for applications requiring high strength and good corrosion resistance at temperatures up to 700 °C. Up to 1000 °C, the oxidation resistance is equivalent to that of AISI-310 stainless. This alloy offers high ductility in notched sections [5]. This alloy is useful in high temperature application such as jet engines, superchargers, blades, and afterburner parts and fasteners.
2. ***Incoloy 800*** is a solid solution strengthened Fe-base super alloy of moderate strength and good resistance to oxidation and carburization at elevated temperatures. It is particularly useful for high temperature equipment in the petrochemical industry, because it does not

Table 3.4 Nominal compositions of Ni-base and Co-base cast super alloys covered in this book and their UNS-designation (Figure 3.3)

Super alloy (trade name)	Composition (wt%)												
	C	Ni	Cr	Co	Mo	Fe	Al	B	Ti	Ta	W	Zr	Others
Ni-base alloys													
B-1900	0.1	64.0	8.0	10.0	6.0	—	6.0	0.015	1.0	4.0	—	0.10	1.5 Hf
CMSX-4	—	Balance	6.5	9.0	0.6	—	5.6	—	1.0	6.5	6.0	—	—
CMSX-10	—	Balance	1.8–4.0	1.5–9.0	0.25–2.0	—	5.0–7.0	—	0.1–1.2	7.0–10.0	3.5–7.5	—	—
Hastelloy X	0.1	50.0	21.0	1.0	9.0	18.0	—	—	—	—	1.0	—	—
Inconel 100	0.18	60.5	10.0	15.0	3.0	—	5.5	0.01	5.0	—	—	0.06	1.0 V
Inconel 718	0.04	53	19.0	—	3.0	18.0	0.5	—	0.9	—	—	—	0.1 Cu, 5 Nb
MAR-M246	0.15	60.0	9.0	10.0	2.5	—	5.5	0.015	1.5	1.5	10.0	0.05	—
René 41	0.09	55.0	19.0	11.0	10.0	10.0	—	1.5	0.01	3.1	—	—	—
Udimet 700	1.0	53.5	15.0	18.5	5.2	—	4.2	0.03	3.5	—	—	—	—
WAX-20(DS)	0.2	72.0	—	—	—	—	6.5	—	—	—	20.0	1.5	—
Co-base alloys													
AiResist 13	0.45	—	21.0	62.0	—	—	3.4	—	—	2.0	11.0	—	0.1 Y
Haynes 21	0.25	3.0	27.0	64.0	5.0	1.0	—	—	—	—	—	—	—
MAR-M302	0.85	—	21.5	58.0	—	0.5	—	0.005	—	9.0	10.0	0.2	—
MAR-M509	0.6	10.0	23.5	54.5	—	—	—	—	0.2	3.5	7.0	0.5	—
NASA Co-W-Re	0.4	—	3.0	67.5	—	—	—	—	0.1	—	25.0	1.0	2 Re
X-40 (Stellite31)	0.5	10.0	22.0	57.5	—	1.5	—	—	—	—	7.5	—	0.5 Mn, 0.5 S

Table 3.5 Composition of mechanically alloyed super alloys covered in this book and their UNS-designation (Figure 3.2)

Alloy	BHN	Order of machinability rating	Composition (wt%)										
			Ni	Fe	Cr	Al	Ti	Ta	W	Mo	Zr	B	Y_2O_3
Fe-base alloy													
Incoloy MA 956	270	1	—	74.5	20	4.5	0.5	—	—	—	—	—	0.5
Ni-base alloy													
Inconel MA 754	277	2	88.60	—	20.0	0.3	0.5	—	—	—	—	—	0.6
Inconel MA 6000	450	3	68.74	—	15.0	4.5	2.5	2.0	4.0	2.0	0.15	0.01	1.1

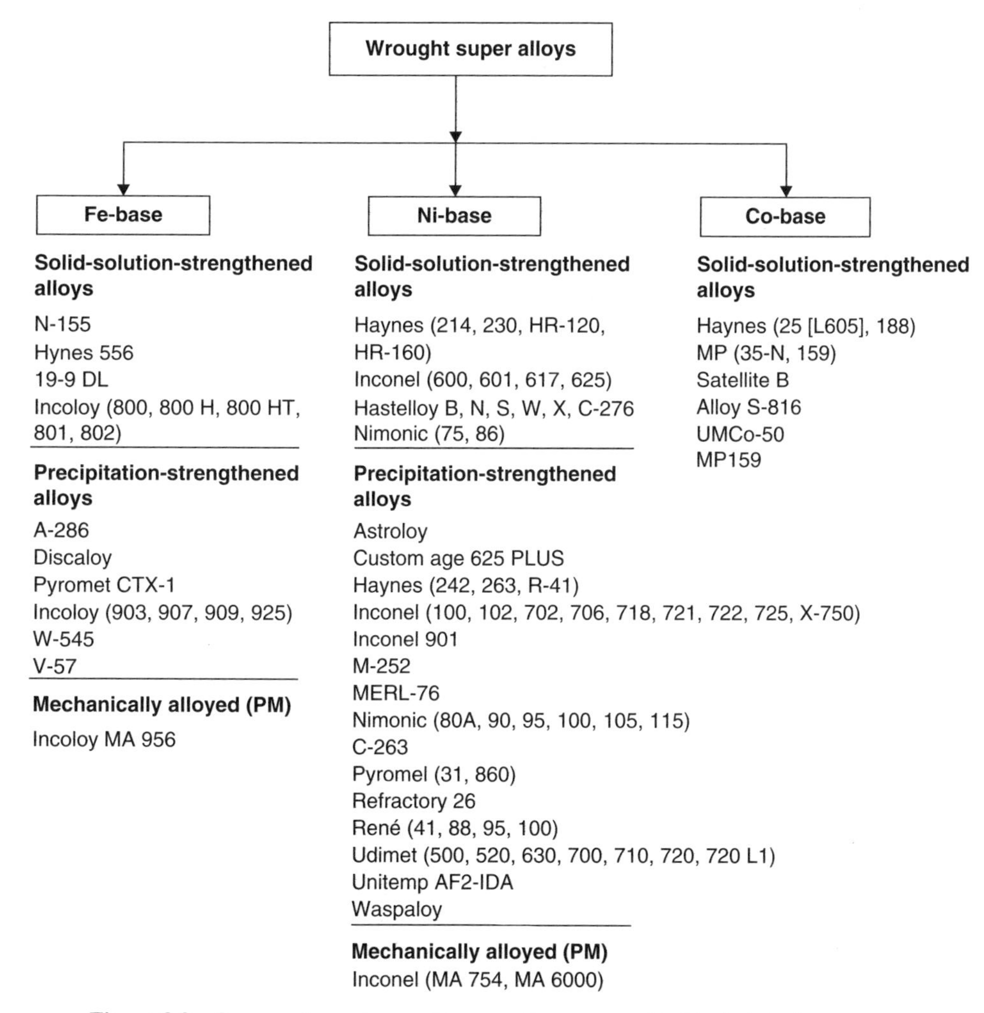

Figure 3.2 Commonly used wrought super alloys, as specified in their trade names.

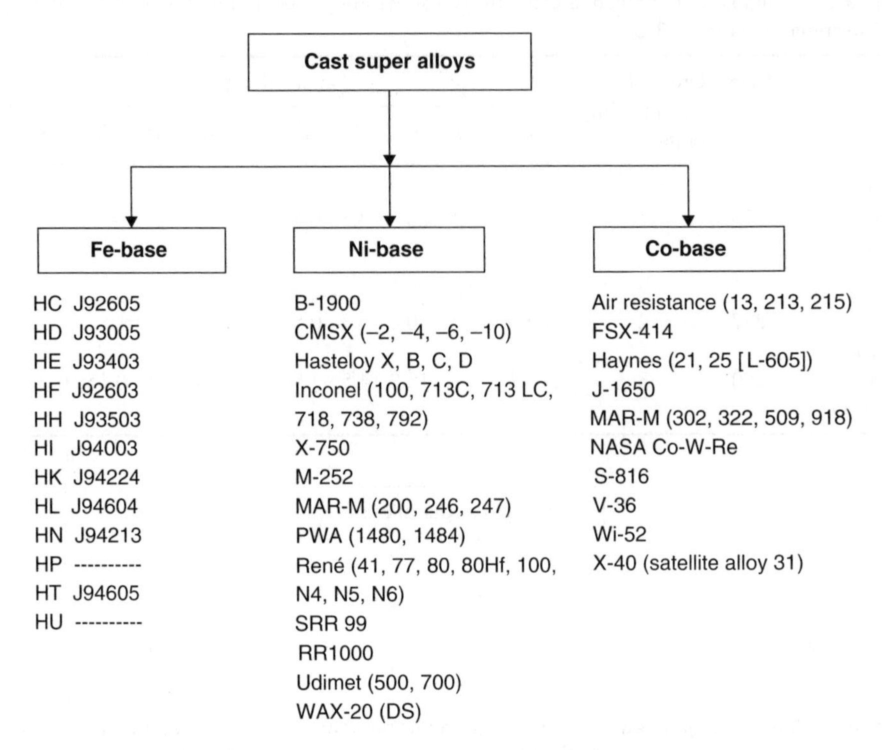

Figure 3.3 Commonly used cast super alloys, as specified in their trade names.

form an embrittling sigma phase after a long time exposure up to 650 °C. Excellent resistance to chloride stress-corrosion cracking is another important feature of this alloy. Therefore, it is applicable for heat exchangers, process piping furnace components, and heating-element sheathing.

3. ***N-155*** is also a solid solution strengthened Fe-base alloy, having good ductility, excellent corrosion resistance, and can be readily fabricated and machined. N-155 is recommended for part necessitating good strength and corrosion resistance up to 850 °C. It is used in numerous aircraft applications such as tail cones and tailpipes, exhaust manifolds, combustion burners, and turbine blades.
4. ***W-545*** is a precipitation hardenable Fe-base alloy. It has exceptionally high creep strength, combined with good ductility, resistance to notching, and excellent oxidation resistance up to a temperature of 750 °C. It is used in gas turbines.

3.2.2 Ni-Base Alloys

These constitute the largest group of super alloys, and are generally very difficult and demanding to machine. They are mainly used in applications requiring high corrosion resistance or high strength at elevated temperatures. They currently constitute over 50% of the weight of advanced aircraft engines. Alloys of this grade contain 38–76% Ni and up to 28%

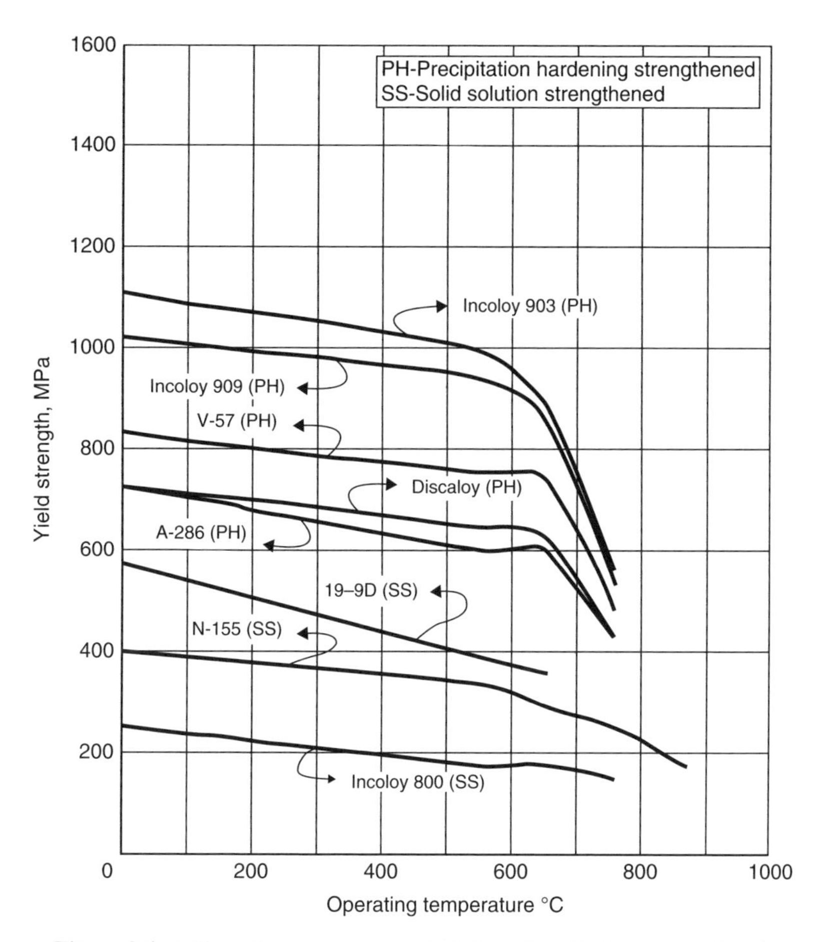

Figure 3.4 Effect of temperature on yield strength of Fe-base super alloys.

of each of the elements Cr, Co, and Mo. Common types of Ni-base alloys include Hastelloy, Inconel, Nimonic, René, Udimet, Astroloy, and Waspaloy series (Figure 3.2).

Ni-base super alloys are basically of four types. Solid solution strengthened (Table 3.1), precipitation hardenable (Table 3.2), cast (Tables 3.3 and 3.4), and oxide-dispersed strengthened (ODS). The solid-solution alloys contain little or no Al, Ti, or Nb. The precipitation hardenable alloys contain several percent Al, and Ti, and a few contain substantial Nb. The ODS' alloys contain a small amount of fine oxide particles (0.5–1% Y_2O_3) and are produced by powder metallurgy techniques. These alloys are expensive and this limits their use to small or critical parts, where the cost is not an issue. These alloys (also called mechanically alloyed super alloys) are listed together with the Incoloy MA 956 in Table 3.5. Inconel MA 745 and Inconel MA 6000 are not containing Fe at all.

As illustrated in Figure 3.5, precipitation-hardening alloys have considerably higher rupture strength values as compared to solid solution strengthened alloys. The same was depicted from Figure 3.4 and 3.6 with respect to the yield strength [6]. For the most demanding

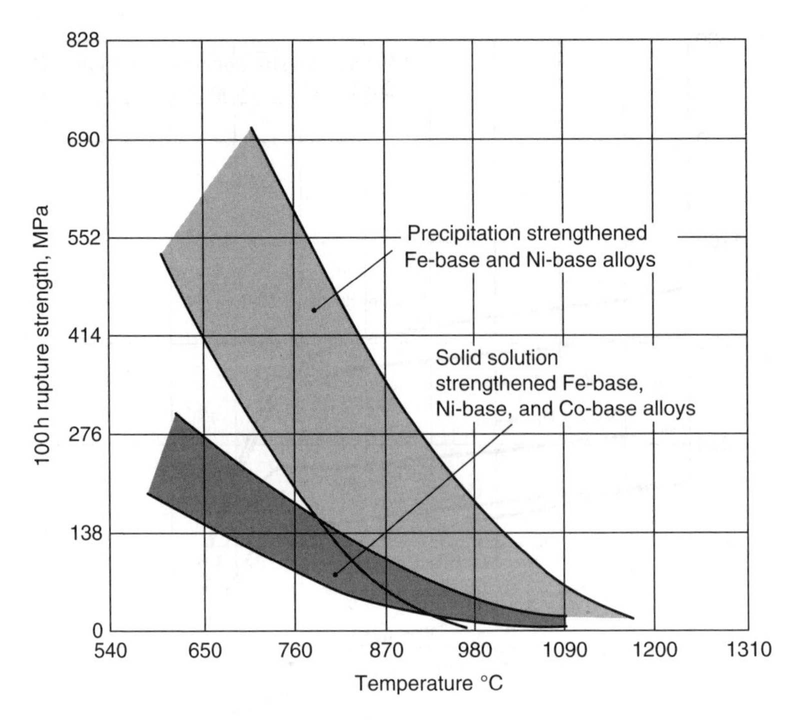

Figure 3.5 Effect of operating temperature on the stress rupture strength of super alloys. (Adapted from: Mathew *et al.* [6].)

of elevated-temperature applications, precipitation strengthened super alloys are preferred. Solid solution strengthened are preferred, when service conditions allow their use because of their ease of weldability and machinability. Figure 3.6 shows the yield strength at different operating temperatures of Ni-base super alloys.

Table 3.6 illustrates how the attractive properties of Ni-base alloys are achieved through the addition of the shown alloying elements [7]. Similar patterns are generally followed as previously illustrated in Section 3.1 for super alloys.

The features and specific application of the commonly used Ni-base alloys are given below:

1. ***Inconel 718*** is a recently developed precipitation hardenable Ni-base alloy, containing significant amounts of iron, niobium, and molybdenum, along with lesser amounts of aluminum and titanium, which is designed to display exceptionally high yield, and creep rupture properties at a temperature up to 700 °C [5]. It has excellent weldability as compared to Ni-base alloys hardened by Al and Ti. As such, it is typically used in applications where properties of temperature resistance, corrosion resistance, and loading resistance are most desired. Examples include plane engines, nuclear activation furnaces, rocket engines, furnaces, and military warships.
2. ***Inconel X-750*** is another Ni-base alloy, precipitation hardened by addition of Al and Ti, having creep rupture strength at high temperatures to about 700 °C. It is widely used for high temperature conditions but is not strong. It is used in nuclear reactors, gas turbines, rocket engines, pressure vessels, and aircraft structures.

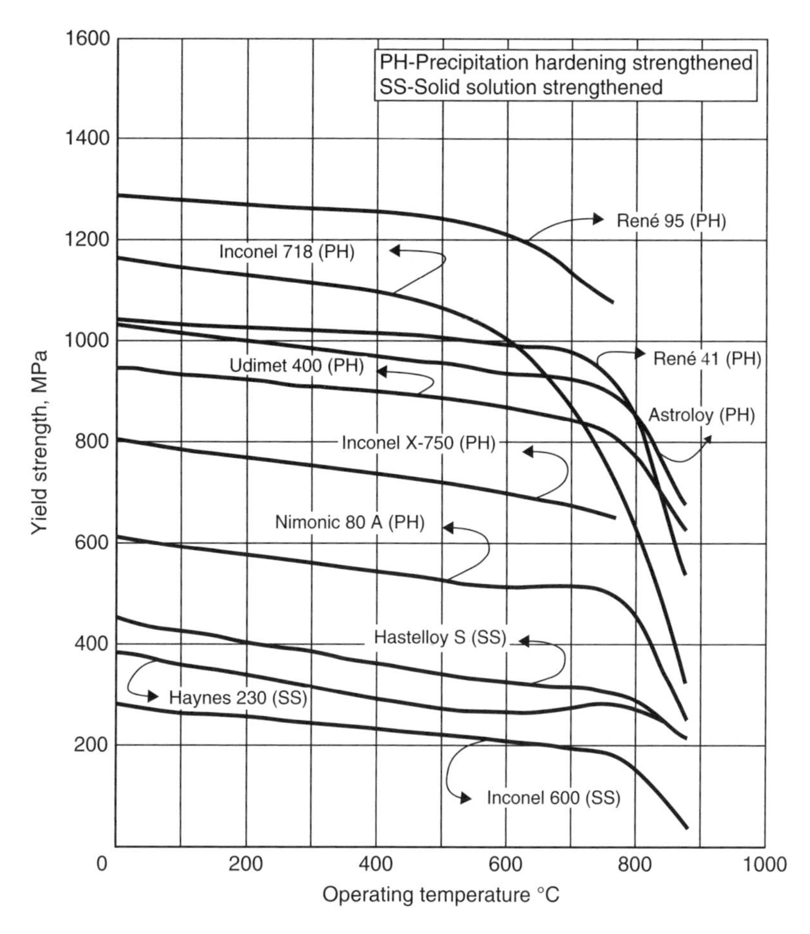

Figure 3.6 Effect of temperature on yield strength of Ni-base super alloys.

Table 3.6 Enhancement of properties of Ni-base super alloys through the addition of the shown elements

Addition of alloying elements	Leading to
Cr, Fe, Mo, W, Ta	Higher strength
Al, Ti	Higher temperature toughness
Al, Cr, Ta	Higher oxidation resistance
B, C, Zr	Higher creep resistance
Hf	For intermediate temperature ductility, and prevents oxide flaking

3. ***Inconel 706*** is a precipitation hardenable Ni-base alloy that provides high mechanical strength, in combination with good fabricability. It has characteristics similar to Inconel 718 except that it is more readily fabricated, particularly by machining. It is used in aerospace applications such as turbine and compressor discs, shafts, and casings.

4. ***Inconel 625*** is a solid solution strengthened, corrosion- and oxidation-resistant, Ni-based alloy of outstanding strength and toughness in the range from cryogenic to 1100 °C. This alloy has excellent fatigue strength and stress-corrosion cracking resistance to chloride ions. It is applicable for heat shields, furnace hardware, gas turbine ducting, chemical plant hardware, and seawater applications.
5. ***Inconel 600*** is a solid solution strengthened Ni-base alloy possessing excellent combination of high strength, hot and cold workability, and resistance to ordinary corrosion. It finds application in carburizing atmospheres, ethylene dichloride crackers, and furnace trays.
6. ***Hastelloy B*** is an additional member of Ni-Mo-family of Ni-base alloys. It is a solid solution strengthened alloy with excellent resistance to hydrochloric, sulfuric, and phosphoric acids at all concentrations and temperatures and other nonoxidizing media. It is intended for chemical processes, vacuum furnaces, and mechanical component in reducing environments.
7. ***Hastelloy S*** is a solid solution strengthened Ni-base alloy with outstanding thermal stability of moderate strength and very good oxidation resistance. It is applicable in gas turbines, seal rings, and casings.
8. ***Hastelloy X*** is a solid solution strengthened Ni-base alloy that possesses exceptional strength and oxidation resistance up to 1200 °C. It is found to be exceptionally resistant to stress cracking in petrochemical applications. The alloy has excellent forming and welding characteristics. It is recommended especially for use in furnace applications because it has unusual resistance to oxidizing, reducing, and neutral atmospheres. Furnace rolls made of this alloy were still in good condition after operation for 8700 hours, at 1180 °C. Hastelloy X has performed well in jet engine tailpipes, after burner components and other aircraft parts. It is excellent for subzero temperature application up to 480 °C [5, 8].
9. ***Hastelloy*** C-276 is a solid solution strengthened Ni-base alloy, which is balanced to provide excellent corrosion resistance to a variety of chemical process environments. It provides resistance to hot, contaminated mineral acids and organic and inorganic chloride contaminated media. It is used in chemical processing components, like heat exchangers, reaction vessels, evaporators, and transfer piping. It is also used in pulp and paper production, waste treatment, and pharmaceutical and food processing equipment.
10. ***Nimonic 90*** is a precipitation hardenable Ni-base alloy of extra-high mechanical properties along with corrosion resistance. It is used in aerospace industries, and in springs operating at high temperatures.
11. ***René 41*** is a high temperature, and high strength Ni-base alloy. It possesses good oxidation resistance at high temperatures up to 800 °C. It is a useful alloy in gas turbine, aircraft, and marine applications.
12. ***Waspalloy*** is a precipitation hardenable, Ni-base alloy, which possesses excellent corrosion resistance; it is used in elevated temperature applications (900 °C), for example, gas turbines, and aircraft jet equipment.
13. ***Haynes 282*** is a wrought strengthened Ni-base super alloy developed for high temperature structural applications in aero-engines and similar gas turbine systems. Examples include casings for compressors, combustors, and turbine sections, along with exhaust and nozzle parts. A key advantage of this alloy over other equivalent Ni-base super alloys, such as Waspaloy, is its unique combination of creep strength stress of 221 MPa to produce 1% creep in 100 hours at 816 °C, thermal stability (UTS (ultimate tensile strength) of 975 MPa after exposure at 870 °C for 1000 hours), and superior weldability.

Table 3.7 Mechanical properties of some common Ni-base super alloys at 870 °C

Trade name of super alloy	Alloy condition	UTS (MPa)	YTS (MPa)	Ratio YTS/UTS	Elongation % 50 mm
Astroloy	PH-wrought	770	690	0.90	25
Hastelloy X	SS-wrought	225	180	0.80	50
IN-100	Cast	885	695	0.79	6
Inconel 625	SS-wrought	285	275	0.97	125
Inconel 718	PH-wrought	340	330	0.97	88
MAR-M200	Cast	840	760	0.90	4
René 41	PH-wrought	620	550	0.89	19
Udimet 700	PH-wrought	690	635	0.92	27
Waspaloy	PH-wrought	525	515	0.98	35

SS = Solid Solution Strengthened and PH = Precipitation Hardened.
Adapted from Kalpakjian and Schmid [9].

Table 3.7 shows the mechanical properties of some very common Ni-base super alloys at elevated temperature (870 °C). It is remarkable that the ratio of yield tensile strength/ultimate tensile strength (YTS/UTS) attains very high value compared to steels and alloy steels. Such ratio is highly evaluated by the designers for applications requiring high strength at high operation temperatures, such as aircraft applications.

3.2.3 Co-Base Alloys

Co-base super alloys have their origin in satellite alloys patented in the early 1900s by Elwood Haynes. Cobalt is an austenitizing element like Ni. High cost tends to limit the use of Co-base alloys to gas turbine applications. They could be a substitute for Ni-base, depending on actual strength needs and type of corrosion a task expected. These super alloys are not as strong as Ni-base alloys, but they display superior hot corrosion resistance, and retain their strength at higher temperatures as compared to Ni-base alloys. Co-alloys show superior thermal fatigue resistance and weldability over Ni-base alloys. Co-base alloys are hardened by carbide precipitation; thus carbon content is critical. Cr provides hot corrosion resistance and other refractory metals (W, Mo, Ta, Nb, Zr, and Hf) are added to give solid solution strengthening.

Compared to Ni-base alloys, the stress rupture curve for Co-alloys is flatter and shows lower strength up to about 930 °C (Figure 3.5). Casting is important for Co-base alloys, and directionally solidified alloys (DS) have led to increased rupture strength and thermal fatigue resistance. Even further improvements in strength and temperature resistance have been achieved by the development of single crystal alloys.

Co-base super alloys generally contain 35–67% Co, 19–30% Cr, and up to 35% Ni. Common alloys of this group include Haynes, and AiResist series. Co-base alloys may be used in lieu of Ni-base alloys, depending on actual strength needs of the component and its corrosion behavior expected. Figure 3.7 visualizes the yield strength at different operation

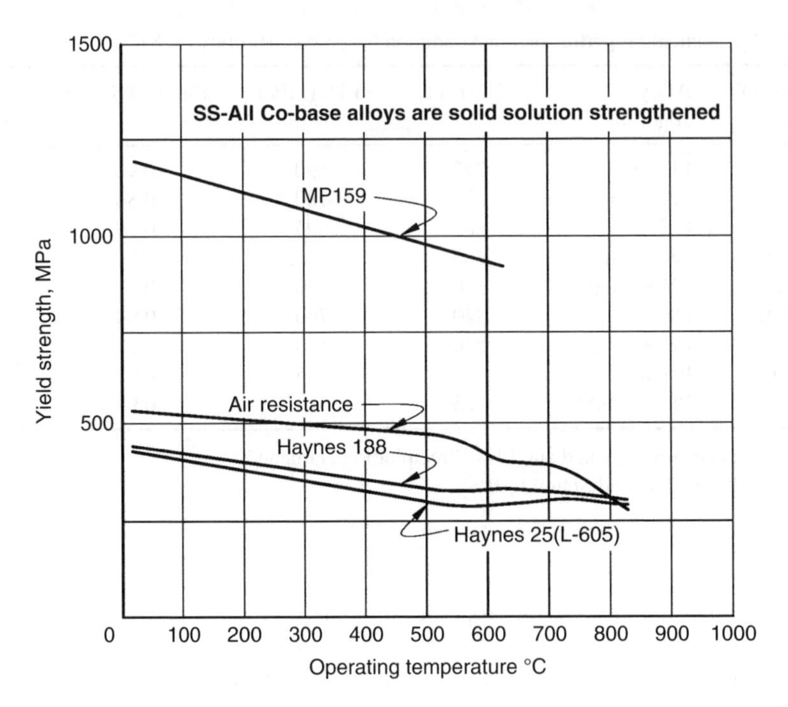

Figure 3.7 Effect of temperature on yield strength of Co-base super-alloys.

temperatures of Co-base super alloys. The features and applications of some selected Co-base alloys are given below:

1. ***Haynes 25 (L-605)*** combines good formability and excellent high temperature properties. This alloy is resistant to oxidation and carburization up to 1050 °C. Haynes 25 has a good service in many jet engine parts. Some include turbine blades, combustion chambers, after burner, and turbine rings. It has been successfully used in variety of industrial furnace applications.
2. ***Haynes 188*** is also solid solution strengthened. It is characterized by excellent high temperature strength, good resistance of oxidizing environment up to 1150 °C, excellent resistance of chloride, and can be rapidly hardened. It is successfully used in military and commercial gas turbines.
3. ***AiResist 213*** is a cast Co-base alloy, which exhibits good castability and excellent resistance to thermal fatigue and oxidation.
4. ***MAR-302*** is also a cast Co-base alloy, which is frequently used as nozzle guidance material in advanced engines.

References

[1] Davis, J.R. (2000) *ASM Speciality HB, Nickel, Cobalt, and Their Alloys*. Davis and Associates.
[2] Sandvik-Coromat(2013) Workpiece Materials – ISO S HRSA and Titanium.
[3] Degarmo, E.P., Black, JT, Kohser, R.A. (1997) *Materials and Processes in Manufacturing*, 8th edn, Prentice Hall Inc., Upper Saddle Rivers, NJ.
[4] IQS Directory Industrial Quick Search http://www.Iqsdirectory.com/alloys/superalloys/ (accessed December 10, 2013).
[5] High Temp Metals www.hightempmetals.com (accessed April 17, 2015).
[6] Mathew, J., Denachio J., Donachie, S.J. (2002) *Super Alloys- A Technical Guide* ASM International, 2nd edn.
[7] Seco (2014) Technical Guide. Turning Difficult-to-Cut Alloys.
[8] MEGA MEX www.megamexcom/monel-500-nickel-alloy.htm.
[9] Kalpakjian, S., Schmid, S.R. (2003) *Manufacturing Processes for Engineering Materials*, 4th edn, Prentice Hall, Inc., Upper Saddle River, NJ.

4

Traditional Machining – Machinability, Tooling, and Cutting Fluids

4.1 Machinability Concept in Metal Cutting

4.1.1 Definition and General Aspects

The machinability of a material is similar to the palatability of wine – easily appreciated but not readily measured in quantitative terms. Hence, it is not an absolute property of a material, but a mode of behavior of the material during cutting. It is a term that was suggested for the first time by Taylor in the 1920s to describe the machining properties of workpiece materials. Since that time, it has frequently been used, but seldom fully explained, as it has a variety of interpretations depending upon the viewpoint of the person using it. Machinability is the relative susceptibility of a material to the machining process. In its broadest interpretation, a material of good machinability requires lower power consumption, with longer tool life, and achieving a good surface finish. The relative importance of three factors (tool life, power consumption, and surface finish) depends mainly on whether the machining is roughing or finishing. In actual production, tool life for rough cuts and surface finish for finish cuts are generally considered to be the most important criteria of machinability (Table 4.1).

An additional machinability criterion sometimes to be highly considered is the chip disposal criterion. Long thin ribbon chips, unless being broken up with chip breakers, can interfere with the operation leading to hazardous cutting area. This criterion is of vital importance in automatic machine tool operation. Chip formation, friction at the tool/chip interface, and BUE (built-up edge) phenomenon determine machinability. A ductile material that has a tendency to adhere to the tool face or to form BUE is likely to produce a poor finish. This has been observed to be true with such materials as low carbon steel, pure aluminum, Cu, and stainless steels (SSs) [1].

Machining of Stainless Steels and Super Alloys: Traditional and Nontraditional Techniques, First Edition. Helmi A. Youssef.

Table 4.1 Relative importance of machinability criterion in roughing and finishing in actual production

Order of machinability criterion	Rough cut	Finish cut
1	Tool life	Surface finish
2	Power consumption	Tool life
3	Surface finish	Power consumption

Mechanical and physical properties also play a role in the magnitude of energy consumption and temperatures generated during cutting. For instance, Ti is not machinable, partly because of the high temperature generated due to its poor thermal conductivity, and partly because of its tendency to adhere to the cutting tool.

Moreover, the machinability is not an absolute material characteristic. It is also more or less related to the selected machining process. A material that is machinable by a certain process may not be machinable by another process. A particular machining process found suitable under given conditions may not be equally efficient for machining the same material under other conditions [2].

From the foregoing, the input parameters affecting machinability (Figure 4.1), are:

1. work material properties and previous history such as modulus of elasticity, yield strength, tensile strength, hardness, work hardening, microstructure, grain size, heat treatment, chemical composition, fabrication, thermal conductivity, and thermal expansion;
2. tool material and geometry;
3. type of cutting operation;
4. machine tool power, rigidity, and accuracy;
5. machining conditions;
6. type and quantity of cutting fluid.

4.1.2 Quantifying and Criteria of Machinability

Work on machinability is needed to bridge the gap between, the material development and manufacturing capabilities before introducing these materials for industrial applications [3]. The usefulness of a means to predict machinability is obvious. Due to the above described complex aspects of machinability, it is really difficult to establish quantitative relationship to evaluate the machinability of a material. The machinability is often assessed on a case-by-case basis, and tests are tailored to the needs of a specific machining operation. It is really an elaborate task to express the machinability quantitatively. For this reason, it is advisable to refer to machining recommendations that are based on extensive testing, practical experience, data collected in manufacturing manuals, and specialized handbooks.

Since there is no unit of machinability, it is usually assessed by comparing one material against another which is taken as a reference. The AISI (American Iron and Steel Institute) determined machinability ratings for a wide variety of materials by running tuning tests at

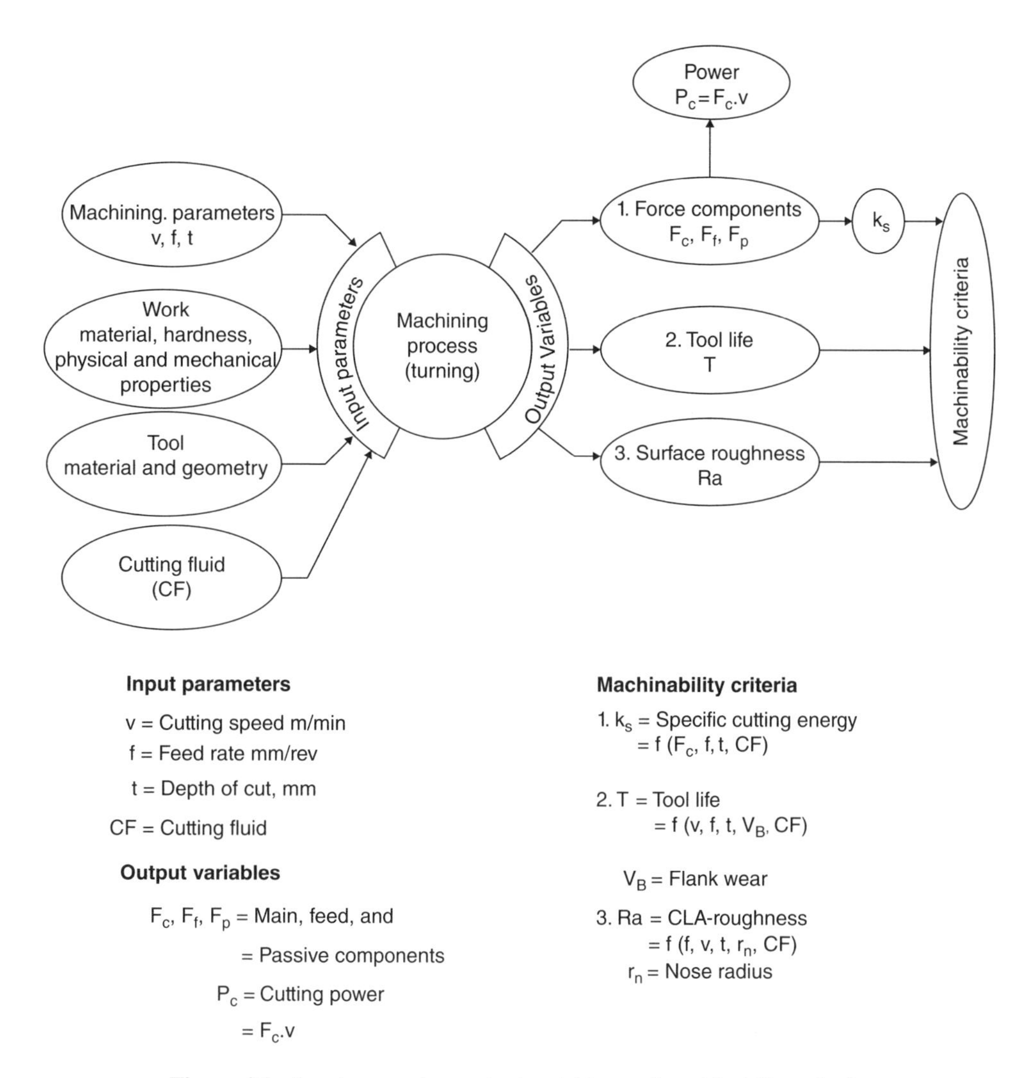

Figure 4.1 Input parameters, output variables, and machinability criteria.

180 sfpm. An arbitrary machinability rating of 100% for a reference material was then assigned. The reference material was AISI-B1112 (resulfurized plain carbon free-machining steel of 160 BHN). The machinability rating is determined by measuring the weighed averages of consumed power, surface finish, cutting speed, and tool life for each material. A material with a machinability rating less than 100% will be more difficult to machine than B1112 and that with a value of more than 100% would be easier to machine.

Table 4.2 lists the relative machinability of some common ferrous and nonferrous alloys in a descending order. The problem associated here is that if different tool materials are used to assess relative machinability, different ratings may occur. Thus tables and data supplied should only be used as guidelines.

Table 4.2 Relative machinability rating for different materials

Machinability rating	Materials
Excellent rating	Mg-alloys, Al-alloys, duralumin
Good rating	Zn-alloys, gunmetal, gray CI, brass, free cutting steel
Fair rating	Low carbon steel, cast Cu, annealed Ni, low alloy steel
Poor rating	Ingot iron, free cutting 18-8 stainless steel
Very poor rating	HSS, 18-8 stainless steel, monel, super alloys
Not machinable	White CI, Stellite, carbides, ceramics

Adapted from Youssef and El-Hofy [1].

4.1.2.1 Tool Life Criterion

Machinability can be based on the criterion of how long a tool lasts (Figure 4.1). This can be useful when comparing machinability of materials that have similar properties and power consumption, but one or more of them are more abrasive, thus decreasing the tool life. The major disadvantage of this approach is that the tool life is dependent on more factors than just the material to be machined. Other factors include cutting tool materials, tool geometry, machining condition, cutting tool clamping, cutting speed, feed rate, depth of cut. The machinability of a material for one tool type cannot be compared to another tool type (i.e., HSS (high speed steel) tool to carbide tool).

For roughing operations, the tool life T (min) or the cutting speed v (m/min) are taken as a yardstick for ranking the material machinability. Both T and v are correlated together in Taylor Equation (4.1) as:

$$vT^n = C \tag{4.1}$$

where

C	=	Taylor constant
n	=	Taylor exponent.

Suppose that, when turning the reference material AISI-B1112, the tool life is 60 min at a cutting speed of 90 sfpm (30 m/min). For the same tool life, if a material has a machinability rating of 70%, then according to the Taylor equation, and assuming the same tool, and otherwise machining parameters, the cutting speed must be 63 sfpm (21 m/min) to realize the same tool life. Figure 4.2a, shows the Taylor relationships for an unknown and a reference material AISI-B1112. These relationships are straight line relations when represented on a double logarithmic coordinate system. The relative machinability R_T is the ratio of the cutting speed of the unknown material to that of the reference material providing the same tool life (T = 60 min), then,

$$R_T = \frac{(v60)_1}{(v60)_2} = \frac{\text{cutting speed of the unknown materal}}{\text{cutting speed of the reference material}} \tag{4.2}$$

To illustrate how a machinability rating might be determined, a series of tool life tests are conducted on two work materials under identical cutting conditions, varying only speed in the test procedure. The first material, defined as the base material, yields a Taylor tool life equation

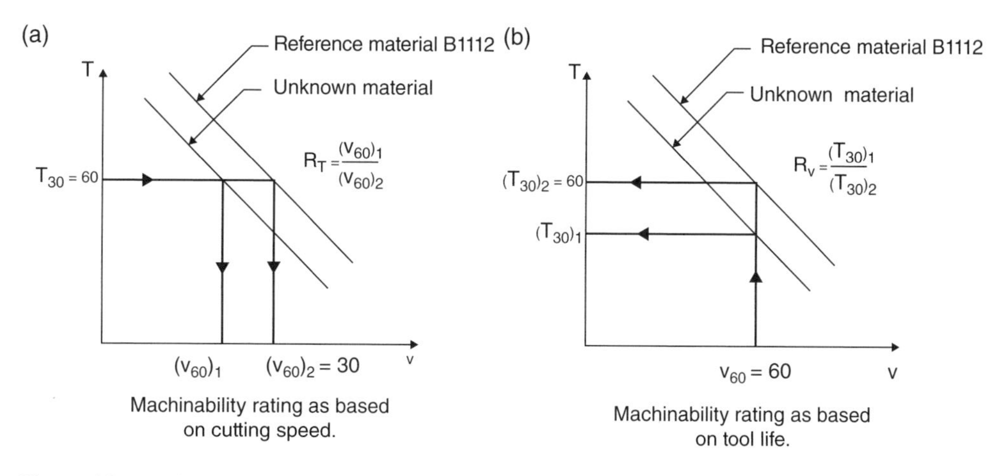

Figure 4.2 Taylor relationships for an unknown and a reference material AISI-B1112. (a) Machinability rating as based on cutting speed. (b) Machinability rating as based on tool life

$vT^{0.28} = 350$, and the other material (test material) yields a Taylor equation $vT^{0.27} = 440$, where speed is in m/min and tool life is in min. The machinability rating of the test material using the cutting speed that provides a 60-min tool life as the basis of comparison. This speed is denoted by v60. The base material has a machinability rating = 1.0. Its v60 value can be determined from the Taylor tool life equation as follows:

$$v60 = (350/60\wedge 0.28) = 111\,\text{m/min}$$

The cutting speed at a 60-min tool life for the test material is determined similarly:

$$v60 = (440/60\wedge 0.27) = 146\,\text{m/min}$$

Accordingly, the machinability rating can be calculated as:

$$R_T\,(\text{for the test material}) = 146/111 = 1.31\,(131\%)$$

On the other hand, the relative machinability R_v of a material, can be expressed as the ratio of the tool life of that material to that of the reference material provided the same cutting speed (v = 30 m/min) (Figure 4.2b), then,

$$R_v = \frac{(T30)_1}{(T30)_2} = \frac{\text{tool life of the unknown material}}{\text{tool life of the reference material}} \tag{4.3}$$

Machining performance is affected by the type of work material. Important mechanical properties include hardness and strength. As hardness increases, abrasive wear of the tool increases so that tool life is reduced. Strength is usually indicated as tensile strength, even though machining involves shear stresses. Of course, shear strength and tensile strength are correlated.

As work material strength increases, cutting forces, specific energy, and cutting temperature increase, making the material more difficult to machine. On the other hand, very low hardness can be detrimental to machining performance. For example, low carbon steel, which has relatively low hardness, is often too ductile to machine well. High ductility causes tearing of the metal as the chip is formed, resulting in poor finish, and problems with chip disposal.

A metal's chemistry has an important effect on properties; and in some cases, chemistry affects the wear mechanisms that act on the tool material. Through these relationships, chemistry affects machinability. Carbon content has a significant effect on the properties of steel. As carbon is increased, the strength and hardness of the steel increase; this reduces machining performance. Many alloying elements added to steel to enhance properties are detrimental to machinability. Chromium, molybdenum, and tungsten form carbides in steel, which increase tool wear and reduce machinability. Mn and Ni add strength and toughness to steel, which reduce machinability. Certain elements can be added to steel to improve machining performance, such as lead, sulfur, and phosphorus. The additives have the effect of reducing the coefficient of friction between the tool and chip, thereby reducing forces, temperature, and BUE formation. Better tool life and surface finish result from these effects. Steel alloys formulated to improve machinability are referred to as free machining steels [4].

4.1.2.2 Cutting Forces and Power Consumption Criterion

The forces required to machine a material are directly related to the consumed power (Figure 4.1). Therefore, the tool forces are often expressed in units of specific cutting energy. This leads to a rating method where higher specific energies mean lower machinability of a material. The advantage of this method is that other factors have a little effect on the rating. Moreover, it takes less time, however, it needs sophisticated apparatus (dynamometers) and setup.

The specific cutting energy k_s is defined as the power Pc (energy/time) to cut a unit volume of a material per unit time Z. Assuming turning operation, then:

$$k_s = \frac{P_c}{Z} = \frac{F_c v}{tfv} = \frac{F_c}{tf} \ \mathrm{N/mm^2} \tag{4.4}$$

F_c is the main cutting force (measured directly by a dynamometer).

Since k_s is expressed in N/mm^2, then it is often called the specific cutting resistance. This parameter depends on the type of work material. However, it is not a material constant, it depends also upon cutting conditions, especially the undeformed chip thickness h which is proportional to the feed rate f (Figure 4.3). Kienzle and Victor [5], expressed the relationship between k_s and h empirically by the exponential Equation 4.5, and Figure 4.4a, which is valid for both orthogonal and oblique cutting.

$$k_s = Ch^{-z} \tag{4.5}$$

When plotted on log–log co-ordinates, it represents a straight line relation (Figure 4.4b). Both equations are shown schematically in Figure 4.4. Through k_s, the machinability of a material, which is machined under given working conditions is fairly estimated.

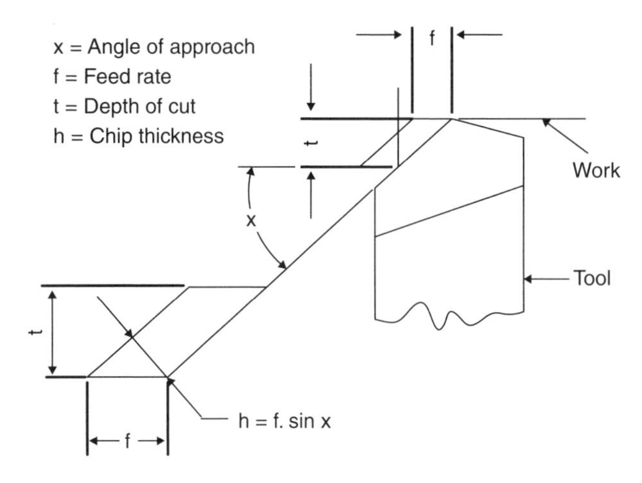

Figure 4.3 Undeformed chip cross-section area in turning

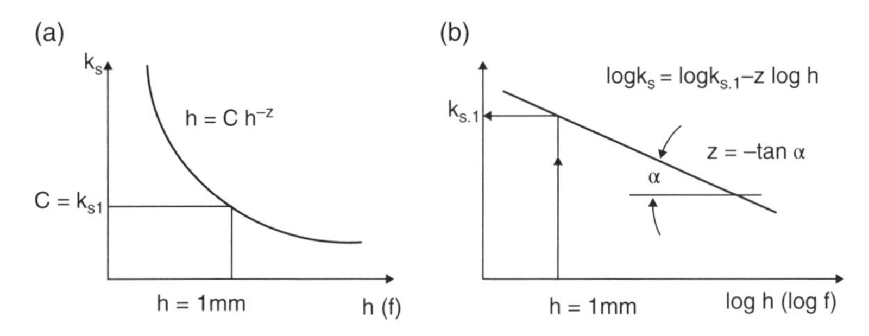

Figure 4.4 Empirical relationship between k_s and h [5], (a) ks–h relationship and (b) log ks–log h straight line relationship

The selection of process parameters can have a significant influence on the specific cutting energy, which reduces by setting the cutting conditions (feed rate, cutting speed, depth of cut) high, thereby shortening the machining time, yet within a value range which does not compromise tool life and surface finish. Figure 4.4a shows the specific energy demand for machining processes as a function of chip thickness, and consequently metal removal rate.

The specific cutting energy can be determined, however, indirectly using a wattmeter, thus avoiding the use of sophisticated setup and force dynamometers. The consumed power in cutting P_c (W) is determined by subtracting the value of the input electrical power P_1 (W) during idle running of the machine from its value P_2 (W) during cutting (Equation 4.6):

$$\begin{aligned} P_c &= P_2 - P_1 \\ &= \frac{F_c.v}{60} \\ &= \frac{k_s.f.t.v}{60} \end{aligned} \tag{4.6}$$

$$\text{Then,} \quad k_s = \frac{60(P_2 - P_1)_\eta}{f.t.v} \ N/mm^2 \tag{4.7}$$

where

v	=	cutting speed (m/min)
f	=	feed rate (mm/rev)
t	=	depth of cut (mm)
η	=	overall efficiency of the machine tool.

This method provides a reasonable estimate of the specific cutting energy k_s, and hence the machinability of a material. However, it is hampered by a drawback, because the drive efficiency under varying loads cannot be accurately estimated.

4.1.2.3 Surface Finish Criterion

The surface finish is sometimes used to measure the machinability of a material (Figure 4.1). Soft ductile materials tend to form a BUE. Stainless steels, super alloys, and other materials with a high strain hardening ability tend to form BUE. Al-alloys, cold worked steels, and free-machining steels do not tend to form BUEs, so these would rank as more machinable. The advantage of this criterion is that it is measured with appropriate equipment. Its disadvantage is that it is sometimes irrelevant. For instance, when making a rough cut, the surface finish is of no importance. This rating method also does not always agree with other methods. Ti-alloys would rate well by this criterion, low by the tool life criterion, and intermediate by the power consumption criterion [6]. This criterion is mainly used in finishing operations, where higher cutting speeds, lower feed rates, and depths of cut are used. Moreover, in finishing, broad nose tools, and suitable cutting fluids are recommended. Less tool wear as compared to roughing should be allowed.

4.1.3 Enhancing Machinability of Difficult-to-Cut Materials

Many techniques have been developed for improving the machinability of difficult-to-cut (DTC) materials, for example, hot machining, ultrasonic-assisted machining (UAM), high speed machining (HSM), using advanced cooling technique, cryogenic treated tooling, and adoption of free-machining alloys (Figure 4.5).

4.1.3.1 Adoption of Free Machining Steels and Alloys

1. *Free machining steels*
   Vast quantities of steels are machined and efforts are directed at improving their machinability mainly by adding lead (leaded steels), sulfur (sulfurized steels), and phosphorus (phosphorized steels) to obtain the so called free-machining steels. These additions produce films of low shear strength, and thus reduce the friction in the secondary shear zone at the tool–chip interface.

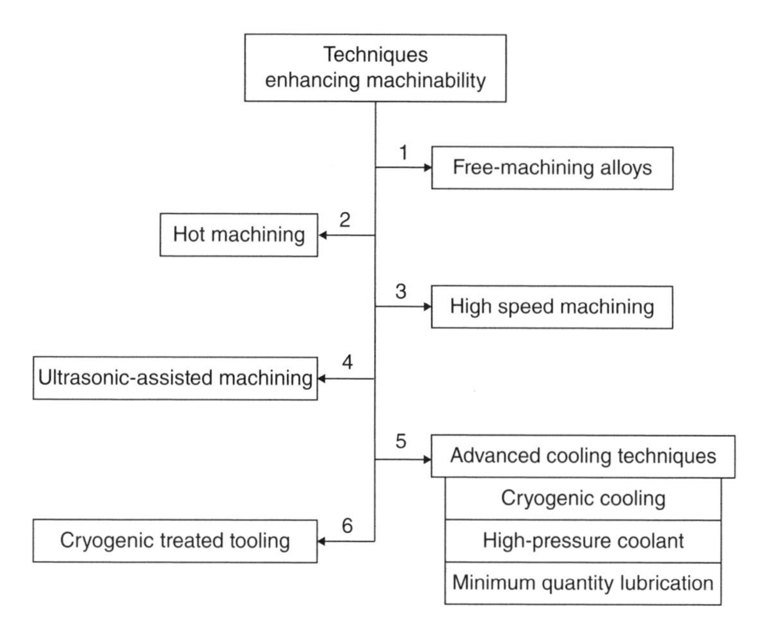

Figure 4.5 Techniques for enhancing the machinability of difficult-to-cut-materials

(a) *Leaded steels*: Lead is added to molten steels and takes the form of dispersed fine lead particles. Lead is insoluble in iron, copper, aluminum, and their alloys. Thus during cutting, lead particles are sheared and smeared over the tool–chip interface, acting as a solid lubricant. It is also believed that lead probably lowers the shear stress in the primary shear zone, thus reducing the cutting forces and power consumption. Because of environmental concerns, the trend now is toward eliminating the use of leaded steels in favor of bismuth and tin (lead-free steels). Leaded steels are identified by the letter L between the second and third numerals of AISI-identification system, for example, 10L45.

(b) *Resulfurized and rephosphorized steels*: Increased sulfur content (resulfurized steels) forms MnS inclusions of controlled, globular shape, which act as stress raisers in the primary shear zone. As a result, the chips produced are small and break up easily, thus improving machinability. An undesirable consequence is reduced ductility, and fatigue strength and slightly reduced tensile strength. Sulfur, however, can severely reduce the machinability of steels because of the presence of iron sulfide, *unless* sufficient Mn is present to prevent the formation of iron sulfide. Phosphorus in steels also improves machinability by increasing their hardness. Rephosphorized steels are significantly less ductile than resulfurized steels.

(c) *Calcium-deoxidized steels*: In these steels, flakes of calcium aluminosilicate (CaO, SiO_2, and Al_2O_3) are formed; therefore the crater wear of cutting tools, especially at high cutting speeds, can be reduced without impairing the mechanical properties of such steels.

2. *Free-machining stainless steels*

The higher strength and lower thermal conductivity of stainless steel result in higher cutting temperatures. The high strain-hardening rate of austenitic stainless steels (AISI 300 series) makes them more difficult to machine. Chatter could be a problem, which necessitates the use of rigid machine tools with high stiffness and damping capacity. However ferritic stainless steels have good machinability (AISI 400 series). Martensitic steels (also AISI 400 series) are abrasive, tend to form BUE, and require tool material with high hot hardness and resistant to crater wear. Precipitation hardening stainless steels are strong and abrasive and thus require hard, abrasive-resistant tool materials.

If necessary, free machining properties can be imparted using alloying elements such as sulfur, phosphorus, selenium, tellurium, lead, and bismuth to enhance machinability. These grades have significantly lower corrosion resistance and they are particularly prone to pitting corrosion attack compared to their standard version. The machining of stainless steels will be presented in detail in Chapter 5.

4.1.3.2 Thermally Assisted Machining (Hot Machining)

Tigham first innovated the process of hot machining in 1889, with the idea of softening the workpiece material whereby part or all of the workpiece is heated. Heating is performed before or during the machining. Researchers utilized various methods for heating the workpiece [7]. The hot machining process prevents cold work-hardening by heating the workpiece above the recrystallization temperature, thereby reducing the specific cutting resistance and hence improving machinability. The limiting highest temperature should be the recrystallization temperature of the workpiece material, as higher heating temperatures may induce unwanted structural changes in the workpiece and even increase the cost of heating. In hot machining, it was observed that the chip thickness ratio increases with the increase in temperature. Hence the machinability of the material improves with increase in temperature and the chip produced at high temperature is of the continuous type, whereas it was of the discontinuous type when machining conventionally. The effect of the temperature of the workpiece is clearly found to be the most significant on tool life [8,9].

The basic concept of thermally-assisted machining (TAM) is the addition of significant amounts of heat to the workpiece immediately prior to the cutting tool so that the material is softened, but the strength of the tool bit is unimpaired (Figure 4.6). A steep temperature gradient is necessary, and a high energy density of the heating source is of prime importance. Two mainly used heat sources are the plasma arc and laser beam of energy densities of 1×10^4 and 1×10^7 W/m m^2, respectively. The plasma arc has a core temperature of 8000°C, and a surface temperature of 3600°C. It can produce 1100°C in the workpiece material in approximately one-quarter of the circumference of the workpiece between the point of its application and the cutting tool. A warm surface is not helpful. Thin-walled workpieces heat more highly than thick-walled ones.

4.1.3.3 High Speed Machining

Concept and Definition

The concept of HSM was introduced by Salomon (German Patent: 523594, 1931) during tests performed on milling of nonferrous metals using high cutting speeds up to 16 500 m/min. Based on his observations, the cutting temperature reached a peak (T_p) at a given cutting speed (v_{cr}). However, according to his model, as the cutting speed was further increased

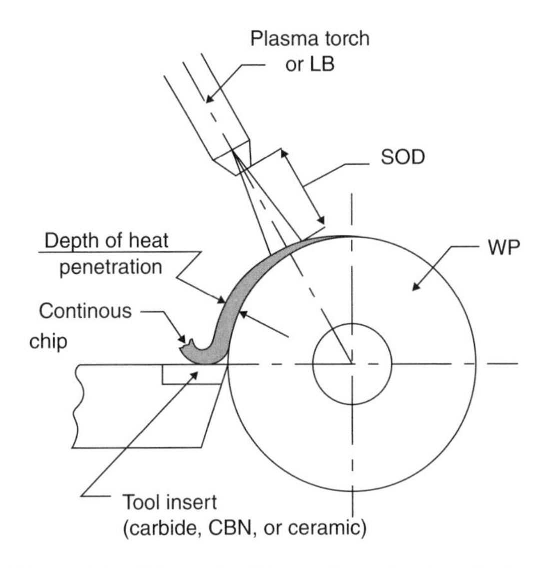

Figure 4.6 Schematic of thermally-assisted machining.

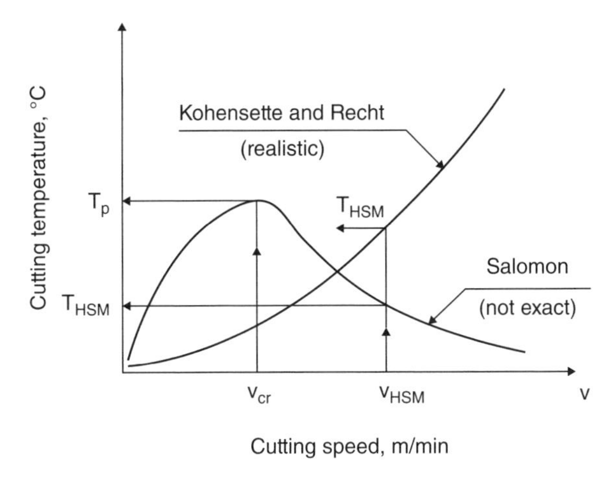

Figure 4.7 Dependence of the cutting temperature on the cutting speed in HSM. According to models of Salomon, and Kottenstette and Recht [10].

(v_{HSM}), the temperature decreased (T_{HSM}) (Figure 4.7). The evidence indicated that the falling trend is not true.

Instead, the temperature increases with the speed, approaching the melting point of the work material, rather than falling off at very high speeds as had been claimed by Salomon (Figure 4.7) [11].

With increasing demands for higher productivity and lower machining cost, more investigations have been carried out since the late of 1950s to increase the metal removal rate (MRR), particularly for applications in the aerospace and automotive industries. HSM for a given material can be defined as that speed above which shear localization develops completely in the primary deformation zone. As a general guide, an acceptable suggestion has been developed by Turkovich [12] as follows:

High speed: 600–1800 m/min
Very high speed: 1800–18 000 m/min
Ultra high speed: above 18 000 m/min.

The studies conducted by Salomon are now of historical interest, since current research is developing more definitive data using more sophisticated techniques. In addition, Salomon's interpretations were responsible for confusion and false expectations concerning HSM. The first systematic investigations on HSM were undertaken by Vaughn at Lockheed Aircraft Corporation [13–15]. In these research works, Vaughn studied a series of variables that became very important for the MRR in HSM such as:

- type and size of the machine;
- power available;
- cutting tool;
- machining parameters (speed, feed, and depth of cut).

Recent advances in computer control systems have provided the capability of accurately manipulating high performance, high speed automatic machine tools. Progress in the field of bearing design, automatic tool changing, and cutting tool materials made contribution toward developing machine tools capable of efficiently machining DTC materials and Al-alloys under conditions of HSM.

In 1970s, tests are performed by the US-Navy with Lockheed Missiles and Space Co. to explore the feasibility of using HSM in production environment, initially with Al-alloys, and later with Ni-Al-bronze, in order to realize a major improvement in production. In the late 1970s and early 1980s, General Electric Company presented a database for machining Al-alloys, Ti-alloys, Ni-base super alloys, and steels [16].

The main purpose of HSM is to increase the metal removal rate (MRR), while keeping the quality of the machined parts regarding the accuracy and surface finish unaffected, or even improved. As previously mentioned, the MRR in turning Z is the product of cutting speed v, feed rate f, and depth of cut t. Therefore, an increase of productivity is affected by increasing of the three parameters v, f, and t. However, increment of f and t is not recommended when machining DTC materials such as stainless steels and super alloys, because increment of f leads to bad surface finish, and increment of both f and t (f.t = cross-sectional area of cut) leads to considerable increase of cutting forces and hence to the deterioration of product accuracy, due to the large deflections that occur in workpiece and cutting tool. Moreover, this increases the risk of catastrophic failure of the cutting tool, demanding more power consumption, and necessitating stiff and rigid machine tools. Machining with higher feed rates and depths of cut may be applicable when machining relatively soft materials such as A1, soft alloys, or low carbon steels.

Chip Morphology in HSM

Depending on the type of work material, and its physical, mechanical, and metallurgical characteristics, there are two types of chips that have been observed in HSM. These are the continuous chips and the shear localized (segmented) chips (Figure 4.8a,b).

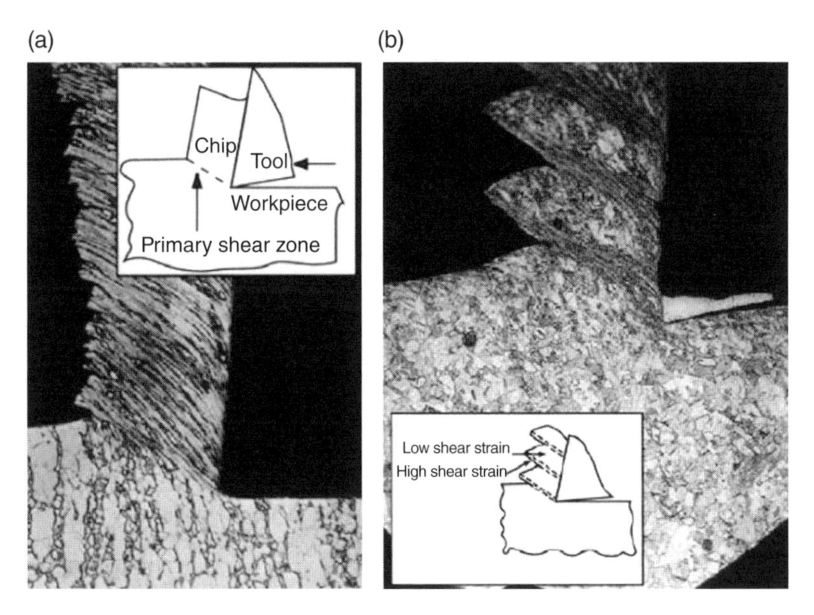

Figure 4.8 (a) Continuous and (b) shear localized (segmented) chips. (Adapted from Kalpakjian *et al.* [17] and Flom [18].)

1. *Continuous chips*
   These are likely to form during HSM of metals and alloys of bcc/fcc structures, of high thermal diffusivity, and low hardness such as Al-alloys, and soft carbon steels (Figure 4.8a); the deformation of the material takes place along a primary shear zone. Although they generally produce good surface finish, continuous chips are not always desirable, particularly in automatics, because they tend to get tangled around the tools; such problem can be alleviated with chip breakers.

   As a result of strain hardening caused by the shear strains, the chip becomes usually harder, less ductile, and stronger than the original workpiece material. As the rake angle decreases, the shear strain increases, and consequently the chip becomes stronger and harder. In this case, the chip tends to behave more like a rigid plastic material [17].
2. *Shear localized (segmented) chips*
   Segmented chips, also called serrated or nonhomogeneous chips, are semicontinuous chips, with zones of low and high shear strain. These occur with such materials as Ti-alloys, Ni-base super alloys, stainless steels, and hardened alloy steels, which are mainly characterized by low thermal diffusivity, hexagonal close-packed crystal structures, and high hardness. This type has a saw-tooth-shaped (serrated) outer profile (Figure 4.8b), consisting [17] of adjacent regions of heavily strained and lightly strained material. This results from the occurrence of rapid plastic (adiabatic) instability taking place in the primary shear zone. This might be attributed, that materials with low thermal diffusivity, previously mentioned, will be more likely to machine with chips containing regions of catastrophic shear than those with a higher thermal diffusivity such as low carbon steel and Al-alloys.

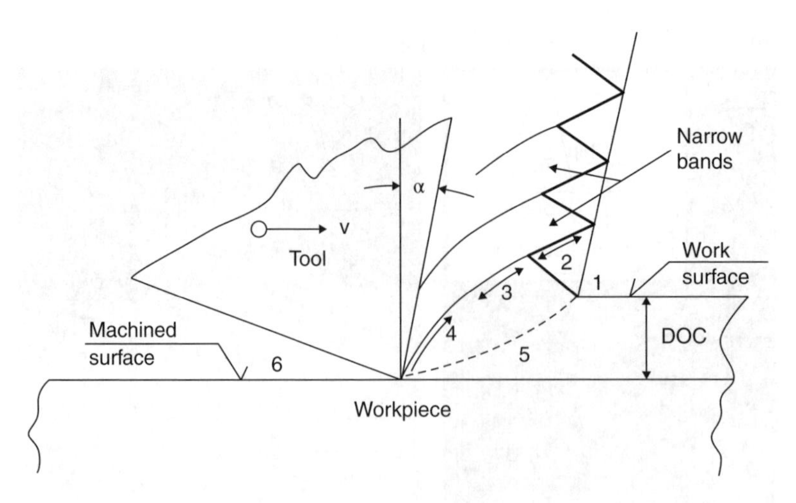

Figure 4.9 Komanduri model [20] of shear localized chip formation. (Adapted from Komanduri [20].). (1) Undeformed surface, (2) part of catastrophically shear-failed surface separated from the following segment owing to intense shear, (3) intense shear band formed by catastrophic shear during upsetting stage, (4) intense sheared surface of segment sliding on the tool, (5) intense localized deformation in the shear zone, and (6) machined surface.

The serrations in Figure 4.8b are inherent to the metallurgical features of the work material and they are caused by the varying strength of the material at the tool–chip interface caused by the occurrence of stick-slip conditions [19].

In his model Komanduri [20] divided the mechanism of chip formation into two stages (Figure 4.9). The first involved plastic instability and strain localization in a narrow band in the primary shear zone leading to catastrophic shear failure along a plane in this zone. The second stage involved the gradual build-up of the segment being formed by an upsetting type of process with negligible deformation of the work material. As the upsetting of the segment progressed, the build-up of stresses caused intense shear between this segment and the one before it, thus repeating the first stage, and so on. No attempt, however, was made to check this model in a quantitative sense by comparing, for example, experimental chip velocities and chip geometry with values given by the model.

Once shear localized chips are formed above a certain speed, they persisted with increases in speed. No further transition into different chip forms occurred, at least up to approximately 30 000 m/min [16]. Because they are easier to dispose of, shear-localized chips are preferable to continuous chips, especially at higher speeds, where individual segments of a chip are completely isolated. Formation of shear localized chip, however, has not been accompanied by rapid reduction of tool wear at high speeds.

With several metals and alloys, the degree of segmentation directly depends on the cutting speed. An example is the AISI 4340, for which continuous chips are formed at 120 m/min (Figure 4.10a). At 975 m/min, however, completely sequenced and detached chips are formed (Figure 4.10b). Similarly, Inconel 718, forms relatively continuous chips below 60 m/min, but within the range 60–120 m/min segmentation begins. At higher speeds, severe detachment occurs, Titanium alloys such as Ti-6Al-4V are unique in that they form segmented chips at all speeds, regardless of their heat treatment conditions.

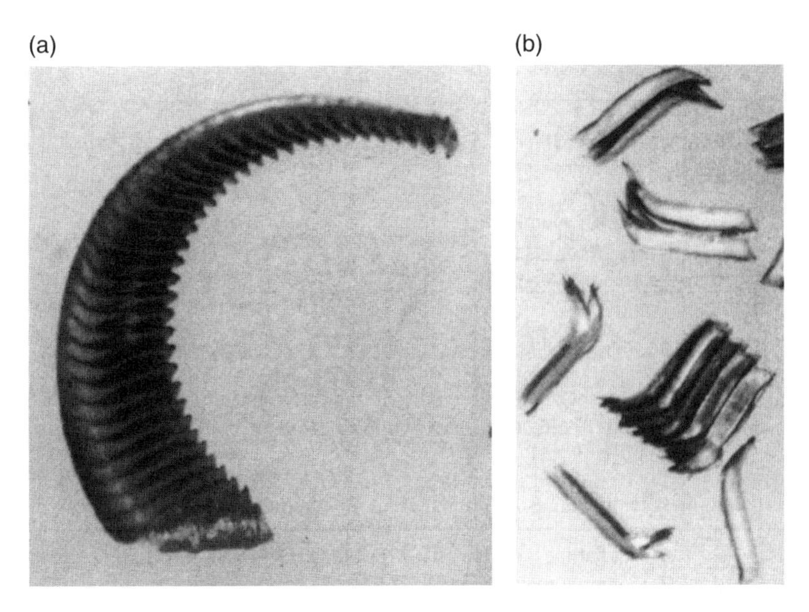

Figure 4.10 Effect of cutting speed on chip morphology of AISI 4340, (a) cutting speed 120 m/min and (b) cutting speed 975 m/min. (Adapted from Flom [18].)

Effects of Cutting Speed on Output Variables in HSM

The range of cutting speeds for machining with acceptable tool life is very broad, depending on the material to be machined. Aluminum can be machined using cutting speeds in the range from 200 to 1000 m/min, and stainless steels are machined at cutting speeds from 50 to 300 m/min, whereas super alloys cutting speeds from 20 to 400 m/min are used [16]. These speeds are valid for roughing (cutting without much consideration to dimensional tolerances and surface finish), und finishing (cutting to obtain final dimensions with acceptable tolerances and surface finish), using carbide and ceramic tooling. Speeds for HSS tools are lower than these ranges. Higher ranges are recommended for coated carbides and cermets. Speeds for diamond tools are significantly higher than any of the upper ranges. For the above conditions, the depth of cut ranges from 0.5 to 12 mm, while the feed rates are ranging from 0.15 to 1 mm/rev. It is claimed that a cutting speed of 8000 m/min can be achieved in case of turning of Al using diamond tool.

The effects of cutting speed on the output variables in HSM are presented:

Cutting force versus cutting speed: Referring to the work of Flom [18], it has been confirmed when HSM of the steel AISI 4340, that the cutting force decreases with increasing speed until a minimum is reached at a speed depending on the workpiece material (1500 m/min for AISI 4340); beyond this speed the force tends to slowly increase (Figure 4.11a). Similar to AISI 4340, aluminum 6061-T6 exhibits a decrease in the cutting force with increasing speed up to 3000 m/min, beyond which also the force increases slightly. In contrast to AISI 4340 and aluminum, the cutting force for Ti remains unaffected all over the total speed range speed (Figure 4.11a).

Accordingly, an improvement of the machinability of DTC materials is realized by the reduction of the specific cutting energy associated with the reduction of the main cutting force, along with the improvement of the machinability due to the improvement of the surface quality and accuracy associated with the high speed.

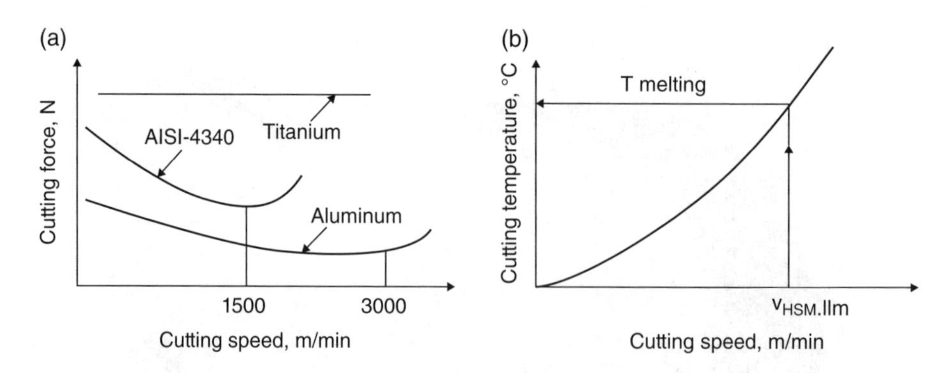

Figure 4.11 Effect of cutting speed in HSM on the cutting force and cutting temperature – schematics. (a) Effect on cutting force and (b) effect on cutting temperature.

Table 4.3 Hot hardness temperatures of tool materials and melting points of some work materials

Tool material	Hot hardness temperature (°C)	Work material	Melting temperature (°C)
HSS	600	Aluminum	600–660
Carbides	1100	Al-alloys	~540
Ceramics	1400	Super alloys	1300–1400
CBN	1500	Titanium	1600–1650
Diamond	1500	Steels	1400–1530

Adapted from Kramer [21].

Cutting temperature versus speed: The vast majority of research works indicated that the chip/tool interface temperature increases with the speed, approaching the melting point of the workpiece material (Figure 4.11b). Because the melting point of Al-alloys is low (~540°C) and well below the temperature limitations of carbide and coated carbide tools, the top cutting speed for Al-alloys appears unlimited from the cutting tool point of view [16].

Table 4.3 lists the hot hardness temperatures of different tool materials, along with the melting points of some common work materials. From Table 4.3, the suitability of a work material to HSM using different tool materials can be estimated.

Surface finish versus speed: There are indications that surface finish tends to improve with increasing speed; however, these results are not conclusive for some reasons including the burnishing action due to tool wear and the dynamic response of the cutting tool [16].

On the other hand, increasing the MRR through increasing the cutting speed improves the surface quality and reduces the cutting forces. The disadvantage of using HSM is the adverse effect on the tool life according to the well-known Taylor relationship. Regardless of this drawback, the correct trend is to use the HSM to improve the machinability and productivity.

Tool wear versus speed: Two major mechanisms are associated with HSM [21]. These are high-speed chemical solution wear and high-speed diffusion-limited wear:

1. *High-speed chemical dissolution wear*
   In the range of cutting speed in HSM, the chemical dissolution of the tool material into the workpiece is the most important contributor to wear. In essence, the tool material dissolves into the flowing chip. The tool material that is the most resistant to dissolution exhibits the least wear.
2. *High-speed diffusion-limited wear*
   As the cutting speed is increased, the cutting temperature rises to a level at which seizure of chip material occurs everywhere on the tool face. The layer of adherent material becomes saturated in the tool surface, serving as diffusion-boundary layer, hence reduces the rate of tool material transport to the chip and consequently the crater wear decreases (interval (a), Figure 4.12). Because the diffusivity increases exponentially with temperature, a further increase in cutting speed beyond the speed for minimum wear produces a rapid increase in crater wear (interval (b), Figure 4.12).

Tool Selection for HSM

HSM is only possible with the application of new tool materials, and improved coating technologies such as the PVD (physical vapor deposition)-technique or the modern technologies that allow us to produce multilayer and nanomultilayer coatings [6]. These coatings have extremely high heat resistance, allowing the coated tools to be used in severe cutting conditions. Based on the considerations regarding tool wear mechanisms in HSM, the following recommendations are to be considered:

- Choose a tool material that is chemically stable with respect to the work material so that chemical dissolution of the tool material does not significantly occur, even at the melting point of the work material.
- Cutting speed is to be selected to approach the diffusion-limited wear.
- Isolate the tool from the workpiece by applying protective layers (coatings). Various lubricants in milling of Ti have been proved successful in this respect.

Tool Selection for Al-Alloys

When machining Al-alloys, the cutting temperature is limited due to their low melting point and high thermal conductivity. Chemical dissolution wear is minimal and wear is primarily a result of the abrasion of the tool material by hard second-phase particles. Abrasion wear decreases with the hardness of the tool material HSS and carbide tooling are suitable for machining most Al-alloys, while PCD (polycrystalline synthetic diamond) is preferred for the highly abrasive cast Al-Si-alloys (10–20% Si) [16].

Tool Selection for Steels

Oxides are the only potential tool materials that are not limited by their chemical stability. The most promising area of development is the enhancement of toughness and strength of these oxide tools. The development of tool materials other than CBN (cubic boron nitride) with high hot strength in the range of 1300–1400°C might allows a transition from dissolution-limited to diffusion-limited wear with a corresponding increase in tool life. Alternatively, HSM of steels of moderate hardness (35–50 HRC) with cutting speed in the range of 1200 m/min and above using CBN was found feasible [16].

Tool Selection for Super Alloys

The recommendations are similar to those for steels. The oxides are quite chemically stable with respect to Ni-and Co-base super alloys, making the development of tough oxide tools a priority. The transition to diffusion-limited wear occurs at high speeds. Consequently, any new high hot strength compositions will likely find application in HSM super alloys. A possible example is the tool material based on Si_3N_4 and alloys of Si_3N_4 and Al_2O_3 referred to as SiAlON. These tool materials are very effective in machining Ni-base alloys at high speeds. Due to the relatively poor chemical stability of these tool materials, it is suspected that they represent the second example (in addition to CBN) of a tool material that has sufficient hot strength to enable very high speed wear transition.

Tool Selection for Ti-Alloys

Ti and Ti-alloys have low thermal conductivity, low specific heat, and a high melting point. These characteristics ensure that cutting temperatures will be high at even moderate speeds. In addition, Ti is highly chemically reactive with all known tool materials, causing rapid wear. It is quite likely that the most wear resistant tool materials (carbides and diamond) have already been identified. Therefore, the most promising area for investigation is the development of effective lubrication techniques to reduce the interaction between the tool and the chip. Moreover, new tool geometries such as the ledge tool are suggested to increase the productivity of Ti-alloys. The size of the ledge (overhang) as an integral part of the tool equals the depth of cut, and its thickness equals the ultimate flank wear to be tolerated.

Rapid tool wear remains a problem in machining of Ti-alloys and other DTC alloys, even though cutting speeds for Ti and Ti-alloys have been recently increased three- to fivefold through judicious choice of cutter grades and geometries fluids, and machining parameters. Partial enhancement of tool life lies in new tool geometries and the use of rotating cutters.

4.1.3.4 Ultrasonic-Assisted Machining

UAM is a technique involving oscillating the cutting tool ultrasonically to improve machinability. An important issue in UAM is the choice of a transducer control system that adapts the different tool holders by tuning the oscillation frequency of the HF-transducer to the natural frequency of the oscillating system. In addition, the dynamic load during cutting causes damping of the US-oscillations. To overcome this difficulty, an auto-resonant control was suggested by Skelton [22] and implemented by Babitskya *et al.* [23] to keep oscillations under resonance during cutting.

Compared to traditional machining, UAM allows significant improvement in noise reduction, tool life, surface finish, and accuracy. The higher accuracy of UAM is attributed to the reduction in the elastic deformation of both cutting tool and workpiece, and the reduction of their temperatures. Significantly lower tool temperature during UAM as compared to traditional machining can be explained by the fact that the cutting tool separates from the chip within each cycle of US-oscillations. Such intermittent contact leads to a reduction in the total time of thermal conduction.

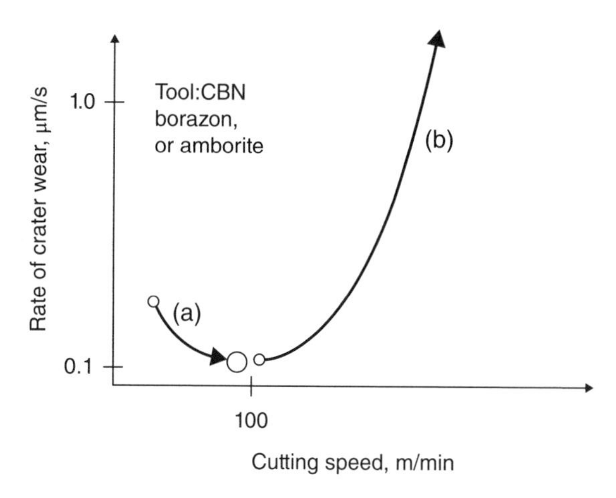

Figure 4.12 Effect of cutting speed on the rate of crater wear of CBN – a schematic.

4.1.3.5 Advanced Cooling Techniques

Cutting fluids improve the performance of machining operations because of their lubrication, cooling, and chip-flushing functions. However, their use has become more problematic in terms of health hazard, and environmental pollution. Their minimization is therefore mandatory; besides it also leads to economic benefits due to saving lubricant costs and workpiece/tool/machine cleaning cycle time.

When machining DTC materials, the heat generation becomes a more important issue than when machining low strength materials. In addition, the thermal conductivity of DTC materials such as Ti-alloys (about 15 W/m°C) and Inconels (about 11 W/m°C) much lower than that of low and medium carbon steels. Accordingly, efficient cooling in metal cutting operations should be considered. Different cooling technologies, especially, for machining DTC materials will be presented.

Cryogenic Cooling

Cryogenic cooling is the efficient way of maintaining the temperature well below the softening temperature of the cutting tool material. It is an environmentally safe alternative to conventional emulsion cooling. In the past, common cryogenic cooling approaches included precooling the workpiece, indirect cooling, general flooding, and enclosed bath. The liquid nitrogen absorbs the heat, evaporates quickly, and forms a fluid gas cushion between the chip and the tool face that functions as a lubricant. Reduction in tool temperature during cryogenic cooling reduces the crater and flank wear [24].

Cryogenic cooling, if properly employed, can provide (besides environmental friendliness) significant improvement in both productivity and improve the overall machining economy even after covering the additional cost of the cryogenic cooling system and cryogen. The beneficial effect of cryogenic cooling by liquid nitrogen may be attributed to effective cooling, retention of tool hardness, and favorable interactions of the cryogenic fluid with the chip–tool and work–tool interfaces [25].

By selectively applying liquid nitrogen to the chip and tool rake face through using a well-controlled jet, tool life can be enhanced. Micro-temperature manipulation with cryogenic cooling is the best means of chip control in the machining of DTC materials. During turning with carbide tools under cryogenic cooling, notching, abrasion, adhesion, and diffusion – wear can be retarded effectively, leading to a remarkable improvement in tool life.

Minimum Quantity Lubrication

The concept of minimum quantity lubrication (MQL) was suggested to embrace the issues of environmental and occupational hazards associated with traditional wet machining. In MQL machining, a small amount of vegetable oil or biodegradable synthetic ester is sprayed on to the tool tip with compressed air. The consumption of oil in industrial applications is in the range of approximately 10–100 ml/h. Machining using MQL is nearly equal to or often better than traditional wet machining in terms of tool life.

During finish turning of Inconel-718 with coated carbide tools (with chip breakers) under MQL (biodegradable synthetic ester as lubricant), the cutting fluid was supplied to the cutting point with compressed gas through oil holes on both the flank and rake faces of the tool. Comparative analysis of oxygen and argon as carrier gas reveals that the poor heat capacity, poor thermal conductivity, and poor lubrication characteristics of argon gas have increased cutting temperature and tool wear. Increasing the quantity of lubricant can only help in improving the surface finish. The cooling efficiency depends on the specific heat of the coolant gas. A coolant of higher specific heat can receive more heat from the tool and workpiece. Thus air acts as a better carrier gas in comparison with argon.

In turning normalized ball bearing steel (100Cr6) using commercial triple-coated carbide tips with a negative rake angle, tool flank wear was studied under MQL (inside nozzle mixing device) and dry cutting. Comparison of dry, rake MQL, and flank MQL for surface roughness and tool wear was made. It is observed that dry cutting and rake face MQL generally have the same behavior. This means that when MQL is applied on the rake surface, lubricant does not reach the cutting area. Under this condition, the tool wear increases. Lubricating the flank surface of a tip by the MQL technique reduces the tool wear and increases the tool life. However, the main drawback of the MQL methodology is that, if the mixture is not properly controlled, it may lead to the formation of mist or dangerous vapors, and thus contamination of the working environment.

High-Pressure Coolant

High-pressure coolant (HPC) delivery is an emerging technology that delivers a high-pressure fluid to the cutting zone. The high fluid pressure allows a better penetration of the fluid into the tool–workpiece and tool–chip contact regions, thus providing a better cooling effect and decrease in tool wear through lubrication of the contact areas.

In finish turning of AISI 1045 steel using coated carbide tools under high-pressure fluid (with high and low flow rates), dry cutting, and conventional fluid application (low pressure, high flow rate), the tool wear was investigated [26]. For HPC, three directions of high-pressure fluid were used:

1. toward the chip-tool interface (tool rake face),
2. toward the workpiece-tool interface (flank face), and
3. toward both flank and rake face.

The longest tool lives were obtained when fluid was applied either simultaneously on the rake and flank faces with high pressure and high flow rate, or when it was applied solely on the flank face with high pressure and low flow rate. When fluid was injected on the rake face, the adhesion between chip and tool was strong, causing the removal of tool particles. When the adhered chip material was removed from the tool by the chip flow, this resulted in a large crater wear. The fluid was not able to penetrate between chip and tool to perform lubrication.

Tool life generally increases with increase in coolant supply pressure. This can be attributed to the ability of the HPC to lift the chip and gain access closer to the cutting interface. This action leads to a reduction of the seizure region, thus lowering the coefficient of friction, which in turn results in reduction in cutting temperature and cutting forces. The drawback of HPC, however, includes the fact that the high pressure generated by the fluid may produce certain subsurface defects, which should be considered machining steps.

4.1.3.6 Cryogenic Treatment of Tool Materials

Cryogenic treatment refers to subjecting materials to very low temperatures. It is believed that the life of cutting tools extends substantially with cryogenic treatment. The basic reason behind enhanced performance of cryogenically treated carbide tools is that carbide formation is facilitated and homogeneously distributed. Cryogenically treated tungsten carbide tools have much greater resistance to chipping compared to untreated ones, especially when used at higher cutting speeds. The high wear resistance of the cryogenically treated HSS is attributable to the reduction of the retained austenite which is transformed into martensite. This treatment includes a combination of deep freezing (boiling point of liquid nitrogen −196°C), maintained for about 24 h, followed by a controlled raising back to room temperature. Subsequently tempering processes follow [27].

Cryogenic treatment has shown good results with die steel, HSS, and carbide tools in terms of improved wear resistance of the tool materials. Coating on cryogenically treated base material is favorable for improved machining performance. It is further suggested that tempering before cryogenic treatment is not recommended as the stabilization of carbides and micro-structural phases during the tempering process inhibits further transformation during cryogenic treatment.

However, there are a few drawbacks of the cryogenic treatment of cutting tools; for example, in continuous cutting the performance of these tools deteriorates and comes down to the performance level of the untreated versions. At higher depth of cut or in rough turning operations, the performance of the cryogenic treated tool can be maintained by supporting the turning process with certain cooling methods, which involves extra cost. Treatment is not effective on coated tools as it results in shorter tool life. The uneven contraction of the coated material and the substrate during cryogenic treatment can cause incipient cracks to appear at the interface.

4.2 Cutting Tool Materials

By the beginning of the twentieth century, Taylor had introduced his HSS instead of tool carbon steel (TCS), thereby increasing the cutting speed to three times. In 1926, Krupp developed a very hard sintered carbide material, used in cutting and forming, called Widia (German acronym for *Wie Diamant*, i.e., like diamond), accordingly cutting speeds of 3–6 times of those of HSS have been realized. Ceramics were then introduced in 1950 as a new tool material, realizing speeds of 6–8 times those achieved by HSS. Latest developments are tool materials such as coated carbides and CBN. Now polycrystalline diamond tools are mainly used in finish turning, achieving cutting speeds of more than 2 km/min [1].

4.2.1 Characteristics of an Ideal Tool Material

Materials utilized in cutting tools have to meet a different set of requirements than those used in general engineering construction applications. The cutting tool is subjected to severe conditions, such as high temperatures of more than 1000°C, severe friction, and high dynamic stresses. The tool should be strong enough to withstand the above mentioned conditions. Accordingly, the ideal tool material should be characterized by:

- High hot hardness (refractoriness), that is, it would retain its hardness at high operating temperatures.
- High strength and toughness to withstand both static and impact loads. Unfortunately, extreme hardness poses limitation on toughness, so a compromise must be struck between adequate hot hardness and improved toughness (Figure 4.13).
- High fatigue strength (endurance limit) to withstand repeated loads.
- High wear resistance to resist mechanical and thermal abrasion.
- Inertness and chemical stability, that is, no metallurgical affinity to workpiece.
- Resistance to thermal shocks, especially if coolant is suddenly turned on during cutting.

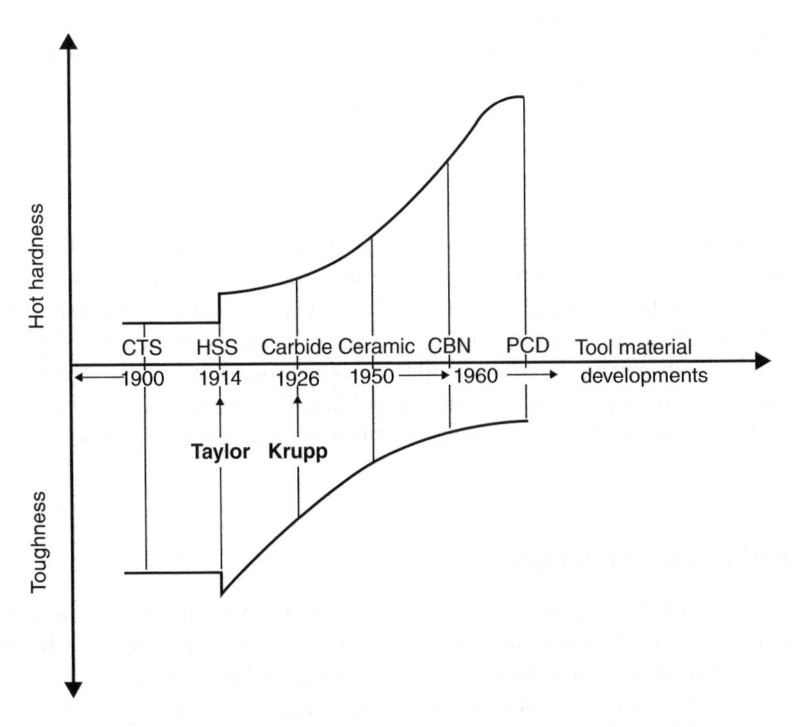

Figure 4.13 Hardness versus toughness of cutting tool materials.

- High thermal conductivity to efficiently conduct away heat from the cutting edge.
- Low coefficient of friction at the tool/chip interfaces to guarantee minimum wear and hence reduced cutting force and consumed power.
- Low tool cost and availability
- Easy to form into the required shape and easy to sharpen.

Of course, the ideal tool material having all the above-mentioned characteristics does not exist. An increasingly wide range of tool materials are, however, available from which the production engineer must select the material that will best perform tasks at the lowest unit cost. Table 4.4 provides the relative costs of typical cutting tool materials in industrial use today [2]. The most important characteristic of any cutting tool material is its hot hardness. Good thermal shock resistance is an important characteristic of suitable cutting materials. Figure 4.14 illustrates the hot hardness of some of the most common tool materials in current use.

4.2.2 Types of Cutting Tool Materials

The cutting tool materials are presented as follows in ascending order of hot hardness and descending order of toughness. They may be ferrous (e.g., tool steels and HSSs) or nonferrous (e.g., cast nonferrous alloys, carbides, cermets, ceramics, and so on.). TCS is the oldest type of tool steels, which is generally used for machining soft materials; it is also used in the manufacturing of hand tools. It is not recommended in machining DTC materials.

4.2.2.1 High Speed Steel (HSS)

Today HSS is misnamed, because it is now considered the general purpose tool material for machining operations performed at low and moderate cutting speeds. HSSs contain mainly carbide forming alloying elements such as W, Cr, and V, besides 0.7–15% C. Therefore, their microstructures consist of a martensitic matrix, in which complex carbides (Fe_2W_2C, $Cr_{23}C_6$, and VC), and cementite Fe_3C are embedded. These carbides help in achieving good hot hardness and sharp cutting edges at operating temperatures up to 600°C but rapidly soften at higher temperatures (Figure 4.14).

Table 4.4 Relative costs of tool materials Youssef *et al.* [2]

Tool material	Relative cost
Tool carbon steel (TCS)	1
High-speed steel (HSS)	1–2
Cast nonferrous alloys	2–10
Cemented carbides	3–10
Sintered alumina oxides, cermets	20
Coated carbides	10–30
Borazon (CBN), diamond	100 and more

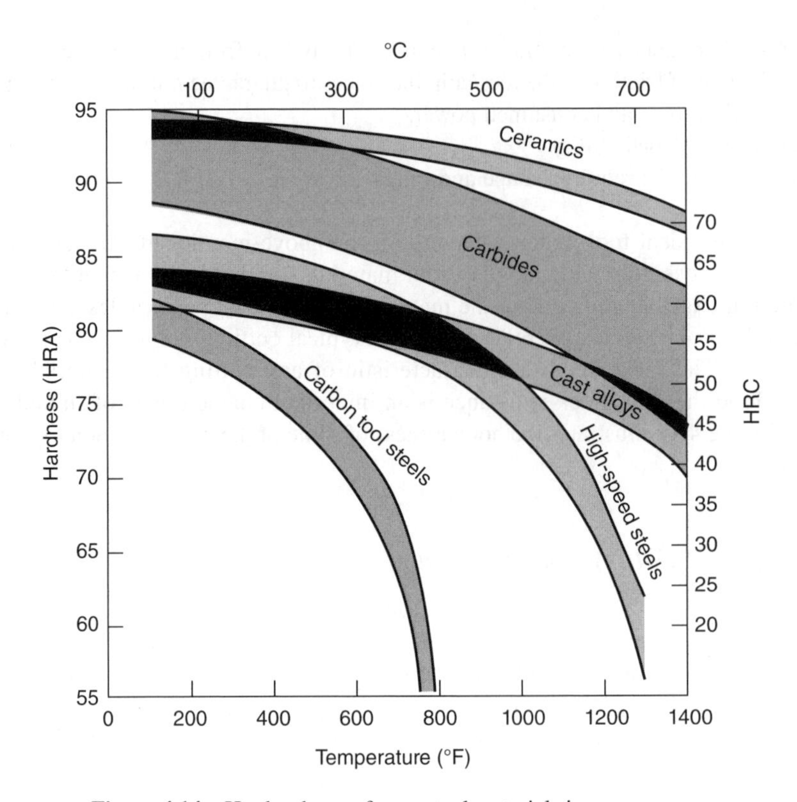

Figure 4.14 Hot hardness of some tool materials in common use.

HSS provides a good balance between hot hardness and toughness (Figure 4.13) to cope with interrupted cuts. Moreover, it can be fully hardened with little danger of distortion and cracking, and the tool can be rapidly sharpened many times during its useful life. Because of its toughness, HSS is especially suitable for high positive rake angles and for low power machine tools with low stiffness. The basic limitation of HSS is the relatively low cutting speeds. For high-production machining, it should be replaced by carbides and other advanced tool material [2].

The first and most common 18-4-1 HSS is designated according to the percentage of its alloying elements W, Cr, and V respectively. During World War II, when there was a shortage of tungsten, it was largely replaced by the strategically available molybdenum. A smaller percentage of Mo (only 5%) behaves in the same manner of W regarding hot hardness [28]. The AISI has classified the commercially HSSs into two groups, namely, the T-group (tungsten HSSs), and the M-group (molybdenum HSSs). The M-group generally has higher abrasion resistance than the T-group, which undergoes less distortion during heat treatment. Today, the M-group constitutes the majority of HSSs produced in the States.

As illustrated in Table 4.5, the majority of HSSs contain 1–2% V. Addition of more V (T-15) allows the formation of complex carbides, which inhibits grain growth at high temperature needed in heat treatment of HSSs; T-15 is commonly used for high strength materials such as Cr-alloy die steels. The pronounced superiority of HSS containing Co (T4, T15, M30,

Table 4.5 Some of widely used HSSs according to AISI

AISI designation		Composition (wt%)						Remarks
		C	W	Cr	V	Mo	Co	
T group	T1	0.73	18	4	1	—	—	Taylor straight W grade
	T4	0.75	18	4	1	0.6	5	W-Co grade
	T15	1.55	12.5	4.5	5	0.6	5	W-Co-V grade
M group	M1	0.80	1.75	3.75	1.15	8.75	—	Straight Mo grade
	M2	0.85	6.25	4	2	5	—	Straight Mo grade
	M7	1.02	1.75	3.75	2	8.75	—	Straight Mo grade
	M10	0.89	0.7	4	2	8	—	Straight Mo grade
	M30	0.80	1.8	4	1.2	8.25	5	Mo-Co grade
	M36	0.85	6	4	2	5	8.25	W-Mo-Co grade
	M42	1.08	1.6	3.75	1.15	9.6	8.25	Mo-Co grade

M36, and M42) is due to the tendency to increase hot hardness by reducing the amount of retained austenite.

The performance of HSS can be improved using different technologies of surface coating. One the most effective methods is the PVD, in which a very thin coating (5–7 μm) metal carbides, nitrides, or oxides is provided. The cost of HSS-coated tools is about twofold that of the normal HSS. TiC, TiN, and $A1_2O_3$ are used as coating materials. Multilayer coatings of these are also possible. TiC coatings are effective for abrasive wear resistance. Al_2O_3 coating is a good thermal barrier due to its lower thermal conductivity. Hence it is used for high-speed and high-feed rate operations. TiN coating is of yellow-gold color; it is the most preferred due to its effectiveness in preventing adhesion at tool–chip interface; this reduces galling and greatly improves the tool life. However, the tools do not perform well at low cutting speeds, because this coating can be chipped off. Proper lubrication during cutting is a must. The wear of HSS-coated tools is generally reduced such that the tool life extends up three to four times that of conventional HSS grade.

The black-colored TiC coating is deposited at high temperature by chemical vapor deposition (CVD). It should be marked that the PVD- and CVD-coated HSS tools must be hardened and tempered after coating. Moreover, a limitation of coated HSS tools is that they cannot be re-sharpened.

Another advancement in HSS-tooling, is that it can be produced by powder metallurgy (PM). It is useful for large-size tools. PM-HSS tools exhibit better grindability, greater toughness, better wear resistance and higher hot hardness, and perform more consistently. However, the PM-HSS is double the cost of regular HSS.

4.2.2.2 Cast Nonferrous Alloys (Stellite and UCON)

Cast nonferrous alloys are cutting tool materials that contain iron except as impurity. They must be cast shape then ground to size. Cast alloys are generally used for simple and large size tooling. These materials serve best in speed range between HSS and cemented carbide cast nonferrous alloys not applicable for machining stainless and super alloys. They are of two basic categories: Co-base (Stellites) and Cb-based (UCON) alloys.

1. Co-base cast alloy is of trade name Stellite, of typical analysis:

$$38-52\%\ \text{Co},\ 30-32\%\ \text{Cr},\ 10-12\%\ \text{W},\ 2\%\text{C}$$

Its structure is composed of Co-matrix, in which complex W-Cr carbides are embedded with volume ratio of 25–30%. It is of a hardness ranging from 58 to 64 HRC.
Stellites are characterized by

- Weak in tension, fragile, resistant to corrosion, and not heat treatable.
    - Not tough as HSS, hence they are only used for jobs free from shocks, intermittent cuts, and vibrations.
    - Maintain hardness at elevated temperatures (up to 750°C) so used at relatively higher speeds (25%) as compared to those of HSS.
    - Available as bars of circular and square sections or as inserts, brazed or attached on lathe tools, and milling cutter bodies.
    - Of less tendency to form BUE.
    - Do not require cutting fluids except if special surface quality required.

Selective applications of Stellites include machining of plain carbon steel, cast iron (CI), and hard bronze. They are generally recommended for deep and continuous roughing operations at relatively higher feeds and speeds. Stellites may be used in the production of slip and limit gauges.

2. Cb-based cast alloy is nitrided refractory columbium-based alloy. UCON is the trade name of Union Carbide Company. Typical analysis of UCON.

$$50\%\ \text{Columbium}\ (\text{Cb}),\ 30\%\ \text{Ti},\ 20\%\ \text{W},\ \text{and free from carbides.}$$

If the tool surface receives a nitriding treatment, UCON attains hardness up to 25% more than ceramics and 50% more than cemented carbides.
UCON is characterized by

- High hardness and reasonable toughness.
- Excellent thermal shock resistance.
- Excellent resistance to diffusion and chip welding.
- Relatively expensive as compared to cemented carbide.
- Edge life is three to five times as compared to cemented carbide.
- Difficult to machine.
- Available as throwaway insert.

UCON is used in machining of common steels at high-speed range of 300–500 m/min. It is not recommended for intermittent cuts. Moreover, it is not practical for machining many materials such as CI, Ti, Ti-alloys, stainless steels, and super alloys [2].

4.2.2.3 Cemented Carbides (Widia)

Cemented carbides (commercially known in Europe as Widia, the trade name of Krupp) are basically composed of finely ground carbides (WC + TiC + TaC). These carbides are extremely hard, but not tough enough to withstand impact loads during cutting. Therefore, they should be bonded by a soft metal, usually Co, just as bricks are cemented together by mortar, hence the name cemented. These carbides are produced by sintering through powder metallurgy technique; hence they also acquired the name sintered carbides. Cemented carbides are composed of hard carbides (85–95% by volume) bonded together in a Co-matrix.

Cemented carbides are classified in two fundamental groups, namely, the straight group WC + Co, and the multicarbide group WC + TiC + TaC + Co. The early tools of the straight group were very effective when machining CI, nonferrous metals, and nonmetallic materials. However, they were not compatible with steels because crater wear is rapidly developed on the tool face. It was found that the crater wear considerably decreased, if TiC and TaC were added to produce the multicarbide group. Such addition promoted wear resistance but reduced the toughness. Therefore, the multicarbide group is best suited for cutting ferrous metals producing long chips like steels.

There are many systems for classification of cemented carbides. However, it appears that the international system ISO is gaining ground, and it is now becoming a world standard. According to ISO 513–91, the carbide inserts for metal cutting applications are grouped into three grades, that are identified by the letters P, M, and K, coded by three different colors, blue, yellow, and red, respectively. Each grade is further subdivided into different types on a scale of 01–40 to indicate hardness and toughness. Types 01 and 10 are used for finishing, while types 30 and 40 are suitable for rough cutting.

Table 4.6 illustrates the physical, mechanical, and thermal properties, as well as the composition, and typical materials being machined by each grade. Referring to this table, grade P is intended for machining of steels. Grade K is mainly composed of WC and therefore tougher than grade P; it is intended for machining materials of broken chips, producing less crater wear. Grade M is an intermediate grade, which is capable of machining materials producing both long and short chips. It is used in cases where it is required to cut several materials with the same tool upsetting [1].

Cemented carbides are also classified according to the C-code used in the United States [5] into:

- *C1 and C2 grades (ISO group K)*: These are tough and composed of WC + Co.
- *C3 and C4 grades (ISO group M)*: These are the general purpose grade, containing up to 10% mixed carbides.
- *C5–C8 grades (ISO group P)*: These contain mixed carbides ranging from 10 to 60% to withstand face crater at high cutting speeds. These are more suitable for machining steels.

Because of their high hardness over a wide range of temperatures (800–1100°C), high modulus of elasticity, high thermal conductivity, and low thermal expansion (Table 4.6), cemented carbides are the most popular cutting tool materials for machining operations. They are capable of machining harder materials at high cutting speeds.

Table 4.6 Cemented carbides for cutting tools

Grade symbol (color)	Type	Trend[a]	Composition (%)			Specific weight (g/cm³)	VH3.0 vickers (kg/mm²)	σ^b_{bend} (Kg/mm²)	σ^b_{comp} (Kg/mm²)	E[b] (Kg/mm²)	$\mu_{1.exp}$ (10^{-6}/°C)	k (cal/cm°Cs)	Materials to be machined
			(TiC + TaC)%	Co%	WC%								
P (blue)	P01	1	64	6	30	7.2	1 800	75	—	—	—	—	Ferrous metals
	P10		28	9	63	10.7	1 600	130	490	53 000	6.5	0.07	with long chips
	P20		14	10	76	11.9	1 500	150	500	54 000	6.0	0.08	(steels)
	P30		8	10	82	13.1	1 450	175	500	55 000	5.5	0.14	
	P40	2	12	13	75	12.7	1 400	190	470	56 000	5.5	0.14	
M (yellow)	M10	1	10	6	84	13.1	1 700	135	—	58 000	5.5	0.12	Ferrous metals
	M20		10	8	82	13.4	1 550	160	500	57 000	5.5	0.12	with long or
	M30		10	9	81	14.4	1 450	180	480	—	—	—	short chips and
	M40	2	10	15	81	13.6	1 300	210	440	54 000	—	—	nonferrous metals (multipurpose grade)
K (red)	K01	1	4	4	92	15.0	1 800	120	—	—	—	—	Ferrous metals
	K10		2	6	92	14.8	1 650	150	570	63 000	5.0	0.19	with short
	K20		2	6	92	14.8	1 550	170	550	62 000	5.0	0.19	chips (C1)
	K30		1	9	91	14.5	1 400	190	480	58 000	—	0.17	nonferrous
	K40	2	0	12	92	14.3	1 300	210	450	57 000	5.5	0.16	metals and nonmetallic materials

Note: $\mu_{1.exp}$ = coefficient of linear expansion and k = thermal conductivity.

[a]Trends: (1) increasing hardness, wear resistance, and cutting speed and (2) increasing toughness and feed.

[b]σ_{bend} = bending strength, σ_{comp} = compression strength, and E = Young's Modulus.

Abstracted from ISO 513-91 [DIN-4990].

Precautions to be observed when using carbides

- The machine tool must be rigid enough and from free vibrations, and the tool and work piece must be rigidly clamped.
- The machine power and speeds must be adequate to allow the carbide tool to cut with high cutting speed.
- The tool must not be allowed to rub after switching off the machine. It is a good practice to withdraw the tool while the feed is still engaged.
- Coolant, when used, must be effective and sufficient.
- Carbide tools should be sharpened dry, and should not be quenched after sharpening.

Cemented carbides are generally available as inserts or tips, which are provided with a number of cutting edges. These inserts are available in many shapes such as square, triangle, diamond, and round. A square insert is stronger (strength depends on included angle only) than triangular or diamond inserts, while the round insert is stronger than all other shapes (Figure 4.15). Inserts are usually clamped on tool shanks made of tool steel in turning, planing, boring, and slotting operations.

Figure 4.16 shows milling cutters equipped with mechanically clamped cemented carbide inserts. Less frequently used are the inserts brazed to the tool shanks. Considerable economic savings are achieved if mechanically clamped inserts (throwaway tips) are used instead of brazed types [2].

Carbide tips are usually provided with flat margins of negative rakes to protect them against chipping caused by mechanical impact. Tips are also provided with grooves formed in the rake face close to cutting edges acting as chip breakers, which curl the chips and clear it from around the tool. This is important for automated production. The ability to cut for long periods without stopping to clear chips has a considerable effect on the machining cost.

Coated Tools

Coated tools constitute the majority of the tools applied in material removal processes, rendering the employment of uncoated ones as an exception. A broad growing market of coated cutting tools has been developed. Moreover, numerous material and manufacturing engineers have joined their expertise, aiming at developing coatings meeting the needs for processing the most DTC materials at the most extreme cutting conditions. The emerging of new workpiece, tool and film materials, the evolution of sophisticated coatings' characterization methods and

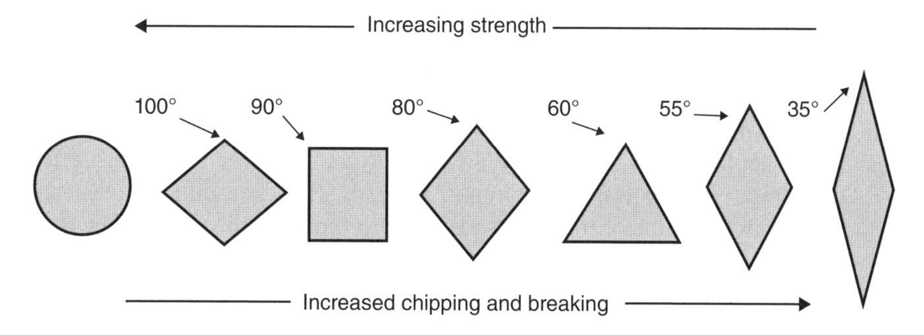

Figure 4.15 Typical shapes of carbide inserts. (Courtesy of Kennametal.)

Figure 4.16 Milling cutters equipped with mechanically clamped carbide inserts. (Courtesy of Krupp, Widia, GmbH, Essen Germany.)

the continuous need for higher productivity rates maintain vividly the industrial and scientific interest for further advancing this field.

In coated carbides, the tool possesses a very hard nonreactive surface, which acts as a diffusion barrier. This is achieved by coating a carbide substrate with a thin coating (typically 5–7 μm) of ceramic such as TiC, TiN, Al_2O_3, or HfN. Several layers may be deposited on each other (TiC is a base layer followed by Al_2O_3 and TiN) [29].

Coatings can be applied by PVD or CVD. PVD is applied to carbide substrate at low temperature to provide a smooth surface that generates less frictional heat, allows lower cutting forces, and resists BUE-formation that can lead to insert chipping.

CVD-coatings are applied to carbide substrate at higher temperatures. It provides interdiffusion of coating with the substrate to assure a strong bond. It also permits the deposition of multilayer coatings that can effectively suppress both crater and flank wear, thereby expanding the range of tool application. CVD is currently the only coating process that can effectively apply Al_2O_3-coating, which permits higher cutting speeds. However, due to higher temperatures associated with CVD, the brittle η-phase is liable to form and this reduces the transverse rupture strength.

When the coating has been worn away, coated tools cannot be re-sharpened, and the wear rate becomes the same as that for uncoated tools. Coated carbides have taken more than half of the cemented carbide market [1].

4.2.2.4 Cemented Titanium Carbides (TiC-Based Tools)

These are recently developed toward enhancing the durability of cemented carbides. Cemented TiC with Mo and Ni bonding material has been available in the form of disposable tips. This material is less tough but more wear resistant than conventional Co-cemented carbides. Developments of TiC-tools have led to superior grades that allow cutting speeds approaching those of ceramic tools; these grades are used in semi rough cutting of both CI and steels. With the relative shortage of tungsten supplies, TiC serves as substitute for conventional carbide grades. Some commercial suppliers now include TiC-based grade in their catalogs. *For machining steels and stainless steels, cemented titanium carbides have gained acceptance.* It is believed, that in the near future, at least one half of steel machining will be carried out using cemented TiC [29].

4.2.2.5 Cermets

These are composites of alumina Al_2O_3, and TiC. Cermets are somewhat more refractory than TiC. The addition of 20–30% TiC promotes toughness and reduces brittleness of alumina. Therefore, performance of cermets is between that of ceramics and carbides. Compared to carbides cermets have higher hot hardness, less toughness, lower thermal conductivity, and greater thermal expansion. So, thermal cracking can be a problem during interrupted cuts; accordingly, cermets are best suited for finishing operations. The enhanced surface quality imparted by cermets is due to its low level of chemical reaction with iron, and hence less craters and BUE are detected. Cermets are successfully used in the high speed finish machining of steels and cast iron. Stainless steels, especially martensitic alloys can be machined by cermets. This tool material is also applicable for machining carbon and polymers [29].

4.2.2.6 Ceramics (Alumina-Based Tools)

Ceramics are fine-grained (<3 μm) pure aluminum trioxide Al_2O_3 particles. As cemented carbides, they are produced through powder metallurgy technique by pressing alumina powder, followed by sintering without binder at high temperature of 1700°C. Alumina is cheap and available; however, the processing is expensive, and therefore; ceramics are not cheap as compared to carbides. They almost completely resist crater wear. Usually, they require no coolant and have the same tool life if operated at double cutting speed of carbides. Ceramics require more rigid tool holders and rigid machine tools to take the benefit of their capacities. They are particularly inert to steel up to high cutting temperatures. They retain their cutting ability for temperatures up to 1200°C, which enables steels, semi-hard steels, and CI to be machined at high speeds (up to 600 m/min) and reasonable feeds (up to 0.25 mm/rev). Higher speeds and feed are realized when machining nonferrous metals.

Ceramic is a multipurpose tool material as compared to carbide tooling because a wide range of materials (ferrous and nonferrous) can be machined at various cutting speeds by one type of ceramic material. However, ceramics are not suitable for machining Al, Ti, and other materials that chemically react with alumina.

Because ceramics have poor mechanical and thermal shock resistance, interrupted cuts, and application of cutting fluids can lead to premature tool failure. It is advisable to chamfer the edge of the workpiece with a carbide or HSS tool before turning with ceramics to avoid severe impact when touching with the ceramic tool. Being nonmetallic, ceramics are less likely to adhere to metal during cutting; hence BUE is not liable to occur, and consequently high surface quality is expected. Better chip flow on the tool face is realized owing to the low coefficient of friction; which leads to reduced cutting forces and machining power when using ceramic tools. It should be considered that the tool/chip interface temperature is high because ceramic tools are a poor conductor of heat [2].

Great advances have been made toward improving the reliability of ceramics. Tools of Al_2O_3 reinforced with 25–40% SiC whiskers, and those made of silicon nitride (Si_3N_4) and Si-Al-O-N ceramics, are tougher and more wear-resistant, and can be used for interrupted cuts. They are extensively used in cutting super alloys and gray CI, [29].

Experience has shown that when ceramic tools properly applied as a replacement to carbide tools, the machining time can be reduced by one-third to half when machining steel and by two-thirds when machining cast iron. Unfortunately, ceramics are more brittle than other sintered inserts. For this reason, their use is still limited in metal cutting. Of all sintered materials in current use, only about 2–3% are ceramics, about 60–70% cemented carbides and cermets, 25% coated carbides, and 5% titanium carbides [30].

4.2.2.7 SiAlON

Si-A1-O-N stands for silicon nitride Si_3N_4-based materials A1 and O additions. It is produced by milling Si_3N_4, aluminum nitride, alumina Al_2O_3, and yttria. The mixture is dried, pressed to shape, and sintered to 1800°C. It is not as tough as cemented carbide; however, it is considerably tougher than alumina and thus suited for interrupted cuts. SiAlON possesses high thermal conductivity and low coefficient of thermal expansion, thus providing resistance to thermal shocks as compared to alumina. SiAlON is mainly used to machine aerospace alloys and Ni-based gas turbine discs at cutting speeds of 200–300 m/min. It is available as inserts of negative rakes and is provided by a chamfer at the cutting edge to strengthen the tool. Improvement in machine tool rigidity enables SiAlON tips to be used effectively [2].

4.2.2.8 Cubic Boron Nitride (CBN)

CBN was developed to overcome the problem of diamond being unstable for machining ferrous metals and Ni-alloys. While not quite as hard as diamond, it is still harder than any other material developed today. Two main commercially available products of CBN are borazon BZN of General Electric Motors (GEM) and Amborite of De Beers. BZN is produced as laminated tool tips of about 0.5 mm thick of polycrystalline cubic boron nitride (PCBN) on WC-substrate. The substrate provides the shock resistance, whereas the PCBN layer provides a very high wear resistance and strength. The Amborite is produced as a tip consisting entirely of consolidated PCBN [2].

The greatest chemical stability of CBN for a long period at elevated temperature of 1000°C makes it possible to machine steel, CI, and Ni alloys at high cutting speeds. The greatest potential of CBN appears in machining hardened steel, HSS, chilled CI at cutting speeds of 60% m/min and feeds of 0.2–0.4 mm/rev. With correct tool geometry and negative rakes. CBN can be employed for taking interrupted cuts on hardened steel. CBN grinding wheels are used for grinding carbides, HSS, Stellites, and so on. CBN is expensive and it is about the same price as synthetic diamond. The material removal rate is several times greater than that possible with carbide tools. It is available in the form of indexable inserts [2].

4.2.2.9 Diamond

Natural diamond is the hardest material known to man. Unfortunately, as a tool material, it is extremely brittle and the slightest impact or fluctuation in cutting force causes it to fracture. Its use is therefore severely restricted and is limited to high cutting speeds and uninterrupted cutting of soft materials such as A1, A1-alloys, bronze, and plastics. It is not applicable to cut ferrous metals and Ni-alloys owing to the adverse chemical reactions between diamond and these materials. Diamond can be applied by CVD as a thin coating of less than 50 μm thick layer to twist drills for drilling nonferrous metals [29].

A recent development (1979s) has led to introducing the PCD. Its main advantage is due to the random orientation of the diamond crystals which prevents the propagation of cracks through the structure, thus making it suitable for interrupted cut. PCD-tool tips are available as sintered inserts or as 0.5 mm thick layers sintered onto a carbide substrate that provides the necessary toughness for shock loads acting on the tool. The composite tip can be clamped as brazed on the tool shank, then ground, lapped, and polished, and resharpened when worn. It takes a longer sharpening time than other tools. It is more expensive, costing typically 20–30 times the equivalent carbide tool [2].

PCD outperforms all other materials on highly abrasive workpieces such as hypereutectic Al-Si alloys. They are also recommended for machining Al, Al-alloys, Cu, and Cu-alloys (e.g., Cu-commutator), and so on. They are applicable for operations of intermittent cutting like milling and threading. However, at high temperatures, it changes into graphite which diffuses into iron and nickel; therefore it is not suitable for cutting steels, stainless steels, cast iron, and super alloys [2,29]. Finally, Table 4.7 provides the physical and mechanical properties of cutting tool materials previously discussed in this section [30].

4.2.3 Tool Materials for Machining of Stainless Steels and Super Alloys

Materials used in the manufacture of aero-engine components generally comprise nickel and titanium base alloys, and stainless steels. These DTC materials provide serious challenges for cutting tool materials during machining due to their unique combinations of properties such as high temperature strength, hardness, and chemical wear resistance. The poor thermal conductivity of these alloys results in the concentration of high temperatures at the tool-workpiece and tool-chip interfaces, consequently accelerating tool wear, and increasing manufacturing cost. Tool materials with improved hardness like cemented carbides (including coated carbides), ceramics, polycrystalline diamond, and PCBN are the most frequently used for HSM of aero-engine alloys. These developments have resulted in significant improvement in the machining of aero-engine alloys without compromising the integrity of the machined surfaces [31])

The recommended types of tool materials for economical machining of stainless steels and super alloys will be discussed in the following sections.

Table 4.7 Physical and mechanical properties of different tool materials

Physical and mechanical properties of tool materials	Tool material (wt%)				
	Carbide (K10) 94%WC + 6% Co	Al_2O_3 90–95% + 5–10% ZrO_2	Whisker-reinforced alumina 75% Al_2O_3 + 25% SiC	SiAlON 77% Si_3N_4 + 12% Al_2O_3 + 1% Y_2O_3	CBN 50–90% + 50 to 10% (TiN-TiC)
Grain size (μm)	1–2	1–2	—	1	1–3
Density (g cm^{-3})	14.8	3.9–4.0	3.7	3.2	3.1
Hardness (HV) at 20°C	1700	1700	2000	1600	3000–4500
Hardness (HV) at 1000°C	400	650	—	900	1800
Fracture toughness (M Nm$^{-3/2}$)	10	1.9	8	6	10
Young's modulus (kN mm^{-2})	630	380	390	300	680
Thermal conductivity (Wm^{-1}°C)	100	8–10	32	23	100
Coefficient of thermal expansion (×10^{-6}°C)	5–6	8.5	—	3.2	5

Adapted from Bhattacharyya *et al.* [30].

4.2.3.1 Cutting Tool Materials for Stainless Steels

The most widely used cutting tool materials for machining SSs are mainly HSSs (wrought or sintered), and cemented carbides.

1. *High speed steels*
   Either tungsten or molybdenum HSS (Table 4.5) can be used. HSSs are particularly useful in machining operation involving high feeds and low cutting speeds. The tungsten types (e.g., T-15) are useful for their good abrasion resistance and hot hardness. The molybdenum HSSs are more widely used. M-42 is useful for applications such as milling, where a good combination of hardness and strength are required at lower cutting speeds. It has better hardness than the more common grade M-2, but may not be as tough however. If the tools are prone to edge chipping, tougher grades such as M-2, and M-10 may be used. If the tools are prone to wearing, the more abrasion resistant grade T-15 must be the proper choice.
2. *Cemented carbides*
   These are used for machining SSs where higher speeds and higher feeds are normally used. The straight tungsten carbides grades are used for machining austenitic and duplex SSs, whereas the complex carbides are used for machining the martensitic and ferritic grades [32].

Table 4.8 provides the recommended HSS and uncoated cemented carbide tooling for machining of stainless steels.

Coated Carbides

They can also be used in machining stainless steels. Tests are carried out with new tool inserts, mounted on a commercial tool holder. Inserts of grade C2 (K40), PVD-coated with 2000 alternate layers of AlN and TiN, each layer 1.25 nm thick, delivered by Sumitomo Electric, have been investigated [33]. The Al/TiN super-lattice film has a hardness of 3900 HV. The wear of PVD-coated and uncoated carbide tools, as well as a precision brazed PCBN in machining ATAVAX, at low cutting speed (44 m/min), depth of cut (3 and 20 μm), and feed rate (5 μm), has been investigated on ultra-precision lathe machine. STAVAX is a specialized martensitic SS-alloy (S42000, of hardness 44 HRC). It was found that, the machined profile with the coated-carbide tool in the presence of natural oil had superior form accuracy and surface finish, even than those of the much more expensive PCBN-tool [33].

ISCAR cutting tools introduced its Al_2O_3 TiAlN, PVD coated submicron carbide substrate with positive rake to machine all types of SSs, especially austenitic SSs, where the machining is influenced by the material's strain hardening, poor thermal conductivity, ductility, and tendency to form BUE [34]. ISCAR provided new grades IC907, which improve machining performance, with good tool life, optimum chip formation, and good surface quality.

Outstanding results have been obtained when machining austenitic SSs (or super alloys) at high cutting speeds and high thermal loads. This is due to the low heat conductivity of the Al_2O_3 layer on the cutting edge on the original TiAl coating. This layer serves also to reduce friction besides insulating the insert from heat conduction, thereby providing a degree of protection against insert wear and plastic deformation. This new combination of an advanced technology of double PVD-coating on a submicron substrate enables manufacturing of coated inserts with sharp cutting edges with positive rakes [34].

4.2.3.2 Tool Materials for Super Alloys

The cutting tool materials for machining super alloys will be considered in the following machining operations.

Turning

Carbide tools are frequently used in turning super alloys, although coated carbides, ceramic, CBN, and even HSS tools are also used. A C2-grade of carbides containing about 94% WC and 6% Co is frequently selected for roughing and C3-grade is used for finishing. Carbide tools permit the highest cutting rates and are recommended for turning operations involving interrupted cuts. Carbide inserts with positive or negative rakes are suitable for both roughing and finishing of super alloys. The cutting force with a standard negative rake insert is always larger than that of a positive rake insert. With carbide inserts, moderate cutting speeds should be used to minimize tip temperatures and encourage longer tool life [16].

Coated carbides yield small increases in metal removal rate (MRR) when cutting the Fe-Ni-base alloys. Generally, metal removal rates can be increased only by about 25%, and higher tool costs limit their application. Coated carbide tools have not yet been proved effective in significantly increasing the MRR of Ni-base super alloys. The limitation is the ability

Table 4.8 Recommended HSS and carbide tooling for machining of stainless steels using different machining operation

Machining operation	Stainless steel group	Stainless steels	
		Free-machining	Nonfree-machining
1. *Turning* (carbides only recommended)	Ferritic	C6/C7	C2/C3/C6/C7
	Martensitic		
	Austenitic	C2/C3	C2/C3
	Duplex	N.A.	
	PH-alloys		C2/C3/C6/C7
2. *Cutting off and form turning* (HSS and carbide are used)	Ferritic	M2/M3-C6	M2/M3-C6
	Martensitic		
	Austenitic	M2/M3-C2/C6	M2/M3/M42/T15-C2
	Duplex	N.A	M2/M3-C2
	PH-alloys		M42/T15-C2
3. *Drilling* (HSS is frequently used)	Ferritic	M1/M7/M10	M1/M7/M10
	Martensitic		
	Austenitic		M1/M7/M10/M42/T15
	Duplex	N.A	M1/M7/M10
	PH-alloys		M42/T15
4. *Tapping and thread cutting* (only HSS is used)	Ferritic	M1/M7/M10	M1/M7/M10
	Martensitic		
	Austenitic		
	Duplex	N.A.	
	PH-alloys		
5. *Milling* (HSS and carbide are used)	Ferritic	M2/M7-C6	M2/M7-C6
	Martensitic		
	Austenitic	M2/M7-C2	M2/M7-C2
	Duplex	N.A.	
	PH-alloys		
6. *Broaching* (HSS is frequently used)	Ferritic	M2/M7	M2/M7
	Martensitic		M2/M7/M42/T15
	Austenitic		
	Duplex	N.A.	M42/T15
	PH alloys		
7. *Reaming* (HSS and carbide are used)	Ferritic	M7-C2	M7-C2
	Martensitic		M7/M42/T15-C2
	Austenitic		
	Duplex	N.A.	M42/T15-C2
	PH-alloys		

HSS: M1/M2/M7/M10/M42/T15, carbides: C2/C3/C6/C7.
PH, precipitation hardenable.

of substrate to resist deformation at substantially higher cutting temperatures regardless of the coating material used [16]. However, PVD-coated carbide tools with positive rakes are suitable for finishing, and super-finishing operations to minimize part deflection and reduce cutting forces and temperatures.

HSS tools are seldom used in turning super alloys, except for interrupted cuts. In such cases, HSS tools are more practical than carbide tools because of their greater shock resistance. The highly alloyed grades such as T15, T36, or M42 are preferred rather than the general-purpose grades such as M2 or T1, despite higher cost.

Ceramic tools, when carefully applied, can improve productivity by allowing higher cutting speeds, ranging from 150 to 1200 m/min, although with feeds of about 80% those of carbides. With ceramics, there is no need for cooling.

Depth-of-cut notching is more pronounced (versus carbides) when cutting super alloys with ceramics because of the relatively low fracture toughness of ceramics, and because the high-temperature strength of the super alloys. The tougher ceramics (SiAlON and SiC-whisker reinforced Al_2O_3) exhibit less depth-of-cut notching than Al_2O_3-TiC ceramics, which are more liable to react chemically with steel or Co-base alloys. Consequently, SiAlON and SiC whisker reinforced Al_2O_3 are more effective when machining wrought Ni-base super alloys. The cast Ni-base alloys, because of their grain structure, chip even the tougher ceramics.

Depth-of-cut notching may also exclude the use of delicate insert geometries (triangles and diamond shaped). Round inserts should be used whenever possible. They are inherently the strongest, although other shapes can be used when the depth of cut does not exceed two-thirds of the nose radius [16].

Rough cuts of wrought Ni-base alloys, when scales and run-out are present, are best turned with the tougher **SiAlON** ceramics at cutting speeds of about 150 m/min. Early experience with SiC whisker reinforced Al_2O_3 ceramics indicated an increase in toughness but a more severe notching problem. Both of these ceramics are characterized by a higher thermal shock resistance than Al_2O_3-TiC ceramics, thus promoting the use of cutting fluids. Semi-finish and finish turning operations can tolerate less insert toughness and can be done with Al_2O_3-TiC ceramics.

CBN-tools are used when turning the harder Ni-base (wrought or cast) and Co-base cast alloys. Tool holding devices must be given high consideration when super alloys are being turned, to significantly increase the tool life [16]. Sharp edge uncoated grades of **PCBN** are used for better surface finishes and close tolerances, whereas coated grades are used to enhance tool life and productivity.

Boring and Trepanning

Super alloys are bored by methods similar to those used for turning; however, speeds and feeds must be reduced because the same cooling and lubricating efficiency cannot be realized as in turning. In addition, the boring tool cannot be held as rigidly as in turning. Otherwise, the selection and application of tool materials are similar to turning but tool geometry varies. The end relief angle of boring tools must be varied inversely with the diameter being bored [16].

Trepanning has not been extensively used in machining super alloys, although Fe-Ni-base alloys have been trepanned in some applications. Experience indicated that speeds, feeds, and tool materials suitable for boring are satisfactory for trepanning under similar working conditions. Several super alloys have been successfully trepanned in the as cast condition (160–210 HB) with HSS tools made of M2 and T5, at cutting speeds of 12–15 m/min and feeds of 0.13 mm/rev.

Planing and Shaping

Planing is generally performed on large super alloys castings but is seldom done on wrought products. Nominal speeds for roughing are 6–9 m/min for HSS tools, and 40–55 m/min for carbide tools. For finishing, speeds of 8–14 m/min for HSS tools, and 45–60 for carbide tools are selected. Planing is usually performed dry, but synthetic emulsions are sometimes used.

Shaping tools with 8° side rake, 0°–3° back rake, 4°–6° relief angle, 1–1.5 mm nose radius are suitable for all types of super alloys. Ram speed must be slow when ***HSS tools*** are used, 2–4 m/min is optimum, using a feed of 0.5–0.75 mm/stroke for roughing, and 0.25–0.4 mm/stroke for finishing. Depths of cut, 1.2–2.5 mm for roughing, and 0.4–0.75 for finishing. Sulfur-free chlorinated oil is recommended as a cutting fluid in shaping [16].

Broaching

HSS is usually used for broaching super alloys. The more highly alloyed grades, such as T4, T5, and T6 are superior in terms of broach wear and life. Although acceptable results can be realized with an M2-broach for some applications involving Fe-Ni-base alloys (A-286), M3-broaches are usually considered more suitable for broaching super alloys [16].

Drilling and Allied Operations

HSS twist drills are used when drilling super alloys, although *carbide drills* are mostly used in the aircraft engine industries. The grades such as T15, M33, or M36 are preferred for drilling many types of super alloys, and their use is often mandatory to achieve acceptable drill life. The higher cost of the drills made from the more highly alloyed HSSs (commonly about four times the cost of their general-purpose grades) and the high cost of resharpening are often warranted by increased tool life. Various surface treatments such as nitriding can also be applied to HSS drills to improve tool life.

For reaming, the unsupported length of reamer shanks should be kept to a minimum. It is preferable to use ***carbide-tipped reamers***, although HSS-reamers are also used to machine wrought alloys and Fe-Ni-base cast alloys. Carbide-tipped reamers should be used when reaming mechanically alloyed (MA) products.

When reaming super alloys, it is generally desirable to use six- or eight-flute reamers; however, four-flute reamers for small holes (less than 6 mm, in diameter) are often used. Optimum angles of reamers may vary somewhat, depending on the super alloy being reamed, size of hole, and number of flutes. Tools used for counterboring and spot facing of all types of super alloys are not necessarily different from those used for similar operations on other metals. However, great emphasis should be placed on the rigidity of the machine and setup [30].

Tapping and Thread Cutting Operations

For tapping (and thread cutting) operations, provided small production quantities, taps made of a general-purpose grade of HSS (such as *M1*) will perform satisfactorily in super alloys, but surface treatment of the taps by liquid nitriding is recommended. In case larger quantities are to be tapped, the higher cost of taps made of one of the more highly alloyed HSSs (such as M4, M36, T15) is usually warranted. The recommended speeds are ranging from 2 m/min (for Ni-and Co-base alloys), to 5 m/min (for Fe-Ni-base alloys).

Milling and Sawing

Because of the interrupted cutting action in milling (and sawing), ***HSS*** is used for cutters in milling super alloys. However, ***carbides*** may be more economical than HSS when milling the more DTC alloys, such as Rene 41 and MA 600. Small solid-carbide end mills have been successfully used in some applications. The more highly alloyed grades of HSS operate better than general-purpose grades [32].

4.3 Cutting Fluids for Stainless Steels and Super Alloys

4.3.1 Functions, Characteristics, and General Considerations

The use of cutting fluids must be economically justified in the form of gains resulting due to increase of quality and productivity. Cutting fluids are commonly applied to machining operations, chiefly to

1. cool the cutting zone to increase tool life, and to improve the accuracy and surface finish;
2. lubricate the area of contact on tool flank and face to reduce friction, and accordingly tool wear, cutting forces, and consumed power;
3. flush away chips and swarf;
4. protect machine from corrosion;
5. control BUE formation on the tool.

The action of cutting fluid, being a lubricant or coolant, depends mainly on temperatures encountered, and consequently depends on the type of cutting operation and cutting speed. Cutting fluids should be applied in a copious stream. The lubricating action is only effective at low and very low speeds, while the cooling action is effective at moderate and high speeds [1].

A proper cutting fluid should be characterized by:

- high heat absorption capacity;
- good lubricating quality;
- high flash point;
- high chemical stability;
- having no emission of fumes while in contact which hot surfaces;
- having less biological and environmental hazards;
- having no corrosive effects on the workpiece and on the machine guide ways;
- being available and nonexpensive.

Since the main purpose of cutting fluids, being either straight or soluble oils, is to remove heat from and to lubricate the tool–workpiece interface, delivery of the right amount of fluid in this area is very critical. The size and form of the nozzle through which the fluid is delivered are of vital importance. For the cutting fluid to be effective, the right volume of it must be delivered to the cutting zone with the appropriate pressure. If the cutting fluid is not applied to the proper area at the tool–workpiece (WP) interface, and not in the correct amount, the fluid will heat up and lose its ability to carry heat away. This in turn will cause degradation of part finish, and decrease the tool life. In general, the temperature of the cutting fluid should not exceed 65°C.

Some sumps (usually small ones) may use chillers to cool the cutting fluid. It is important to make sure, that chillers work properly if they are installed on the machine. Emulsifiable fluids lose water due to evaporation over time. This water must be replenished to keep the proper concentration of coolant in the machine and to eliminate excessive heat buildup in the sumps.

Another important consideration is the cleanliness of the cutting fluid. Swarf, chips, grit, and dust enter the cutting fluid, and if delivered to the tool–WP interface, can ruin the finish of the workpiece. Therefore, it is highly recommended to filter and periodically change the cutting fluid in cooling tank. Before introducing new product to the tank, the tank and all pipes, pumps, and sumps must be properly cleaned and flushed. Swarf and chips attract dust and bacteria to multiply more quickly. This can further degrade the emulsion.

The general characteristics of cutting fluids and their recommendations for specific machining operations performed on stainless steels and super alloys are presented throughout this chapter. In selecting an appropriate cutting fluid and the method of its application, considerations should be given to the possible effects on the workpiece material (e.g., corrosion, stress corrosion cracking, probable staining), the components of the machine tool, health hazards, environmental effects, and recycling and disposal possibilities [17].

4.3.2 Types of Cutting Fluids

Cutting fluids may be in the form of liquids, gases, chemical solutions, lubricating solids.

4.3.2.1 Water-Base Liquids

These may be emulsion (oil dispersed in soft water), or synthetic fluids (also called chemical fluids, containing no oil, only soluble wetting agents, lubrication inhibitors, and salts).

Water-base emulsions have high thermal capacity (mass × spec. heat) compared to mineral oils. Emulsions are generally used, where the cooling action is the most important. For this reason, emulsions are frequently used in high speed cutting operations (about 90% of all cutting fluids). Emulsions are a nonexpensive type of coolants, in which oil is mixed with water in certain proportion. A small amount of soap is added as an emulsified agent; thus the emulsion has a milky white color. If the mixture is weak (oil is low), it may cause corrosion, and is of low lubricating properties. Typical concentrations (oil/water) are:

- 1/40–1/60 for grinding operations;
- 1/20 for other machining operations.

It is important to soften hard water before mixing by addition of 2 g of soda/1 l of water. The main causes of oil separation after mixing, is the incorrect mixing procedure or the excessive water hardness [1].

Synthetic fluid lubricants have a controlled molecular structure with predictable properties. Synthesized hydrocarbons such as polyalphaolefin have been used to replace mineral oil in some cutting fluid applications. Long-chain alcohols have also been used for special applications. However, a significant market was not developed for these synthetic fluids [16].

4.3.2.2 Neat Oils

These are only used when the lubricating action is the most important consideration. So they are only confined to operations of low cutting speeds, such as gear cutting, threading, tapping, reaming, broaching, and honing, to promote superior finish of machined surfaces, and to suppress the cutting forces. For operations such as tapping and reaming, greases which melt at the tool edge may be more convenient than oils. Paraffin is sometimes used on Al-alloys rather than neat oil because of its superior wetting properties [1].

Despite their much higher price, straight cutting oils find another important application in automatics, because water-based coolants, which are quite satisfactory for most turning operations, are likely to find their way to the head stock, causing oil contamination, and hence serious deterioration of indexing mechanisms [1].

Straight cutting oils may be blended from two basic types of oils:

- Mineral oils, for example, paraffin and other petroleum oils. They are cheaper and more stable than fatty oils. For these reasons, they are generally blended with fatty oils.
- Fatty oils are organic, of animal or vegetable origin (whale, lard, and rapeseed). They are expensive, and emitting an unpleasant odor, however, they are environmentally safe. They possess very high lubricating properties and promote good finishes, especially if lard oils are used for high tensile steels. On the other hand, they are less stable than mineral oils, and may decompose if used for a long time.

Sulfurized and Chlorinated EP-Oils

Extreme pressure (EP) additives are added when cutting forces are practically high (tapping and broaching), or in operations performed at high feeds. EP-additives provide a tougher and more stable form of lubrication at the chip/tool interface. These additives include sulfur, chlorine, or phosphorus compounds that react at higher temperatures in cutting zones to form metallic sulfides, chlorides, and phosphates. Most cutting oils are sulfurized by introducing chemically combined sulfur into the oil. Its main advantage is that it prevents pressure welding of chips by forming a film on the tool/chip interface that possesses anti-weld properties, thus inhibiting BUE-formation. These oils, however, may cause dark staining on some alloy steels, and super alloys, will be considered later on.

4.3.2.3 Liquid Gas or Cryogenic Coolants

These are recently used as coolants in cutting and grinding, mainly, to eliminate the adverse environmental impact caused by conventional mineral oils and emulsions. Nitrogen in liquid state (–200°C) is injected through small nozzles into the cutting zone to reduce drastically its temperature, then directly evaporates. Consequently, the tool hardness and tool life are maintained, and high cutting speeds are allowed. Because the hot chips are severely quenched, they become more brittle, hence easily broken without the need of chip breakers; they are then easily disposed and recycled. The outcome of cryogenic cooling is an improved machining economy without environmental effects.

A specially designed tool holder, as shown in Figure 4.17, is used, in which liquid nitrogen is converted into the gaseous state before coming in contact with the tool. Nitrogen is made to flow just beneath the insert through a small hole. In this design, the gas is directed toward the

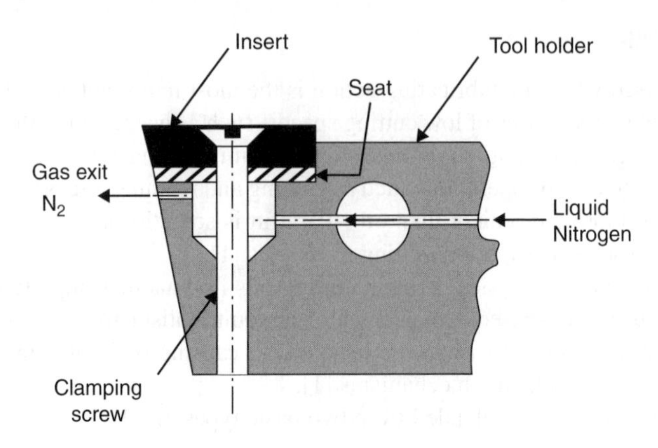

Figure 4.17 Specially designed tool holder for cryogenic cooling.

cutting edge, to cool the newly generated chips. This will enhance the chip brittleness for easy chip breaking. The cryogenic fluid exit orifice is found closer to the tip of the cutting tool.

4.3.2.4 Solid Lubricants

These are MoS_2/Ti-composite coating or WC/C-coatings and show useful results when machining Inconel 718 [35].

4.3.3 Application Methods

The method of applying cutting fluid is as important as its selection. The methods of application are [29] as follows.

1. *Manual application*: This method is not really acceptable even in job-shop situations, because it is not effective and unclean.
2. *Flooding application*: This is the commonly used method of cutting fluid application. Most machine tools are equipped with a recirculating system incorporating filters. The fluid is usually applied at a rate of up to 15 l/min for each engaged cutting edge. For convenience, the tool is flooded from the chip side, although better cooling is secured by application into the clearance cavity.
3. *Coolant-fed tooling*: There are drill and other tooling available in which holes are provided so that pressurized cutting fluids can be pumped to the cutting edges, ensuring access of cutting fluids and facilitating chip disposal [29].
4. *Mist application*: Fluid droplets suspended in air provide effective cooling through their atomization, although separate flood cooling of the workpiece may be required. Measures must be taken to limit air-born mist by the use of demister [29].
5. *Minimum quantity lubrication (MQL)*: Also called Near Dry Machining (NDM), which is defined as the dispensing of cutting fluids at optimum (very low) flow rates, tinny quantities of cutting fluid are sprayed to the cutting zone directly. The MQL provides a comparable performance in terms of tool wear, surface finish, and BUE-formation as compared to flood

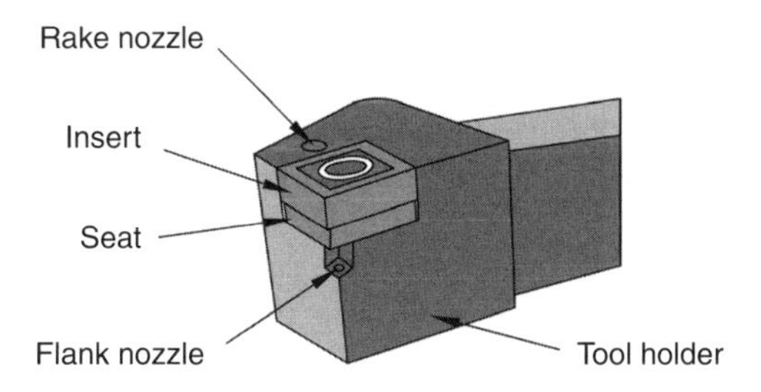

Figure 4.18 Holes in the tool holder for MQL.

application [36]. In MQL, the cutting fluid is supplied to the cutting point of a turning tool with compressed air through oil holes on both the flank and rake faces of the tool, as shown in Figure 4.18.

However, dry machining has the following advantages:

- Nonpollution of atmosphere, which reduces health hazard, in particular, skin and respiratory decreases.
- No residues are left on machined components, which reduces cleaning costs.

4.3.4 Cutting Fluids for Stainless Steels

The use of cutting fluids is more desirable for machining stainless steels than for machining carbon or alloy steels for two reasons. First, stainless steels are generally less machinable than carbon or alloy steels, and second, the lower heat conductivity of stainless steels increases the need for cooling. Restricting the flow of heat away from the cutting zone leads to burning the tool and overheating the machined surface. Overheating stainless steels surfaces, characterized by the formation of heat tinting colors, can impair corrosion resistance and so must be avoided. Overheating can also result in distortion that can be difficult to compensate for or correct.

For machining stainless steels, either mineral oils or water-soluble emulsifiable oil can be used. Mineral oils are more suited to severe machining operations with heavy loads at low speeds or where HSS tools are being used. Emulsifiable oils are used for machining at higher speeds with carbide tooling.

4.3.4.1 Sulfo-chlorinated Cutting Oils

Modern technology stainless steels cutting oils comprise premium quality, heavily de-waxed paraffinic base stocks with active EP and natural and synthetic fatty oil additives. Some of the additive packages available in cutting oils today are keyed in for specific operations or for specific stainless steels grades. Due to the complexity of these additives, stainless steel cutting oils tend to be more expensive than cutting oils used for other metals.

Two types of sulfo-chlorinated oils are generally recommended for machining stainless steels:

1. Sulfo-chlorinated mineral oil containing active sulfur, and about 8–10% fatty oil (viscosity ≈ 200 SUS at 38°C). It is generally used for the nonfree-machining stainless steels.
2. Sulfo-chlorinated mineral oil containing active sulfur, without addition of fatty-base oil (viscosity ≈ 130 SUS at 38°C). It is usually suited for the free-machining alloys of stainless steels.

Paraffin is used to dilute both oils, depending on machining condition. Normally, a 1:1 mixture of sulfo-chlorinated oil to paraffin oil is used initially. If excessive tool wear occurs, more paraffin oil should be added. If the cutting edge is tending to blunt (or burn), and the chips weld to the tool, consider reducing the dilution.

When using sulfo-chlorinated oils for machining SSs, the following guidelines should be considered (data provided by Clark Oil and Chemical, Cleveland, Ohio):

1. When starting a new stainless steel job, it is important to remember that the more difficult the job is, the more highly sulfurized the cutting oil should be.
2. When using automatic screw machines at high speed and light feeds to machine free-machining stainless steels, a mixture of one part sulfo-chlorinated oil to five parts paraffin oil is suggested. When using these machines at average speeds to perform operations that include drilling, threading, tapping, and milling of free-machining alloys, more viscous mixture of one part sulfo-chlorinated oil to three parts paraffin oil with a higher percentage of EP additives is suggested.
3. For a coarse nut tapping, bolt- and tube-threading, and broaching, a mixture of one part sulfo-chlorinated oil to two part paraffin oil is suggested. It is not advised however, to use sulfurized oils on high Ni-alloys.
4. When machining austenitic nonfree-machining stainless steels, a mixture of one part sulfo-chlorinated oil (with addition of 8–10% fatty oil) to one part of paraffin oil is suggested.
5. Generally, for machining the free-machining steels, the mixture should contain a larger percentage of paraffin oil.
6. When drilling smaller diameter holes, care must be exercised to ensure that the viscosity of the cutting oil is sufficiently low to allow it to reach the tool tip.
7. It is important to work with the oil supplier to maximize the effectiveness of the cutting oil as it relates to feeds and speeds, type of stainless steels being machined, critical tooling, and part finish. The oil supplier can accomplish this by fine tuning of their product to achieve the maximum performance for the specific application.

4.3.4.2 Emulsifiable Fluids

Water-emulsifiable (water-soluble) cutting fluids may also be used for machining stainless steels, especially in operations where greater cooling capability is necessary, such as cutting with carbide tooling at high speeds. To machine stainless steels, these fluids should contain similar EP-additives that are used in straight stainless steels cutting oils. However, water-emulsifiable fluids may only be used in machines where mixing of the cutting fluid and the lubricating oil will not happen. Many emulsifiable fluids, no matter how well formulated, may not provide the same benefits as straight oil products. Moreover, they will not withstand some of the more severe cutting operations. Nevertheless, using these fluids may in some cases result in better machined surface, and better heat removal ability from the cutting zone. The use of these cutting fluids is also more economical than using a straight oil product.

4.3.4.3 Selection of Cutting Fluid for Stainless Steels

Further information on the selection of cutting fluids for machining various types of stainless steel alloys using different machining operations is available in Table 4.9.

Codes of Cutting Fluids

A Heavy duty, active sulfur, fats, and chlorinated compounds / heavy anti-weld properties;
A2 Heavy duty, active sulfur, fats, and chlorinated compounds / heavy anti-weld properties in lighter viscosity under 120 SUS@ 35°C;
B Heavy duty, active sulfur, fats, and chlorinated compounds / heavy anti-weld properties with viscosity 190/220 SUS@ 35°C;
C Heavy duty, active sulfur, fats, and chlorinated compounds 150/170 SUS@ 35°C;
D Chlorine free, noncorrosive, heavily fortified sulfurized fatty acids, and EP additives;
E High EP, heavily fortified anti-weld, and load carrying additions with active sulfur;
F Heavy duty, inactive sulfur, fats, and chlorinated compounds / heavy anti-weld properties with viscosity 150/170@ 35°C;

Table 4.9 Suggested guidelines for cutting fluid selection for austenitic, ferritic and martensitic stainless steels

Machining operations	Austenitic SSs		Ferritic and martensitic
	Free	Nonfree	
Turning Milling Reaming Drilling	D,M,N	F,L	D,M,N
Deep hole drilling Gun drilling Trepanning	A2,L	A2	A2,L
Tapping Threading Thread chasing	C	L,D	B
Form tapping Thread rolling	E,L	B,E	E,L
Vertical broaching	L	L,A	L
Horizontal broaching	A	A	A
Sawing	L	L	L
Centerless grinding	O	O,L	O

Data are provided by Clark Oil and Chemical, Cleveland, Ohio, [36].

L Super heavy duty soluble oil with high EP-additives;
M General purpose, highly fortified synthetic with no chlorine or sulfur, with bio-stable, low foam good rust preventative additives;
N General purpose highly fortified semi-synthetic EP-additives;
O Heavy duty synthetic (made for centerless grinding).

4.3.5 Cutting Fluids for Super Alloys

Almost any cutting fluid, or none, can be used in machining super alloys. In many applications, Ni-base alloys respond well to ordinary sulfurized mineral oil. Sulfur imparts improved lubricity and anti-weld properties. The recommended cutting fluids used for machining various types of super alloys will be considered for the following machining processes:

4.3.5.1 Turning, Planing, Shaping, and Boring

Water-soluble oils in mixtures of oil: water = 1:(20–40) are frequently used in turning super alloys. Water-base chemical emulsions have also proved to be acceptable [17]. Supplying a constant flood to the cutting zone is more important than the fluid composition. In some turning applications, sulfurized and chlorinated cutting oils, applied straight or diluted 1:1, with low viscosity mineral oil are used. Diluting with mineral oil permits better mobility and cooling. Active oils are preferred to soluble oils where surface finish is critical and when HSS-tools are used [17].

If sulfochlorinated oil is used, the temperature of the oil and workpiece becomes high enough to cause brown sulfur staining of the work, which can be readily removed with a cleaning solution of sodium cyanide or chromium-sulfuric acid. This should be done before any thermal treatment, including welding and head treatment, because during further exposure to high temperature the staining may cause intergranular surface attack. To avoid this, the parts should be immersed in the above-mentioned cleaning solution only long enough to remove the stain. HSM that creates high temperatures might preclude the use of sulfurized oil because of the sulfur embrittlement of carbide tools, because sintered carbides have Ni-or Co-matrix, which is sensitive to sulfur attack at high temperatures [17].

4.3.5.2 Broaching

The main difficulty of broaching super alloys is the low cutting speed used, where BUE is liable to form on the cutting edge; consequently rapid wearing of the broach occurs. A good cooling can help considerably, because the cutting fluid inhibits the formation of for acceptable results [17]. Oils containing about 1% active sulfur, along with chlorine, and synthetic additions, are often used. A plentiful supply of cutting fluid during broaching is of equal importance, if not greater than fluid composition. It should be supplied under pressure up to 35 kPa. The cutting oil should be diluted 1:1 with plain mineral oil. Cutting oils with viscosity higher than 300 SUS are not recommended for broaching. Also, thorough cleaning of sulfochlorinated oils after broaching is extremely important before welding or heat treatment service to prevent staining damage of part surfaces.

4.3.5.3 Drilling and Reaming

Active cutting oils (sulfurized, chlorinated, or sulfochlorinated) usually provide better drill life and productivity than soluble-oil emulsions or other water-base cutting fluids. The superiority of active oils becomes more significant as the hardness of the work material and the hole depth increase. When active oils are used, all traces of oil must be removed before any treatment at high temperature. Blind holes require special attention [16].

4.3.5.4 Tapping and Thread Cutting

Sulfochlorinated oils should be used for tapping super alloys. It is recommended to be supplied in plentiful amounts, and forced under pressure of about 35 kPa through a nozzle directly in the hole being tapped. If the oil is too viscous, it can be diluted with a thinner mineral oil without seriously impairing its lubrication characteristic. Also the oils must be care fully removed after tapping or threading before any subsequent high temperature service [16].

4.3.5.5 Milling

When milling super alloys, sulfochlorinated oil is introduced in copious amounts at the exhaust side of the cutter, although soluble-oil emulsions are also used, since they provide better cooling for tools and workpieces than do straight oils. The latter are often diluted with thinner mineral (up to 50%) to provide fluidity with no large sacrifice in promoting cutting action and good surface quality.

4.3.5.6 Sawing

Chemical emulsions are generally recommended as cutting fluids for sawing super alloys. However, plain soluble-oil emulsions have also provided satisfactory results. Regardless of cutting fluid used, a plentiful supply at the area being sawed is important.

4.3.5.7 Grinding

Because super alloys have low thermal conductivity, grinding fluids must be applied at grinding area in plentiful amounts to limit heating, and distortions of the work surface. Grinding fluids for super alloys can be classified into four principal groups (Table 4.10). For efficient removal of heat, highly sulfurized water-base soluble-oil emulsions are the best for any super alloy. Sulfurized oils are appropriate; however, they remove heat less rapidly than water-base soluble-oil emulsions.

Chlorinated oil (about 1% chlorine) is particularly useful for wet dressing of form grinding wheels to a tolerance of 5 μm. Synthetic solutions and water-base soluble-oil emulsions do not have this capacity [17].

A disadvantage of chlorinated fluids is that any residual or entrapped fluid will react with super alloys during high temperature service of the workpiece. For this reason, most

Table 4.10 Identification and classification of grinding fluids for super alloys

Grinding Fluid	Remarks
Soluble-oil emulsions (regular)	
S1	Contains soap
S2	Contains soap and fatty materials
S3	Emulsified kerosene
Soluble-oil emulsions (heavy-duty)	
H1	Contains sulfur and chlorine
H2	Contains sulfurized fats, designed for SS
H3	Contains sulfur and extreme-pressure additives, high percentage of fats
H4	Contains fatty materials, synthetic soaps, designed for SS
Chemical (synthetic) solutions	
C1	Contains 35% potassium nitrite (KNO_2) before dilution for use
C2	Contains a moderate percentage of sodium nitrite
C3	Based on synthetic wax
Grinding oils	
G1	Transparent sulfochlorinated grinding oil containing fats, 4% sulfur, and 2% chlorine (both active), viscosity SUS at 40°C
G2	Dark sulfochlorinated grinding oil containing fats, 3% sulfur, and 0.5% chlorine (both active), viscosity 190 SUS at 40°C
G3	Inactive grinding oil containing fats, no added sulfur or chlorine, viscosity 300 SUS at 40

Adapted from ASM Handbook [16].

users of super alloys do not prefer using chlorinated grinding fluids [17]. As in other TMPs (traditional machining processes), the selection of cutting fluids is mainly based on cost considerations. Emulsions are the least expensive, while grinding oils are the most expensive grinding fluids.

The type of grinding fluid has an effect on the grindability. Table 4.11 shows the effects of water-base fluids on the grinding ratio (G-ratio) of two selected super alloys, U-500 and J-1570. An alumina grinding wheel A-60-H-8-V is used. The G-ratio is the volume of metal removed per volume of wear of the grinding wheel. The higher this index is, the easier the material to grind. In Table 4.11, all the grinding fluids were of 10% concentration, which is higher than normal. Grinding dry (air) proved to be less satisfactory than grinding with most of the water-base fluids. Grinding with plain water resulted in a very low G-ratio. The highest G-ratios were obtained with synthetic fluids. Detailed information on grinding fluids can be found in reference [17].

Table 4.11 Effect of fluid on the grinding ratio of two super alloys (Grinding wheel A-60-H8-V7)

Grinding fluid	Grinding ratio obtained for	
	U-500	J-1570
C1	3.5	2.8
H1	3.3	—
H2	1.5	—
C4	—	1.7
H3	—	1.3
C7	—	1.3
S1	—	4.1
S3	—	4.4
Air	—	0.9

Adapted from ASM Handbook [16].

References

[1] Youssef, H., El-Hofy, H. (2012) *Principles of Traditional and Nontraditional Machining*, Dar Elfath Press, Alexandria.

[2] Youssef, H.A., El-Hofy, H., Ahmed, M.H. (2011) *Manufacturing Technology – Materials, Processes, and Equipment*, 1st edn. CRC Press, Taylor & Francis Group, Boca Raton, FL.

[3] Axinte, D.A., Andrews, P. (2007) Some considerations on tool wear and workpiece surface quality of holes finished by reaming or milling in Ni-base super alloys. *Proc. Inst. Mech. Eng., Part B* **221**, 591–603.

[4] Groover, M.P. (2010) *Fundamentals of Modern Manufacturing, Materials, Processes, and Systems*, 4th edn. John Wiley & Sons Inc.

[5] Kienzle, O., Victor, H. (1957) Spezifische Schnittkräfte bei der Metallbearbeitung. Werkstattstechnik und Maschinenbau, S. 222–225.

[6] Wikipedia. Machinability http://en.Wikipedia.org/wiki/Machinability (accessed March 12, 2011).

[7] Uehara, K., Skuurai, M., Takeshita, H. (1973) Cutting performance of coated carbides in electric hot machining of low machinability metals. *CIRP Ann.* **32**(1), 97–100.

[8] Maity, K.P., Swain, P.K. (2007) An experimental investigation of hot machining to predict tool life. *J. Mater. Process. Technol.* **198**, 344–349.

[9] Anderson, M., Patwa, R., Shin, Y.C. (2006) Laser-assisted machining of Inconel 718 with an economic analysis, Center for Laser-based Manufacturing, School of Mechanical Engineering, Purdue University, USA, November 2005, *Int. J. Mach. Tools Manuf.* **46**, 1879–1891.

[10] Kottenstette, J.P., Recht, R.H. (1982) Ultra-high-speed machining experiments. *Proceedings of NAMREC, 1982*, pp. 263–270.

[11] Sharma, V.S., Dogra, M., Suri, N.M. (2008) Advances in turning process for productivity improvement – A review, *Proc. Inst. Mech. Eng, Part B*. **222**, 1417–1442.

[12] Turkovich, B.F. (1979) Influence of very high cutting speed on chip formation mechanics. *Proceedings of NAMRC-VII, 1979*, pp. 241–247.

[13] Vaughn, R.L. (1960) Ultra high speed machining, *Am. Mach.* **107**(4), 111-126.

[14] Vaughn, R.L. (1960) *Recent Developments in Ultra High Speed Machining*. Technical paper 255, Vol. **60**, Book 1, Society of Manufacturing Engineers.

[15] Vaughn, R.L. (1960) *Ultra High Speed Machining – Feasibility Study*, Final report, contract AF 33 (600) 36232, PED, Lockheed Aircraft Corporation, June 1960.

[16] Flom, D.G., Komanduri, R. (1989)*ASM Handbook: Machining*, Vol. **16**, ASM International, Materials Park, OH

[17] Kalpakjian, S., Schmid, S.R. (2003) *Manufacturing Processes for Engineering Materials*, 4th edn. Prentice Hall, Inc., Upper Saddle River, NJ.

[18] Flom, D.G. (1985) High speed machining, in *Innovations in Materials Processing* (eds G. Bruggeman, V. Weiss), Plenum Press, pp. 417–439.
[19] Manyindo B.M., Oxley, P.L.B. (1986) Modelling the catastrophic shear type of chip when machining stainless steels, *Proc. Inst. Mech. Eng. Part C*, **200**.
[20] Komanduri, R. (1982) Some clarification on the mechanics of chip formation when machining Ti-alloys. *Wear* **76**, 15.
[21] Kramer, B.M. (1984) *On Tool Materials for High Speed Machining, in High Speed Machining*, ASME, pp. 127–140.
[22] Skelton, C. (1969) Effect of ultrasonic vibration on the turning process. *Int. J. Mach. Tool Des. Res.* **9**, 363–374.
[23] Babitskya, V.I., Kalashnikov, A.N., Meadows, A., Wijesundara, A.A. (2003) Ultrasonically assisted turning of aviation materials. *J. Mater. Process. Technol.* **132**, 157–167.
[24] Wang, Z.Y., Rajurkar, K.P., Murugappan, M. (1997) Wear of CBN tools in turning of silicon nitride with cryogenic cooling. *Int. J. Mach. Tools Manuf.* **37**(3), 319–326.
[25] Paul, S., Dhar, N.R., Chattopadhyay, A.B. (2001) Beneficial effects of cryogenic cooling over dry and wet machining on tool wear and surface finish in turning AISI 1060 steel. *J. Mater. Process. Technol.* **116**, 44–48.
[26] Diniz, A.E., Micaroni, R. (2007) Influence of the direction and flow rate of the cutting fluid on the tool life in turning process of AISI 1045 steel, *Int. J. Mach. Tools Manuf.* **47**, 247–254.
[27] Yong, A.Y.L., Seah, K.H.W., Rahman, M. (2006) Performance evaluation of cryogenically treated tungsten carbide tools in turning. *Int. J. Mach. Tools Manuf.* **46**, 2051–2056.
[28] Shaw, M.C. (1982) *Metal Cutting Principles*.Clarendon Press, Oxford.
[29] Schey, J.A. (2000) *Introduction to Manufacturing Processes*, 3rd edn. McGraw-Hill, International Editions.
[30] Bhattacharyya, S.K., Pashby, I.R., Ezugawu, E., Khamsehzadeh, H. (1987) Machining Inco 718 and Inco 901 super alloys with SiC-whisker reinforced Al_2O_3 composite ceramic tools. *Proceedings of the 6th International Conference on Production Engineering, Osaka, Japan, 1987*, pp. 176–181.
[31] Ezugwu E.O. (2004) High speed machining of aero-engine alloys. *J. of the Braz. Soc. of Mech. Sci. & Eng.* **26**(1): 1–11.
[32] http://www.bssa.org.uk/ (accessed June 12, 2015).
[33] Liew, W.Y.H. (2008) Ultra-precision machining of stainless steel using coated carbide tools, *Arch. Mater. Sci. Eng.* **31**, 2, 117–120.
[34] ISCAR-Cutting Tools Ltd (2002).
[35] Dudzinski, D., Devillez, A., Moufki, A., Larrouquère, D., Zerrouki, V., Vigneau, J. ((2004) A review of developments towards dry and HSM of Inconel 718, *Int. J. Mach. Tools Manuf.* **44**, 439-456.
[36] Byrnel, G.D. Dornfeld, B. Denkena (2003) Advancing cutting technology, *CIRP Ann.* **52**(2) 483–507.
[37] Clark Oil and Chemical Cleveland, Ohio (2002) Guide Lines Suggested.

5

Traditional Machining of Stainless Steels

5.1 Machinability of Stainless Steels

Ordinary stainless steels (SSs) described in Chapter 2, are difficult-to-machine because of their work hardening properties and their tendency to seize during cutting. For these reasons, special free-machining alloys have been developed, within each family of the basic alloys by the addition of free-machining elements such as S, Se, Te, and so on, thus forming inclusions in these stainless steels that significantly improve their overall machining characteristics. However, the benefit of improved machinability due to the addition of these elements is not of course obtained without degrading other important properties of stainless steels such as:

- corrosion resistance;
- strength, ductility, and toughness;
- hot workability;
- cold formability;
- weldability.

The improvement in machinability must be balanced against the possible reduction of the degraded properties, especially corrosion resistance.

Machining of Stainless Steels and Super Alloys: Traditional and Nontraditional Techniques,
First Edition. Helmi A. Youssef.

5.1.1 Free-Machining Additives of Stainless Steels

Free-machining additives form inclusions which increase tool life, allow high cutting rates, and may also affect disposability, and surface finish. Important free-machining additives are:

1. *Sulfur (S)*: Its use to improve machinability of stainless steel dates back to the early 1930s. The amount of sulfur addition is limited by the allowable degradation of other properties. Increase in sulfur continuously improves machinability, although at a decreasing rate (Figure 5.1). Increasing sulfur also produces relatively large increases of tool life, even within the typical limit of sulfur content (0.03%) allowed for the non-free-machining austenitic alloys (AISI-304).

   According to Figure 5.1, it is depicted that the machinability of free-machining stainless steels decreases gradually again if sulfur ratio is increased to above 0.3–0.35%. The graph validates published literature that more sulfur, especially at levels beyond 0.3% shows a diminishing return to machinability. The main culprit is the loss of lubricity at the elevated machining temperatures due to high cutting speeds. At that point, sulfur acts mainly as a chip breaker that stops lubricating the tool [2]. Sulfur is present in stainless steels as discrete sulfide inclusion (MnS) (Figure 5.2). Sulfides that are larger and more globular provide greater machinability than smaller and elongated ones [3–5]. Although larger sulfides enhance tool life and disposability (ease of cut), they decrease surface finish [5]. When cutting the free-machining austenitic SS AISI 303, the mean surface roughness (Ra) increases linearly with the increase of sulfide inclusion area as schematically shown in Figure 5.3. Moreover, the increase of the sulfide area deteriorates again the machinability.

   Except at low level of Mn, the sulfides in free-machining stainless steels are basically manganese sulfides. Higher Mn (or Mn/S ratio) will further improve machinability of free-machining stainless steels [6] (Figure 5.4). The use of higher Mn to increase machinability has a deteriorating effect on the corrosion resistance. SSs containing Mn-rich sulfides can be less corrosion resistant compared to those containing sulfides with higher Cr-content. Free-machining austenitic alloys are not affected to the same extent as free-machining ferritic and martensitic alloys [7].

   Manganese can also be intentionally limited in AISI-303 or AISI-416 to give these free-machining alloys improved corrosion resistance. Another approach is to produce alternative sulfides such as titanium sulfides. An example of this type is the free-machining ferritic stainless steel (S18235) (Table 2.1).
2. *Selenium (Se)*: This is another commonly used free-machining agent after S in stainless steels, forming inclusions of MnSe analogous to sulfides. Sulfur is used in Europe, whereas Se is frequently used in USA. Se is less effective than an equivalent weight percent of S in improving the overall machining characteristics of stainless steel (better drillability) [8] (Figure 5.5). However, Se-bearing alloys provide a better machined surface finish than S-bearing alloys [9] (Figure 5.6). In addition, Se-bearing stainless steels may offer improved cold formability and somewhat improved corrosion resistance compared to the corresponding S-bearing alloys. Se additions have also been used in S-bearing alloys to promote sulfides that are larger and more globular, and which are more beneficial to machinability.
3. *Tellurium (Te)*: This is like Se and S, because it forms inclusions similar to sulfides (MnTe). Tellurium is more effective in improving the machinability of austenitic stainless steel [10]

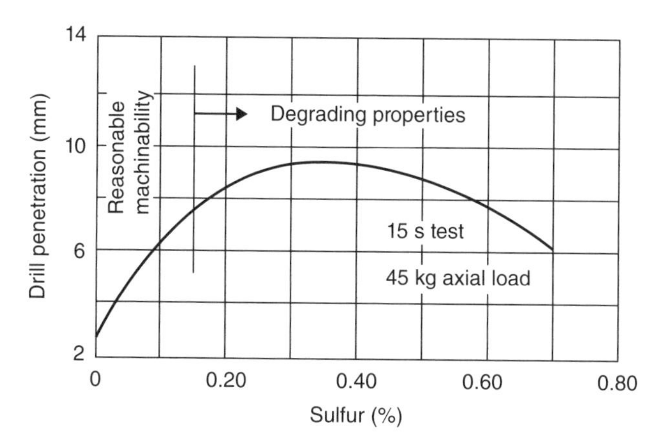

Figure 5.1 Effect of sulfur content on machinability in a drill penetration test for an AISI-304(18Cr-9Ni) austenitic SS. (Adapted from Metals Handbook [1].)

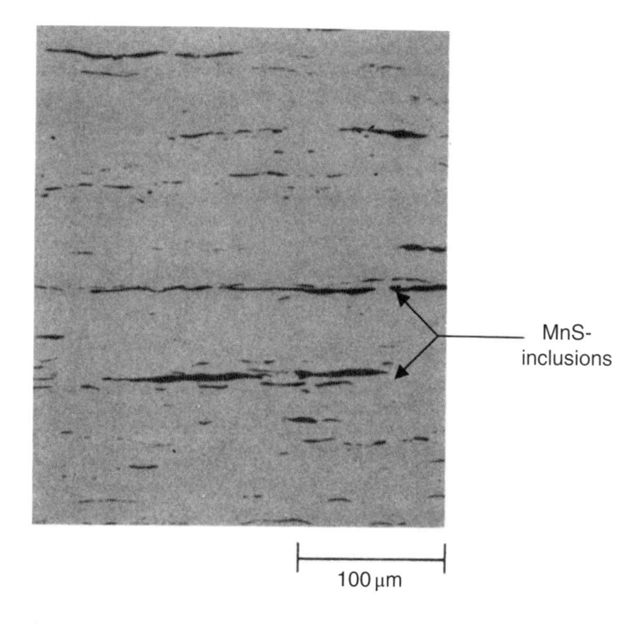

Figure 5.2 Photo of sulfide inclusions in an AISI-303 free-machining austenitic SS-alloy.

(Figure 5.7). Like Se, tellurium can be used to promote sulfides that tend to retain their globular shape [11]. The use of Te is limited, primarily because of its adverse effect on hot workability, particularly in case of austenitic alloys. However, Te has recently been used in ferritic alloys in conjunction with sulfur and lead [9].

4. *Lead (Pb) and bismuth (Bi)*: Both Pb and Bi have low solubility in stainless steel, forming metallic inclusions that improve machinability [10]. Lead is more beneficial to machinability of austenitic stainless steels than other free-machining additives [10] (Figure 5.8). It is also reported that the use of Pb, with or without limited S, results in better machined

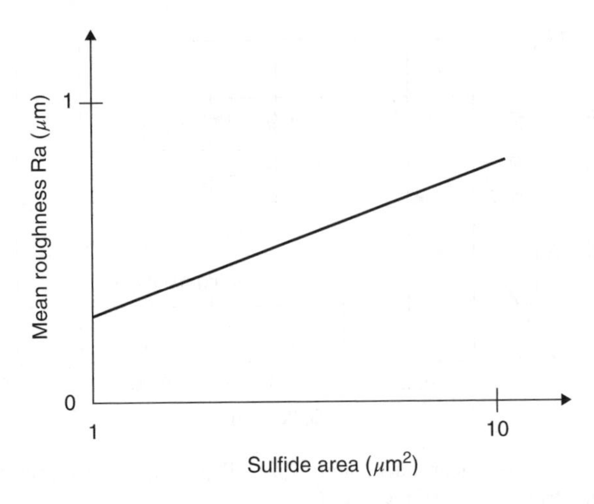

Figure 5.3 Schematic relationship showing the effect of the sulfide inclusion area on surface roughness Ra of an AISI-303 free-machining austenitic SS-alloy.

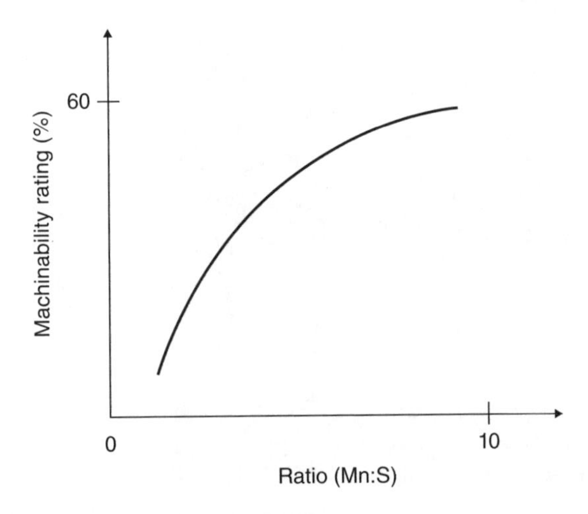

Figure 5.4 Effect of Mn:S ratio on the machinability of free-machining martensitic stainless steel – schematic relationship.

surface finish, corrosion resistance, and cold formability than the use of S alone [13]. Similar effects have been attributed to the use of Bi [14, 15]. Despite the benefits realized by Pb and Bi, stainless steels containing these elements are commercially available to a limited extent.

Problems with the use of Pb include toxicity, reduction in hot workability, and erratic machinability associated with the difficulty in obtaining a uniform dispersion in the steel [10, 14, 15]. Boron addition has been used to alleviate hot workability problems [16].

5. *Phosphorous (P)*: This is added in conjunction with S or Se to enhance machinability, not by forming inclusions, but by modifying the matrix properties of the alloy. The purpose in adding P is to embrittle tough austenitic alloys.

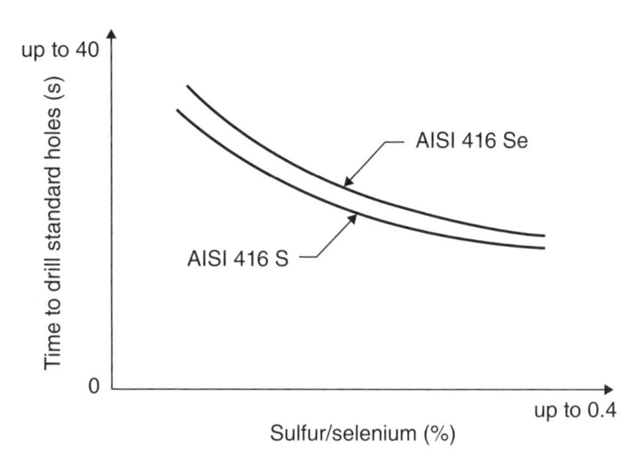

Figure 5.5 Machinability comparison test of S- and Se-bearing free-machining martensitic stainless steels 416 and 416 Se. (From Clarke [8].)

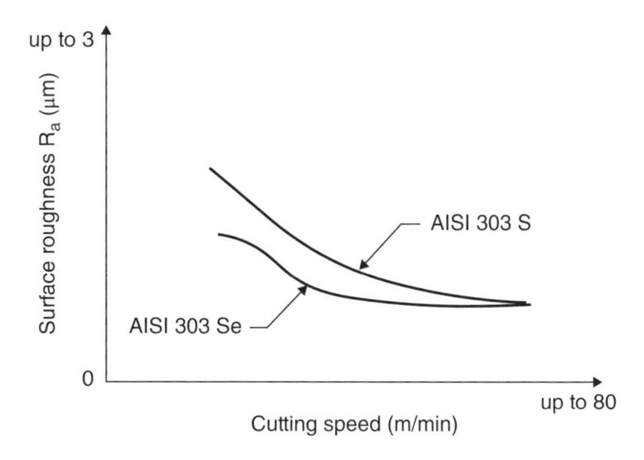

Figure 5.6 Surface finish comparison test of S- and Se-bearing free-machining austenitic steels 303 and 303 Se. (From Tipnis [9].)

5.1.2 Machinability of Free- and Nonfree-Machining Stainless Steels

The free-machining alloys are found in applications where mechanical properties are second in level of importance. Their properties are specifically designed to facilitate the cut. Tables 2.1–2.3 list the free-machining alloys as derived from the three basic stainless alloys. Free-machining alloys are currently unavailable in both the duplex, and the PH-SSs. Duplex alloys are noted for excellent corrosion resistance but have somewhat limited hot workability. The addition of free-machining elements, which would likely degrade both properties, is undesirable. Similarly, PH-alloys are noted for good toughness at high strength levels, making it undesirable to add free-machining elements, which degrade toughness.

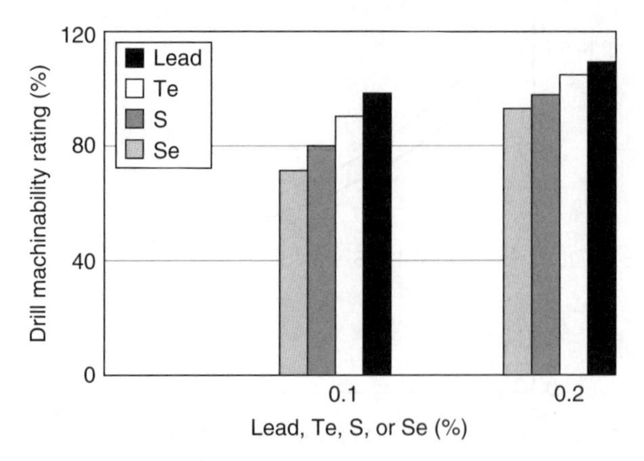

Figure 5.7 Machinability of lead-, Te-, S-, and Se-bearing austenitic Stainless steel 18Cr/9Ni – a comparison. (From Kovach and Eckenrod [10].)

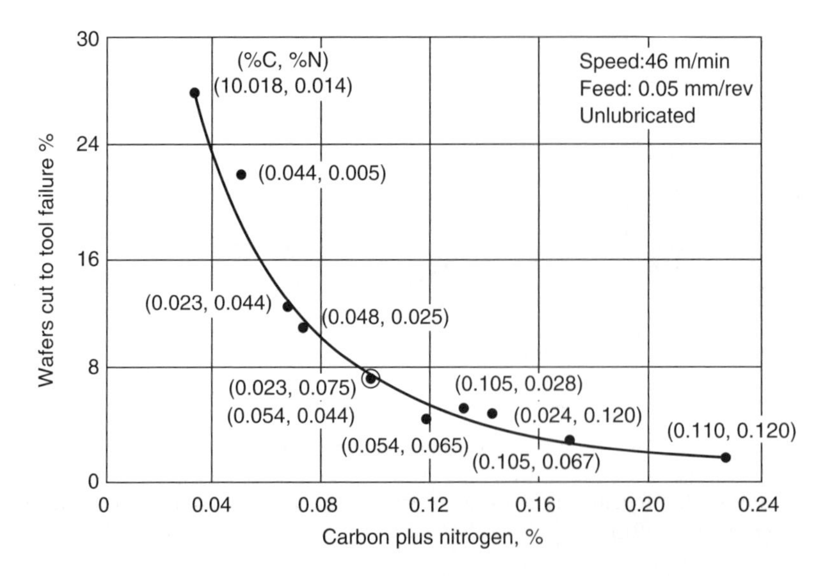

Figure 5.8 Effect of C- and N-contents on machinability (expressed in tool failure) for a free-machining austenitic steel S30310 [XM-5]. (Cited: Eckenrod *et al.* [12]. U.S. Patent 4,613,367 – 1986.)

Table 5.1 lists some nonfree- and free-machining (S-and Se-bearing) alloys within the three basic families (ferritic, martensitic, and austenitic). The best-known nonfree-machining alloys in the basic families are AISI-430 (ferritic), AISI-410 (martensitic), and AISI-304 (austenitic). They have corresponding free-machining alloys (Table 5.1). In addition, the more corrosion-resistant Mo-bearing alloys include AISI-444 (ferritic), and AISI-316 (austenitic), having free-machining versions, and the higher carbon, higher strength alloys AISI-420 (martensitic), and AISI-440C (martensitic), having free-machining versions in martensitic family.

Table 5.1 Nonfree-machining basic alloys and derived free-machining (S- and Se-bearing) alloys

Nonfree-machining	Related free-machining	
Alloys	Selenium-bearing	Sulfur-bearing
Ferritic		
S43000	S43023	S43020
S44400	—	S18200[a]
Martensitic		
S41000	S41623	S41600, S41610[b]
S42000	S42023	S42020
S44004	S44023	S44020
Austenitic		
	—	S20300
S30200/S30400	S30323	S30300
		S30310[b]
		S30345[c]
		S30360[d]
S30430	—	S30330
		S30431[e]
S31600	—	S31620
S34700	S34723	S34720

[a] Does not contain titanium.
[b] Contains high manganese.
[c] Contains aluminum.
[d] Contains lead.
[e] Contains lower copper.
Source: Metals Handbook [1], with permission.

Significant differences in machinability exist between different types of stainless steels including free-machining alloys. These will be presented in the following sections.

5.1.2.1 Ferritic and Martensitic Alloys

Free-machining ferritic alloys such as S43020 (AISI-430F), and annealed, low carbon martensitic free-machining alloys such as S41600, (AISI-416) (Tables 2.1 and 2.2), are the easiest to machine among stainless steels [9, 17, 18]. Their machinability ratings are close to those of certain free-machining carbon steels [17–19].

The nonfree-machining of lower Cr-ferritic alloys (S40500, S43000) (Table 2.1), and annealed low carbon, straight-Cr martensitic alloys (S40300, S41000) (Table 2.2), are generally easier to machine than most other nonfree-machining alloys [17]. The higher Cr-ferritic nonfree-machining alloys, such as S44600 (Table 2.1), are somewhat more difficult to machine than the lower Cr-alloys, because of their gumminess and stringy chips.

Besides the presence or lack of a free-machining additive, the machinabilily of martensitic alloys are influenced by:

- hardness level
- C-content
- Ni-content
- phase balance, that is the percentage of the free- or the δ-ferrite in the martensitic matrix.

In martensitic alloys, the machinability decreases as the carbon content increases from: S41000 (AISI-410) to S42000 (AISI-420) to S44004 (AISI-440C) for the nonfree-machining alloys, and from S41600 (AISI-416)/S41623 (AISI-416Se) to S42020 (AISI-420F)/S42023 (AISI-420FSe) to S44020 (AISI-440F)/S44023 (AISI-440FSe) (Table 2.2). With higher C- levels, there is a tendency for smaller difference in machinability between the corresponding free- and nonfree-machining versions. This trend is primarily due to the large quantities of abrasive chromium carbides present as C-level increases in these alloys. Ni-content also influences machinability by increasing the annealed hardness levels. Consequently, alloys such as S41400 (AISI-414) and S43100 (AISI-431) will be more difficult to machine than S41000 (AISI-420) (Table 2.2), in the annealed condition [19, 20]. It has been found that increasing free- or δ-ferrite content results in improved machinability, tool life, and surface finish [3, 8, 21]. The introduction of a higher ferrite-content also results in decreasing the hardness of the alloy.

5.1.2.2 Austenitic Alloys

The difficulties in machining stainless steels are, in general, more specifically attributed to the austenitic stainless steels. Compared to ferritic and martensitic alloys, typical austenitic alloys have a higher work-hardening rate, a wider spread between yield tensile strength (YTS) and ultimate tensile strength (UTS) (Figure 2.3), and higher toughness and ductility. The high ductility of these steels works against them in machining. Poor chip breaking and built up edge (BUE) at cutting face can easily occur. The thermal conductivity of austenitics are low, compared to other types of stainless steels (Table 5.2), so heat can easily build up at the tool face. Distortion or poor tolerance control during machining can be affected by the higher thermal expansion rates of these steels [3]. In their annealed condition, these steels are not ferromagnetic.

This means that magnetic clamping devices cannot be used. The combination of these effects makes austenitics appear difficult to machine. When machining austenitic stainless steels, particularly, the nonfree-machining alloys are more characterized by:

- Tools become hotter, tending to form a large BUE.
- They form stringier chips are, with a tendency to angle, making their disposal more difficult.

Table 5.2 Thermal conductivities of austenitic SSs as compared to ferritic SSs and carbon steel

Work material	Thermal conductivity (W/m K)
Carbon steel	44
Austenitic SS (S30200)	16
Ferretic SS (S43000)	23

- Chattering occurs if tool rigidity is inadequate or marginal.
- They produce work-hardened surfaces, which are more difficult to machine, especially if the feed rate is too low.

Accordingly, the general precautions for machining stainless steels are particularly important when machining austenitic alloys. However, a moderate amount of cold working may be regarded as beneficial to the overall machining characteristics of austenitic alloys [21]. The cold working reduces the ductility, resulting in cleaner chips, and less tendency for a BUE formation. This, in turn, produces a better machined surface finish but with some loss of tool life due to higher workpiece hardness [20].

Carbon and nitrogen can affect the work-hardening rate and increase the strength and hardness of austenitic stainless steels. Higher levels of either or both elements decrease machinability [12] (Figure 5.8). Consequently, high-N austenitic alloys such as S20910 and S28200 (Table 2.3) are more difficult to machine than the standard lower-N austenitic alloys [18]. Strong (carbide/nitride)-forming elements, including Ti and Nb are used in stainless steels such as S32100 and S34700 (Table 2.3), to prevent the formation of carbides on grain-boundaries, which can reduce intergranular corrosion resistance. However, the (carbide/nitride) – inclusions are of abrasive nature, thus tending to increase the tool wear [22].

5.1.2.3 Duplex Alloys

The machinability of duplex stainless steels is generally poor. It is limited due to their high annealed strength levels. Machining of duplex causes chip hammering, and generates a lot of heat, which causes plastic deformation and severe crater wear. Small entering angles are preferable to avoid notch wear and burr formation. Good tool clamping, and work fixation are essential.

Table 5.3 compares the machinability (in terms of drilling depths) of duplex alloy S32950, with that of high-N austenitic alloy S20910, and a conventional austenitic alloy S31600 of standard S-content of 0.004%, and enhanced-machining version of S-content of 0.03% (Table 2.3). The duplex S32950 has a hardness level comparable to that of the high-N austenitic alloy S20910, but provides higher machinability. However, it is not machinable as either the standard or the enhanced-machining alloy S31600. Other N-bearing duplex

Table 5.3 Comparison of machinability in terms of drill penetration depth (mm), for a duplex SS (S32900), a high-N austenitic SS (S20910), standard austenitic SS (S31600), and enhanced machining a lower nitrogen austenitic SS (S31600)

Work material	Drill penetration depth (mm)
High N-austenitic SS (S20910), 98 HRB	1.9–2.4
Duplex SS (S32950), 100 HRB	2.4–3.0
Standard austenitic SS (S31600), 79 HRB	3.2–3.7
Enhanced lower N-austenitic SS (S31600), 76 HRB	3.7–4.2

Adapted from Metals Handbook [1], (25 seconds, 45kg thrust load).
HRB = Hardness Rockwell B.

alloys are expected to machine similar to S32950. No enhanced-machining versions of duplex alloys are available.

5.1.2.4 PH-Alloys

The machinability of PH-SSs depends on the type and hardness level of the alloy. Martensitic PH-SSs (Table 2.7), are often machined in the solution-treated condition. Therefore, only a single-aging treatment is required afterward to reach the desired strength level. In this condition, the relatively high hardness limits the machinability. Most of these alloys machine comparable to, or somewhat worse than, a standard austenitic alloy such as S30400. The martensitic PH-alloy S17400 (Table 2.7) is available in enhanced-machining version that allows machining at higher cutting speeds with a significantly reduced tendency toward chattering. Martensitic PH-SSs can also be machined in the aged condition, so that heat treating can be avoided to maintain closer tolerances. In the annealed austenitic condition, semi-austenitic alloys (Table 2.7) can be expected to machine with difficulty, somewhat worse than an alloy such as S30200 (Table 2.3), which has a high work-hardening rate. Semi-austenitic alloys S35000 and S35500 (Table 2.7), provide best machinability if supplied in over-tempered condition. As with martensitic PH-SSs, the machinability decreases with aged hardness level. Austenitic PH-alloys such as S66286 (Table 2.7) machine poorly, requiring lower speeds than even the highly alloyed austenitic stainless steels [18]. Machining in the aged condition requires even lower speeds.

5.1.3 Enhanced Machining Stainless Steels

In some cases, compositional changes and processing of SS-alloys can be modified within broad limits to provide optimum level of machining performance. This approach is valid for nonfree- and free-machining alloys, resulting in enhanced-machining versions without affecting the desired properties of the standard alloys. It should be emphasized that the enhanced-machining versions of nonfree-machining alloys provide machining performance superior to that of the corresponding standard alloys, but still do not provide the machinability of comparable free-machining alloys. However, other properties are superior to those of the corresponding free-machining alloys. Whereas free-machining alloys are only available in ferritic, martensitic, and austenitic alloys, the enhanced machining versions are available in all types of stainless steels.

5.1.4 Machinability Ratings of Stainless Steels

From the previous discussion about 50 grades of stainless steels, comprising nonfree- and enhanced machining alloys, are considered. They are grouped according to their machinability levels. Table 5.4 provides a proposal for ranking these grades through a 10-level machinability in a descending order, based on a reference material (resulfurized and rephosphorized plain carbon steel AISI-1212). It shows also the recommended cutting speeds used. The machinability rating of the Ni-based heat resisting super alloy (Hastelloy X), and some plain carbon and alloy steels have been considered for comparison purposes.

Table 5.4 Machinability ratings for stainless steels and some selected steel alloys and related cutting speeds in descending order

Grade (AISI-designation)	Approximate cutting speed (m/min)	Machinability rating based on 1212 (R%)
Resulfurized and rephosphorized free cutting plain carbon steel AISI 1212	60	100
Stainless steels: in annealed condition unless otherwise stated		
1. Mart. 416	66	110
2. 203/303/Ferr. 430F	51	85
3. Mart. 420F/Ferr. 430	40	67
4. Mart. 403/Mart. 410	32	54
5. Mart. 416 High Tens./Mart. 420//Mart. 422/Mart. 17-4 PH-H1150 (age hardened)/Standard Duplex 2205	28	50
6. 303 High Tens./Mart.431/Mart.440FSe/ Mart.15-5 PH/Mart. 15-7PH-(Custom 450)/Mart. 17-4 PH/Semi-aust. 17-7 PH	30	45
7. 302/304/304L/Pyromet,Semi-aust.AM350PH/ Semi-aust.AM355 PH/Mart. 440A/Mart. 440C	24	40
8. 309/310/316/316L/317/317L/317LM/321/347/ Ferr.446/Mart. 13-8M PH	21	36
9. 302″B″/304″B″/Mart. Custom 455 PH/Aust. Pyromet A286 PH/Duplex 7-Mo Plus	18	30
10. Mn-aust.: Nitronics 40,50,60/316″B″	13	22
Superalloy: Hastelloy X	10	17
Plain carbon and alloy steels: In annealed condition		
1020,1015	43	72
1050	32	54
1141	48	81
L 1412	102	170
52100 (ball bearing steel)	24	40

Collected data from Falcon Metals Group [23], BSSA [24], and Carpenter [25].

5.2 Traditional Machining Processes of Stainless Steels

In this section, the machining of stainless steels using different machining processes will be treated. The machining parameters, the tool geometries, and so on are specifically considered. The cutting tool materials and the cutting fluids have been previously considered in Chapter 4.

5.2.1 Turning

Traditionally, high speed steel (HSS)-tools have been used for most turning operations of SSs, but carbides and carbide-coated tips are also used to realize higher removal rates. The choice of tool material depends on machining parameters, production rate, and the available power and rigidity of machines.

Figure 5.9 illustrates the suggested tool geometry of single-edge tool, intended for turning stainless steels, whereas Table 5.5 suggests the recommended cutting speeds, and depths of cut for turning SSs, using HSS- and carbide tools. Too high speed can result in tool tip burning, and too low speed can result in BUE-formation. Cutting fluids must be used.

Referring to Table 5.6, tools with positive side rake will generate less heat and cut freely with a cleaner surface. It is beneficial to select as large a tool as possible to provide a greater heat sink as well as a more rigid setup. To ensure adequate support for the tool, the front clearance angle should be kept to a minimum that is 7–10°.

Because of their toughness, and work-hardening characteristics, austenitic stainless steels require HSS-tools ground with side rakes of 10–15° to control chips, and may require increased side relief angles of 5–6° (Table 5.6), to prevent rubbing and localized work-hardening.

Nonfree-machining stainless steels tend to produce long stringy chips that can be very troublesome. This difficulty can be alleviated by using chip curlers, or chip breakers, that, in addition to controlling long chips, and reduce friction on tool face. Chip breakers or curlers for free-machining stainless steels do not need to be as deep as those needed for the nonfree-machining alloys. If a chip curler or breaker cannot be ground into the tool, it is advisable to have a steep side rake angle.

Carbide tools can also be used. They allow higher speeds than HSS-tools. However, they require greater attention to the rigidity of the tooling and workpieces; moreover, interrupted cuts should be avoided. Speeds for coated carbides are approximately 30% higher than those for carbides.

5.2.1.1 Form Turning and Cutting Off

In form turning, the side relief and clearance angles should be between 1° and 5°. The deeper the cut, the larger are these angles. The above-center distances should be about 3 mm. Sufficient depths and feeds should be allowed to avoid surface work-hardening problems. This applies to

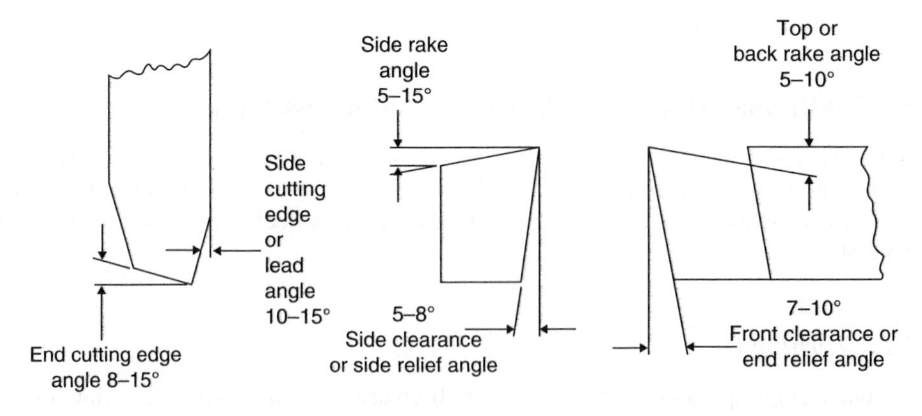

Figure 5.9 Suggested geometry of single-edged turning tools for machining stainless steels.

Table 5.5 Recommended speeds and feeds for stainless steels using single edge turning tool

Stainless steel alloy	Treatment and HB	Rough-R or finish-F	Speed (m/min)			Tool/code[a]	
			HSS	Carbides		HSS	Carbides
				Brazed	Insert		
Wrought martensitic							
Free-machining	Ann.-160	R	45	150	170	M2/M3	C6/C7
grades		F	52	170	190	M2/M3	C6/C7
Lower C/lower Cr	Ann.-175	R	33	120	145	M2/M3	C6/C7
grades		F	40	145	175	M2/M3	C6/C7
Lower C/lower Cr	Q&T-300	R	20	80	100	M2/M3	C6/C7
grades		F	25	98	120	M2/M3	C6/C7
Higher C/higher	Ann.-240	R	20	95	105	M42/T15	C6/C7
Cr grades		F	24	105	120	M42/T15	C6/C7
Higher C/higher	Q&T-320	R	15	67	80	M42/T15	C6/C7
Cr grades		F	21	80	90	M42/T15	C6/C7
Wrought ferritic							
Free-machining	Ann.-160	R	45	150	165	M2/M3	C6/C7
grades		F	50	165	185	M2/M3	C6/C7
12–17% Cr grades	Ann.-160	R	33	137	155	M42/T15	C6/C7
		F	40	155	180	M42/T15	C6/C7
Wrought austenitic							
Free-machining	Ann.-160	R	31	125	140	M2/M3	C2/C3
		F	36	140	155	M2/M3	C2/C3
Other grades (304,	Ann.-160	R	23	85	95	M42/T15	C2/C3
316, 321, etc.)		F	28	97	112	M42/T15	C2/C3
Wrought duplex	Ann.-230	R	24	53	55	M42/T15	C2/C3
		F	30	60	70	M42/T15	C2/C3
Cast plain Cr	N&T-210	R	21	83	100	M42/T15	C2/C3
		F	27	105	130	M42/T15	C2/C3
Cast austenitic	Ann.-270	R	15	68	83	M42/T15	C2/C3
		F	20	83	100	M42/T15	C2/C3

Roughing (R): Depth of cut = 3.8 mm, feed = 0.38 mm/rev.
Finishing (F): Depth of cut = 0.75 mm, feed = 0.2 mm/rev.
These conditions depending on the machine power and capability.
Treatment: Ann = annealed; N&T = normalized and tempered; Q&T = quenched and tempered.
HB = Brinell hardness; HSS = high speed steel.
Speeds: For coated carbides are approximately 30% higher than those for uncoated.
[a] *Tool code*, see Section 5.2.
Source: Adapted from BSSA-www.bssa.org.uk/topics,php?Article=194-2007/2012 [24].

both rough and finish cuts in multicut operations. Feed must be maintained constant as the tool enters the workpiece. For form turning deep or complex shapes, slower speeds should be considered. The flow of cutting fluid must be carefully controlled to ensure that a consistent, large flow volume is delivered continuously to the cutting edges during form turning of stainless speeds.

Table 5.6 Suggested cutting point angles for single-edged turning tools used for SSs

Material		HSS		Carbides			
		BRA	SRA	Brazed		Disposable	
				BRA	SRA	BRA	SRA
Martensitic	160 HB	0°	10°/15°	0°	6°	0°/–5°	–5°
Martensitic	300 HB	0°	15°	0°	6°	–5°	–5°
Free-machining grades	160 HB	5°	8°	0°	6°	–5°	–5°
Wrought ferritic and cast plain Cr	160 HB	5°	8°	0°	6°	0°	5°
Wrought and cast austenitic	165 HB	0°	10°/15°	0°	6°	0°	5°

BRA = back rake angle;
SRA = side rake angle
ERA = end relief angle = 5°
SRA = side relief angle = 5–6° } These angles apply in all cases
ECEA = end cutting edge angle = 5°
Nose radius = 0.5–3 mm. Use as large as possible radius consistent with part being turned
HB = Brinell hardness;
HSS = high speed steel

Source: MHB-M16, adapted [1].

Cutting-off requires an end relief angle, usually between 10° and 15°. This angle should be reduced as the depth of cut increases on larger diameter workpieces to about 5°, to avoid tool deflection. The above-center heights should be about 3 mm.

5.2.2 Drilling

In drilling, the material is mechanically pushed from the center of the drill tip (chisel edge) to the outer cutting edge. This pushing action work-hardens the material at the drill point, in turn hardens the material, causing excessive drill wear and hard spots in the stainless. This work-hardening effect especially of the austenitic grades, such as 304 and 316, is the main cause of the problem associated with drilling stainless steels. In most cases, the steel hardens progressively as it is cold worked; however, the metal may be full annealed (softened), where deep or small diameter holes are to be drilled, and consequently the drillability is partially improved [24]. On the other hand, however, it is well-known that higher hardness (or tensile) of austenitic alloys allows for better surface finish. Therefore, the right level of hardness of these alloys can provide for a good combination of drillability and surface finish. Excessive hardness on the other hand can increase tool wear when parts are machined.

Center punching with conical shaped punches can result in much localized work-hardening to make drill entry difficult. This can make the drill tip deflect or wander, glaze the surface and/or blunt the drill tip and result in drill breakages, particularly where small holes are being drilled. If a punch mark is needed to help get the hole started, a light mark using a three-cornered pyramid punch may be a better idea.

Chips must be allowed to get away from the hole freely. Build up of chips in the hole leads to hole-wall roughness, or in cases of severe clogging to drill breakage. When deep holes are being drilled, the drill should be retracted with minimum dwell as follows:

Drilling sequence	Maximum drilling depth
First retraction	(3–4) × drill diameter
Second retraction	2 × drill diameter
Third retraction and	1 × drill diameter
Subsequent retractions	

5.2.2.1 Important Hints When Drilling Stainless Steels

The following are some of guide lines that improve the drillability of stainless steels.

1. Select the most appropriate drill considering the type of stainless steel, the capacity of the machine, and the coolant to be used.
2. Select the shortest possible drill and make sure of its proper fixation and alignment. The work must be firmly clamped. Whenever necessary, use drilling jigs.
3. Both too slow and too fast drill speeds can cause drill breakage. It is important to observe that small diameter drills should not be used at too slow speeds.
4. Dwell must be avoided, particularly with austenitic alloys. Allowing the drill to dwell causes glazing of the hole bottom. Entry and re-entry should be done at full speed and feed rate.
5. Proper feed increases drill life and productivity.
6. Heavier feeds and lower speeds may be necessary to reduce work hardening, which is a main problem in machining stainless steels.
7. Stepless drives are preferred than stepped gear-boxes for drilling stainless steels.
8. When drilling deep holes, reduce the speed and feed as the drill moves into the metal; however, with depths up to three times the hole diameter, no reduction is necessary as mentioned before.
9. It is essential that drills should be sharp and properly ground, and make sure that any wear has been removed.
10. When drilling through-holes, a backing plate should be used to help avoid drill breakage as the drill comes out of the blind side of the hole.
11. When using coolants or lubricants, ensure a copious supply, which is directed to the drill point.

Table 5.7 lists the recommended speeds and feeds, for drilling different grades of stainless steels. Although standard HSS-drills are suitable for drilling stainless steel, they should preferably be shorter to reduce deflection and breakage in use. Carbide-tipped drills can also be used to realize higher productivity.

Figure 5.10 illustrates the proper tool geometry of a HSS-drill used for drilling stainless steels. The point angle should be 140°, although smaller angles (120°) can be used for drilling free-machining stainless steels. Wear of the drill point is an indication that a larger point angle

Table 5.7 Recommended speeds and feeds for drilling stainless steels using HSS-drills

Stainless steel alloy	Treatment and HB	Speed (m/min)	Feed (mm/rev) for hole diameter of							HSS-grade[a]
			1.6 mm	3 mm	6 mm	12 mm	20 mm	25 mm	50 mm	
Wrought martensitic										
Free-machining grades	Ann-160	36	0.025	0.075	0.15	0.255	0.33	0.4	0.635	M1, M7, M10
Lower C/lower Cr grades	Ann-175	18	0.025	0.075	0.125	0.205	0.3	0.4	0.61	M1, M7, M10
Lower C/lower Cr grades	Q&T-300	17	0.025	0.075	0.1	0.175	0.255	0.3	0.455	M1, M7, M10
Higher C/high Cr grades	Ann-240	15	0.025	0.05	0.075	0.125	0.2	0.255	0.38	M42, T15
Higher C/high Cr grades	Q&T-320	12	0.025	0.05	0.05	0.1	0.15	0.2	0.3	M42, T15
Wrought ferritic										
Free-machining grades	Ann-160	40	0.25	0.075	0.15	0.255	0.355	0.455	0.635	M1, M7, M10
12–17% Cr grades	Ann-160	20	0.25	0.05	0.1	0.175	0.255	0.3	0.455	M42, T15
Wrought austenitic										
Free-machining grades	Ann-160	30	0.25	0.075	0.15	0.255	0.355	0.455	0.635	M1, M7, M10
Other grades (304, 316, 321, etc.)	Ann-160	17	0.25	0.05	0.1	0.175	0.255	0.3	0.455	M42, T15
Wrought duplex	Ann-230	14	0.25	0.05	0.075	0.125	0.2	0.255	0.38	M42, T15
Cast plain Cr	N&T-210	14	0.25	0.05	0.075	0.125	0.2	0.255	0.33	M42, T15
Cast austenitic	Ann-170	10	0.25	0.05	0.075	0.15	0.2	0.255	0.28	M42, T15

[a] HSS – grade see Section 6.2.

Ann = annealed; Q&T = quenched and tempered; N&T = normalized and tempered.

HB = Brinell hardness; HSS = high speed steel.

Adapted from: BSSA-www.bssa.org.uk/topics.php?article = 194-2007/2012 [24].

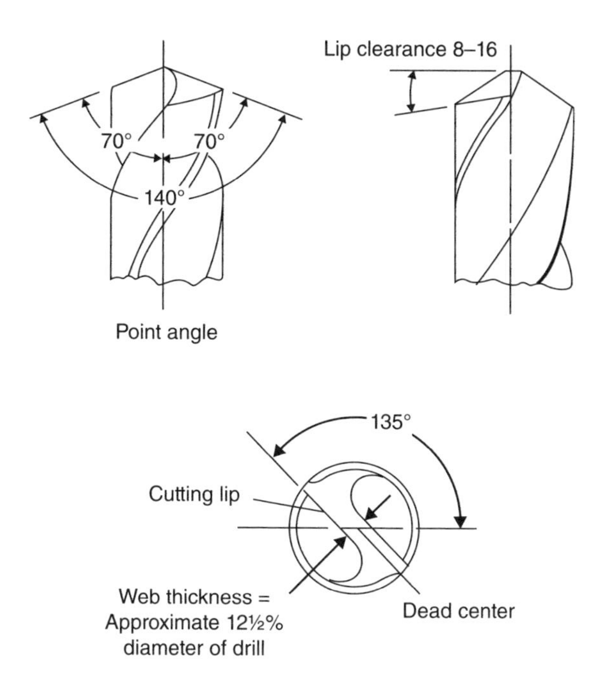

Figure 5.10 Recommended geometry of HSS-drills for drilling stainless steels.

should be used. The lip clearance should be 8–16°. It decreases as the drill diameter increases as is indicated in the following table.

Drill diameter (mm)	3	6	12	20	25 and over
Lip relief angle (°)	16	14	12	10	8

Excessive feed force, poor drillability in spite of sharp edges, or drill breakage indicates that the drill tip is of insufficient lip relief. Conversely, if the drill is tending to dig-in, then a reduced relief may be better.

The web thickness at the point (normally increasing along the length of the drill) should be generally around 1/8 of the drill diameter, except for small diameters where thicker webs improve the strength and rigidity of the drill. This can however impede chip flow. Conversely, with a thinner web, although the drill has less rigidity, chip flow is easier; additionally, it is easier to start the hole. A thinner web reduces the feed force, heat generation, and work hardening of the hole bottom; in other words, it enhances the drillability.

Some jobs require exceptionally deep drilled holes of aspect ratio of 8–10. In such cases, short chucking is impossible, and special drills known as crankshaft drills may be useful. These drills were originally designed for drilling oil holes in forged crankshafts and connecting rods, but have found widespread use in drilling deep holes [1]. They possess a very heavy web and a higher spiral (helix) angle to aid chip removal. They usually have a notched point type of web thinning which is done using a sharp-cornered grinding wheel (GW). For best results in sharpening HSS-drills, medium grain size, soft grade, dry grinding is performed. Burning and quenching should be avoided to eliminate cracking of the drill. Both lips must be symmetrically ground, that is, they should have the same angles to the center line of the drill, and of the same length. A mixture of chip thicknesses and sizes indicates that the drill is not symmetrically ground.

5.2.3 Reaming

The difficulties involved in reaming stainless steels are most often caused by previous operations, particularly with the nonfree-machining austenitic alloys. For example, if the feed in the previous drilling operation is too light, the hole wall can be severely work hardened and can resist cutting by reamer. It is also very important that ample material be left from the previous operation to allow a positive reaming cut to be made to undercut the new work-hardened layer produced.

5.2.3.1 Tool Geometry of Reamers for Stainless Steels

Straight- or spiral-fluted designs of reamers are used for cylindrical or tapered holes. Spiral fluted are preferable for reaming stainless steels as they produce less chatter, can better dispose chips from deep holes, and are capable of producing a better finish than straight-fluted reamers. For normal clockwise tool rotation, right hand spiral tools cut more freely than left hand spiral tools, but tend to self-feed into the hole. Left hand spiral tools have a lower tendency to self-feed which can be beneficial when precise feed rate control is needed.

Figure 5.11 illustrates a HSS-spiral-fluted reamer. The rake angles should be between 3° and 8°, larger angles suiting austenitic stainless steel grades (304, 316, and so on). Margin width should be 0.13–0.38 mm for HSS-tools and 0.05–0.125 mm for carbide-tipped reamers. The margin width increases within this range with increasing reamer diameter. Insufficient primary relief angle or too wide a land can cause chattering. Table 5.8 illustrates the recommended tool geometries of HSS and carbide-tip reamers for machining stainless steels.

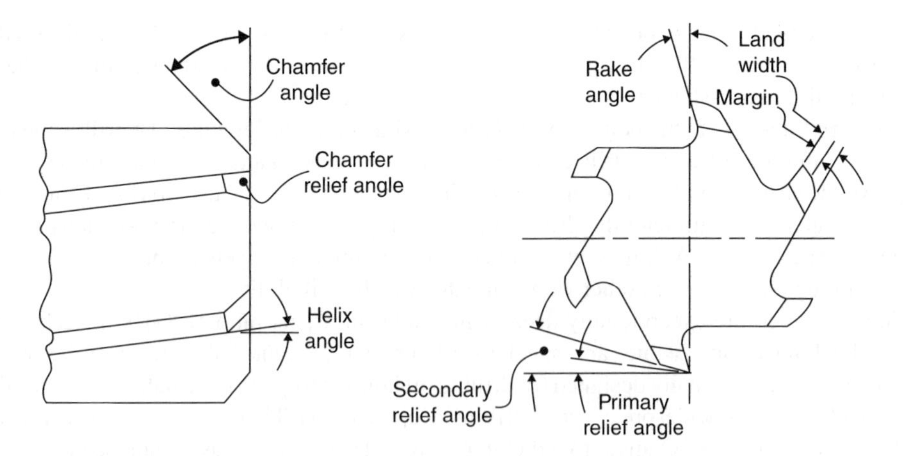

Figure 5.11 HSS-spiral-fluted reamer for machining stainless steels.

Table 5.8 Recommended geometries of HSS and carbide-tipped reamers for machining stainless steels

Geometrical feature	Values	
	HSS-reamer	Carbide-reamer
Primary (working section.) relief angle (°)	4–5	6–12
Margin (land) width (mm)	0.13–0.38	0.05–0.125[a]
Chamfer angle (°)	30–35	2
Chamfer length (mm)	1.5	4.8
Chamfer relief angle (°)	4–5	NA
Rake-angle (working sec.) (°)	3–8	7–10
Helix angle (°)	0–10	5–8

[a] For reaming the nonfree-machining ferritic and austenitic grades and the PH-grades, using carbide-tipped reamers, the margin width should be increased to 0.125–0.25 mm.
Source: MHB-M16, compiled [1].

5.2.3.2 Reaming Parameters

Recommended speeds and feeds for reaming stainless steels using HSS and carbide-tipped reamers are listed in Table 5.8. Smooth finishes require significantly lower speeds as compared to drilling. To improve the surface finish of finally reamed holes, the suggested speeds in Table 5.7 should be reduced by 50% and the feeds by 25%. Rough holes or burned tools usually indicate that too high speeds are used. To obtain smooth surfaces, the cutting fluid must also kept clean. Reaming produces slivers and very fine chips which can float in the cutting fluid and damage the surface finish, especially if the machine is equipped with a circulating system.

5.2.4 Milling

HSS-cutters are used in milling stainless steels, although tooling with carbide inserts can also be used, particularly for alloys that are more difficult-to-machine. The smoothest finishes are obtained with helical or spiral HSS-cutters running at high speed, particularly for cuts over 20 mm wide. Helical cutters cut with shearing action and hence cut smoothly and with less chatter than straight tooth cutters. Coarse-tooth (heavy duty) cutters work under less stress and permit higher speeds than fine-tooth (light duty) cutters. For heavy, plain milling work, a 45° left-hand spiral heavy-duty cutter is preferred. In wide slab milling, such cutters produce smooth finishes and avoid chattering.

Milling deep slots in stainless steel, sometimes presents problem of chattering, and jamming of wide chips; such difficulties are eliminated by using a staggered-teeth cutter. End milling of stainless steels is recommended using a solid-shank end mill because of its high strength.

Figure 5.12 illustrates the suggested geometry of a side-milling cutter. Excessive vibration indicates that the cutter has insufficient clearance, provided that the rigidity of tooling, fixtures, and machine is adequate. Hogging-in, generally, indicates too much rake or possibly too high cutting speed.

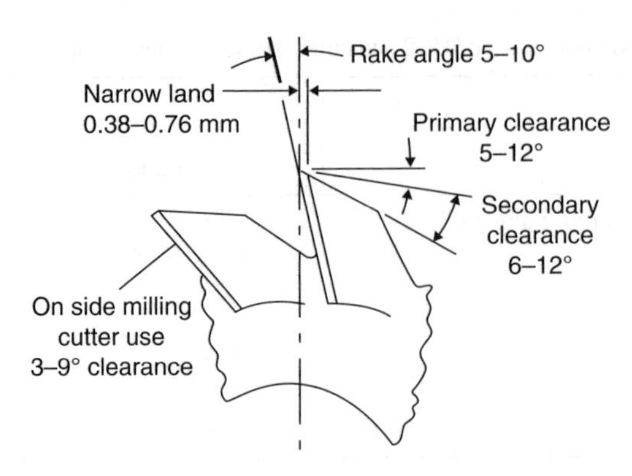

Figure 5.12 Recommended geometry of milling cutters for machining stainless steels.

Table 5.9 lists speeds and feeds in case of peripheral end milling stainless steels using either HSS or carbide cutters. If the feed is too light, the tool will burnish the work; if too heavy, the tool life will be shortened. A roughing cut runs with heavier feeds and slower speeds than those used for finishing cuts.

Once a milling cut has been started, it should not be stopped unless absolutely necessary as the tool will undercut when starting again. When it is necessary to back out and start again, the tool should be retracted two or three turns behind the workpiece before starting again to eliminate the danger of backlash and guards against under cutting.

5.2.5 Broaching

Broaching is a fast way to remove metal either externally or internally, and produces a job to close tolerances. Related machines fall into two general classes, vertical or horizontal. Either can be used for push or pull broaching. For internal broaching, a properly drilled or reamed hole is satisfactory. For external (surface) broaching, preliminary machining operations are seldom required. It is essential that chip not be allowed to build up between the teeth, otherwise damage to the broach may result from chip packing. Damage may also occur if the broach is not properly aligned, leading to excessive localized load on the tooth.

Broaches for stainless steels are usually made of HSS or PM-HSS (powder metallurgy-high speed steel). Basically, a broach can incorporate a roughing, a semi-finished, and a final precision cuts (Figure 5.13a). In designing the broach, the manufacturer provides maximum tooth strength and a sufficient pocket between teeth for chips. When a broach becomes dull, it should be re-sharpened only on a broach grinder or returned to the manufacturer to be ground.

For internal broaches, the back-off angle should be held to minimum (preferably 2° and not exceed 5°) (Figure 5.13b). Too much back-off angle will shorten broach life due to size reduction from resharpening. Any nicks on the cutting edges of the broach will score the surface of the work. Therefore, careful handling is very important.

Table 5.10 lists the nominal feeds and speeds for broaching free- and nonfree-machining stainless steels using Micro-Melt powder HSS-tools (Carpenter-Trademark of AK steel Corp.). When higher hardness is required, pieces should be first broached and then heat treated. Sulfo- chlorinated oils diluted with paraffin oil, rather than emulsifiable fluids are suggested.

Table 5.9 Nominal end-milling parameters for machining wrought stainless steels using HSS and carbide tipped tools

Stainless steel alloy	Treatment and HB	Speed (m/min)	Feed (mm)/tooth for cutter diameter				Tool material HSS: M2, M7 carbides: C2, C6
			6 mm	13 mm	19 mm	25–50 mm	
Wrought martensitic							
Free-machining grades	Ann-160	30	0.03	0.06	0.11	0.12	M2, M7
		90	0.03	0.06	0.13	0.17	C6
Lower C/lower Cr grades	Ann-175	34	0.05	0.075	0.12	0.15	M2, M7
		107	0.025	0.05	0.12	0.15	C6
Lower C/lower Cr grades	Q&T-300	27	0.025	0.05	0.12	0.15	M2, M7
		82	0.025	0.05	0.12	0.15	C6
Higher C/higher Cr grades	Ann-240	23	0.025	0.05	0.12	0.15	M2, M7
		72	0.025	0.05	0.12	0.15	C6
Higher C/higher Cr grades	Q&T-300	20	0.025	0.05	0.12	0.15	M2, M7
		68	0.025	0.05	0.12	0.15	C6
Wrought ferritic							
Free-machining grades	Ann-160	44	0.03	0.06	0.1	0.12	M2, M7
		122	0.03	0.06	0.13	0.17	C6
12–17% Cr grades	Ann-160	32	0.05	0.075	0.13	0.15	M2, M7
		100	0.025	0.05	0.1	0.15	C6
Wrought austenitic							
Free-machining grades	Ann-160	32	0.03	0.06	0.11	0.14	M2, M7
		100	0.03	0.06	0.11	0.14	C7
Other graded (304, 316, 321, etc.)	Ann-160	24	0.05	0.075	0.13	0.15	M2, M7
		82	0.025	0.05	0.075	0.13	C6
Wrought duplex	Ann-230	23	0.05	0.075	0.13	0.15	M2, M7
		79	0.025	0.05	0.075	0.13	C2
Wrought PH-alloys	Ann-210	22	0.025	0.05	0.07	0.1	M2, M7
		62	0.025	0.05	0.07	0.12	C2

Ann = annealed; Q&T = quenched and tempered; N&T = normalized and tempered.
Optimum values of speeds and feeds may be higher or lower, depending on machining conditions.
HB = Brinell hardness; HSS = high speed steel.
Source: Adapted from MHB-M16 [1].

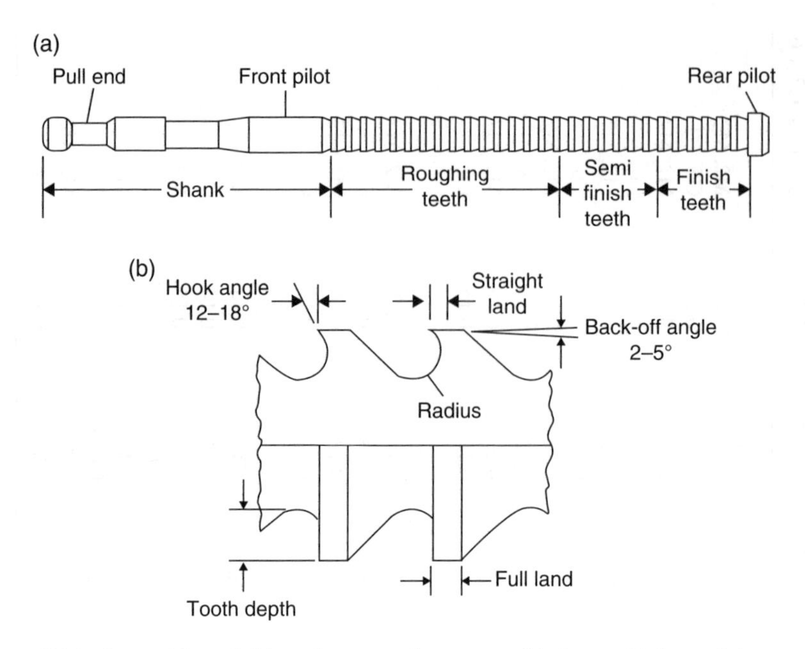

Figure 5.13 Internal broach (a), and suggested geometry (b), for machining stainless steels.

5.2.6 Grinding

Corundum (Al_2O_3) wheels are most commonly used for stainless steels. However, Carborundum (SiC) wheels can also be used for special applications, but at a reduced wheel life. Medium density wheels of hardness grades H to L are generally selected for grinding stainless steels, although harder wheels are more suited for thread grinding. Grit sizes commonly used are 46, 54, or 60; finer grits can be used to produce a finer finish. Vitrified and resinoid bonded wheels are mostly used. GWs used previously to grind other metallic materials should not be used to grind SS, because particles of the other materials may be embedded in the SS, affecting its corrosion resistance.

Typical wheel speeds are 1500–2000 m/min. For surface grinding, table speeds are 15–30 m/min with a down feed of up to 0.05 mm/pass for rough grinding, and 0.013 mm/pass for finishing, and a cross-feed of 1.3–13 mm/pass. Because of lower thermal conductivity of stainless steels, an efficient coolant is necessary. Conventional water-soluble fluids generally provide lower GW-life than heavy-duty sulfo-chlorinated oils.

5.3 Surface Treatments of Stainless Steel after Machining

Stainless steels are designed to be naturally self-passivate whenever a clean surface is exposed to an environment that can provide enough oxygen to form the chromium rich oxide surface layer, on which the corrosion resistance of these alloys depends. To achieve maximum corrosion resistance of machined stainless steel parts, they should be chemically cleaned to remove grease, oil, fine metal debris, and fingerprints from the machined surface. The average shop degreasing solution is usually sufficient for that purpose. After cleaning (pickling), acid passivation, under some circumstances should be considered [26].

Table 5.10 Nominal broaching parameters of free- and nonfree-machining stainless steels using HSS-broaches

Stainless steel alloy		Speed (m/min)	Super-elevation (mm/tooth)	HSS* broach
Martensitic alloys				Recommended broaches are M42 or T15
Nonfree	410	8	0.1	
Free-machining	416	10	0.1	
Nonfree	420	6	0.075	
Free-machining	420F	8	0.075	
Nonfree	431	6	0.075	
Nonfree	440A, 440B	6	0.05	
Nonfree	440C	4	0.05	
Free-machining	440F	6	0.05	
Ferritic alloys				
Nonfree	430	8	0.075	
Free-machining	430F	12	0.1	
Nonfree	443	8	0.075	
Austenitic alloys				
Nonfree	302, 304, 316	6	0.1	
Free-machining	303	8	0.085	
Free-machining	203	8	0.085	
Nonfree	321, 347	6	0.075	
Nonfree	S24110	4	0.075	
Nonfree	S24904	4	0.075	
Nonfree	S20910	4	0.075	
Duplex alloy: Annealed, 230 HB		5	0.075	
PH-alloys				
Martensitic – Custom 455-Annealed		3	0.05	
Custom 455-aged		4	0.05	
Semi-Austenitic-Pyromet 350-355-over temperature		4	0.05	
Pyromet .350-355 Aged		3	0.05	

* The listed speeds and feeds are conservative recommendations. Higher values may be attainable depending on machining environment.
HB = Brinell hardness; HSS = high speed steel.
Source: Compiled from MHB-16 [1] and Carpenter [25].

5.3.1 Chemical Cleaning (Pickling)

The chemical cleaning or pickling of SS-surfaces is an important operation which must be performed directly after machining to allow these surfaces to be passivated and heat treated if required. Parts made of martensitic stainless steels or PH-alloys may be hardened or solution treated at high temperatures. The parts must be thoroughly cleaned with a degreaser to remove any traces of cutting fluids. During machining, parts can pick up minute particles of iron from the cutting tool which can form rust spots on the surface. Pickling usually involves nitric/hydrofluoric acid mixtures, whereas traditionally passivation has been done using only nitric acid. For pickling, the finished parts are immersed in a nitric-Hf pickling solution (10% nitric, 2% Hf) either warm or at ambient temperature.

5.3.2 *Passivating*

Passivation treatments are sometimes specified, but it is important to consider whether this is strictly necessary or not. After the parts have been properly cleaned by pickling, they should be immediately passivated by exposure of the cleaned surface to air to form an extremely thin transparent film, which is tenacious, uniform, stable, and passive. It imparts the SS-surface the property of passivity, normally associated with noble or inert metals, making them corrosive resistant. This oxide film repair itself spontaneously if damaged both in air (O_2) or when immersed in solutions. This passivation can be, however, chemically enhanced.

Good cleaning will prevent contamination of the passivation bath and avoid reactions that may lead to *flash attack* or a heavily etched or darkened surface. The traditional acid passivation practices for stainless steels are illustrated in Table 5.11.

The A-A-A (Alkaline-Acid-Alkaline) method outlined for passivating free-machining stainless steels, illustrated in Table 5.12, prevents corrosion that may otherwise occur from residual trapped in pits remaining after the free-machining inclusions that have been removed by the passivation bath.

Passivation is not a scale-removal method. Any particle of oxide or heat tint must be removed by pickling before passivating. When passivating stainless steels, the following must be noted:

- Baths should be replaced on a regular schedule to avoid a loss in passivating potential that can result in flash attack.

Table 5.11 Nitric acid passivation of stainless steels

Stainless steel grades	Passivation practice
• Cr-Ni grades (series 300) • Grades with 17% Cr or more, except series 440	20% by vol. nitric acid at 49–60°C for 30 min
• Straight Cr-grades (12–14% Cr) • High C/high Cr grades (series 440) • PH-stainless	20% by vol. nitric acid + 229/l sodium dichromate at 49/60 °C, for 30 minutes, or 50% by vol. nitric acid at 49/60 °C, for 30 minutes

Parts should be thoroughly cleaned and degreased prior to nitric acid passivation. Parts should be rinsed after immersion in acid and sodium dichromate.
Adapted from: ASTM A967 [26].

Table 5.12 Alkaline-acid-alkaline passivation of free-machining alloys

Free-machining stainless including AISI types 420F, 430F, 440F, 203, 182-FM and Carpentacer project 70+ types 303 and 416
Alkaline
1. 5% by weight sodium hydroxide at 71/82°C, for 30 min.
2. Water rinse
Acid
3. 20% by vol. nitric acid + 22 g/liter sodium dichromate at 49/60 °C, for 30 min
4. Water rinse
Alkaline
5. 5% by weight sodium hydroxide at 71/80 °C, for 30 min
6. Water rinse

Parts should be thoroughly cleaned and degreased prior to nitric acid passivation.
Adapted from: ASTM A967 [26].

- The bath should be maintained at proper temperatures, shown in Table 5.12, since lower temperatures may allow localized attack.
- Carburized or nitrided part should not be passivated, since their reduced corrosion resistance may result in attacking the bath.
- High carbon martensitic stainless steels should be in hardened condition, in order to provide sufficient corrosion resistance.
- Water used for passivation baths should have relatively low chloride content (preferably <50 ppm).

Citric Acid Passivation as an Alternative to Nitric Acid Treatments: Citric acid passivation treatments can also be considered as an alternative to nitric acid as both provide the oxidizing conditions necessary for passivation. They are becoming more popular in avoiding the use of mineral acids or solutions containing sodium dichromate $Na_2 Cr_2 O_7 \cdot 2H_2O$. Citric acid is a less hazardous method and has environmental benefits in terms of "NOx" fume emission and waste acid disposal. Solution strengths of 4–10% citric acid are specified for passivation treatments in ASTM A967. Citric acid passivation treatments are useful for several grades of SS. They are illustrated in Table 5.13.

Table 5.13 Citric acid passivation of stainless steels

Stainless steel grades	Passivation practice
Types 316/316L Project 70 + types 316/316L Types 304/304L Project 70 + types 304/304L Custom Flo 302 HQ Type 305 Nitrogen-strengthened austenitics Type 430 17Cr-4Ni Project 70 + Custom 630 15 Cr – 5 Ni Project 70 + 15 Cr – 5 Ni Custom 465	10 wt% citric acid, 66 °C, 30 minutes
Type 409 Cb	10 wt% citric acid, 82/93 °C, 30 minutes; after passivation and water rinse, neutralize in 5 wt% sodium hydroxide, 77 °C, 30 minutes
Type 303 Project 70 + type 303	10 wt% citric acid, 66 °C, 30 minutes; after passivation and water rinse, neutralize in 5 wt% sodium hydroxide, 77 °C, 30 minutes
Type 410 Type 420 Trim Rite	10 wt% citric acid, 49/54 °C, 30 minutes; after passivation and water rinse, neutralize in 5 wt% sodium hydroxide, 77 °C, 30 minutes
Type 409 Cb-FM Type 416 Project 70 + type 416	10 wt% citric acid (adjusted to pH 5 with sodium hydroxide), 43°C, 30 minutes; after passivation and water rinse, neutralize in 5 wt% sodium hydroxide, 77 °C, 30 minutes

Parts should be thoroughly cleaned and degreased prior to citric acid passivation.
Parts must be water rinsed after immersion in acid and sodium hydroxide baths.
Adapted from: ASTM A967 [26].

References

[1] Editor Committee of ASM International (1989) *Metals Handbook* Machining 19th edn, Vol. *16*, ASM International Materials Park, OH.

[2] Drab, B. (2010) Making Stainless Steel More Machinable, Schmolz & Bickenbach, USA, http://www.Productionmachining.com/articles/making-stainless-more-machinable(2) (accessed April 24, 2015).

[3] Richmond, F.M. (1967) A decade of progress in machinability, finishing and forming, *Met. Prog.* Aug.: 85–86.

[4] Kovach, C.W. (1975) *Sulfide Inclusions and Machinability of Steel*, American Society of Metals (ASM), pp. 459–479.

[5] Sparre, C. (1972) Stainless free-cutting steel, *Wire*, April: 56–60.

[6] Kovach, C.W., A. Moskowitz (1969) Effects of manganese and sulfur on the machinability of martensitic stainless steels, *Trans. Met. Soc., AIME*, *245*, 2157–2164.

[7] Henthorne, M. (1970) Corrosion of resulfurized free-machining stainless steels, *Corrosion*, *26* (12), 511–528.

[8] Clarke, W.C. (1964) Which free-machining stainless?, *Metalwork. Prod. 9,* 68–71.

[9] Tipnis, V.A. (1971) Machining of stainless steels, *Wire*, August: 153–161.

[10] Kovach, C.W., Eckenrod, J.J. (1971) Free-machining austenitic stainless steels. *13th Mechanical Working and Steel Processing Conference, American Institute of Mining, Metallurgical, and Petroleum Engineers, 1971*, pp. 300–325.

[11] Tipnis, V.A. (1976) Methods of making stainless steels having improved machinability. US Patent 3,933,480.

[12] Eckenrod, J.J., Kenneth E.P., Geoffrey O.R., William E.R. (1986) Low carbon plus nitrogen free-machining austenitic stainless steel. US Patent 4, 613,367.

[13] McClaymonds, N.L. (1964) Machinability of leaded stainless, *Met. Prog.*, Aug.: 166–168.

[14] Clarke, W.C. (1964) Which free-machining chromium stainless? *Metalwork. Prod.* May: 43–45.

[15] Kimura, A. (1986) Super STARCUT stainless steel 304 BF with Bithmus, *Bull. Bismuth Inst.* 1–5.

[16] Ono, K.M. Ynagida, T. Kawano. M., N. Shibata (1983) Development of leaded free-machining austenitic stainless steel, *Denki Seiko*, *54* (4), 265–274.

[17] Committee of Stainless Steel Producers, AISI (1975) *Free-Machining Stainless Steels*, American Iron and Steel Institute.

[18] Carpenter Technology Corporation (1985) *Guide to Machining Stainless Steels and other Speciality Metals*, Carpenter Technology Corporation, 1985.

[19] Blott, D.M. (1977) *Machining Wrought and Cast Stainless Steels, HB of Stainless Steels*, McGraw-Hill, pp. 24-2–24-30.

[20] Tipnis, V.A. (1974) Stainless steel having improved machinability. US Patent 3, 846, 186.

[21] Divine, Jr., C.A. (1968) What to consider in choosing an alloy, *Met. Prog.*, Feb.: 19–23.

[22] Wright, P.K., Bagchi, A. (1981) Wear mechanism that dominate tool life in machining, *J. Appl. Metalwork. 1*: 15–23.

[23] Falcon Metals Group, http://www.falcon-metals.com (accessed June 15, 2015).

[24] British Stainless Steel Association (BSSA) www.bssa.org.uk/topics.php.?article=194-2007/2012 (accessed April 24, 2015).

[25] Carpenter www.Cartech.Com/techarticles.aspx?id=1578 (accessed April 24, 2015).

[26] ASTM A967. (2003) *Standard Specification of Chemical Passivation Treatments for Stainless Steel Parts*, ASTM.

6

Traditional Machining of Super Alloys

6.1 Machinability Aspects of Super Alloys

Super alloys are generally classified as having poor machinability. Much of high machining cost is due to the fact that allowable cutting speeds are only 5–10% of those used for steels. The Fe-base alloys, which have descended from stainless steels, usually machine more easily than the Ni-base and Co-base super alloys under similar conditions of processing and heat treatment. However, the Fe-base alloys do present chip braking problems, which often require special tool geometries. The Ni-base and Co-base alloys have several characteristics in common, that contribute to high machining costs.

All super alloys have, in general, high strength at high temperatures and produce segmented chips during cutting, hence creating high dynamic forces. An increase in high temperature strength of these alloys makes them harder and stiffer at the cutting temperature, thus increasing forces at the cutting edge during machining and consequently promoting chipping or deformation of the tool edge. Poor heat conductivity and high hardness generate high temperatures during machining. The strength and work hardening characteristics of super alloys create notch wear at large depths of cut and extremely abrasive environment of the cutting edge.

Accordingly, when machining super alloys, carbide inserts should have good edge toughness, and when coated, a good adhesion of coating to the substrate should be secured to provide good resistance to ablation and plastic deformation. In general, tool inserts of sharp edges and positive rakes are to be used.

The machining behavior varies depending on the prior treatment of the material. When machining super alloys in the soft condition (prior to heat treatment), the heat generated and cutting forces are considerably higher than for normal steel. By treating super alloys at

Machining of Stainless Steels and Super Alloys: Traditional and Nontraditional Techniques,
First Edition. Helmi A. Youssef.

elevated temperatures, small intermetallic particles are precipitated, thus hindering the movement in the crystal structure, and as a result the material will be more difficult to deform and machine. After aging, the heat generated during cutting is so high that only grades with highest hot hardness and abrasion resistant are practical. Furthermore, the machined surface is prone to work hardening, which means that notch wear is a critical issue.

To summarize, the main factors affecting the machining characteristics of super alloys are that they:

- exhibit austenitic matrix which promote rapid work hardening during machining;
- retain strength at high temperatures, where less efficient high speed steel (HSS)-tools begin to soften;
- possess usually high dynamic shear strength;
- contain in their micro-structure hard carbides that make them abrasive;
- possess low thermal conductivity, which leads to high cutting edge temperature;
- form a tough continuous chip, which leads to BUE-formation;
- produce abrasive carbides in their microstructures;
- are reactive with cutting tool materials under atmospheric conditions.

As previously described in Chapter 3, super alloys constitute a wide spectrum of alloys. With such a wide spread of alloys under the generic heading of super alloys, it is impossible to quote one set of cutting recommendations for the entire class, and the machining behavior can vary greatly even within the same alloy group. In fact, the same material can have numerous machining recommendations depending on its heat treatment conditions.

The super alloy as a raw material could be provided as cast, wrought, and forged (bar and plate stocks), and in sintered (powder metallurgy (PM)) forms. Forged materials usually have a finer grain size than in castings. Forgings generally possess higher strength, improved grain flow, better fatigue, and fracture resistance, as compared to cast super alloys. They are, however, more abrasive with greater tendency to deform the tool during machining. Machining with reduced speeds and increased feeds leads to reduce the potential of work hardening and notching of the cutting edge. In casting, the opposite applies, that is, applying higher speeds and lower feeds can be beneficial. Components produced from casting techniques exhibit excellent creep strength combined with toughness. These characteristics create machinability problems due to poor chip segmentation. Casting typically features a hard orange peel mottled surfaces, which makes machining more difficult and can cause notch wear on the insert. This necessitates harder and wear resistant insert grades to be used than for forgings. Casting alloys are intended for parts requiring less strength and are suitable for the production of near net shape (NNS) components such as turbine blades.

More complicated and NNS components can be produced using the PM technique. Components from this manufacturing route exhibit extremely low machinability and are very abrasive.

6.2 Machinability Rating of Super Alloys

As previously discussed in Chapter 4, one of the main criteria to assess machinability was the tool life and the related cutting speed to affect a predetermined allowable wear mark that terminates the tool life. Another important criterion for assessment of the machinability was the power consumption. A material of good machinability requires lower power consumption, and consequently lower specific cutting energy.

6.2.1 *Machinability as Based on Tool Life and Nominal Cutting Speeds*

As usual, turning process has always been selected to determine the machinability of materials. The machinability of a large number of super alloys as based on tool life and nominal cutting speeds, suggested by Metcut [1], are arranged in Table 6.1, in a descending order. Accordingly, these super alloys fall into 13 different categories, that may be classified into three main groups, namely, easy-, medium-, and hard-to-machine groups.

The easy-to-machine group consists of five categories, listed as shown in Table 6.1. These are the cast Fe-base categories (Fe-C1, Fe-C2, Fe-C3), and the wrought Fe-base category (Fe-W), along with the wrought Ni-base category (Ni-W5), representing only one alloy (TD-Nickel-90% Ni and 2% ThO_2). The medium-to-machine group comprises also five categories. These are the wrought Ni-base categories (Ni-W4, Ni-W1, Ni-W2) and the cast Ni-base category (Ni-C1), along with the wrought Co-base category (Co-W). The hard-to-machine group comprises 3 categories which are the cast Co-base category (Co-C), the cast Ni-base category (Ni-C2), and finally, the wrought Ni-base category (Ni-W3).

The mechanically alloyed versions (PM) of super alloys are listed in Table 3.4, Y_2O_3 has no significant effect on machinability rating (MR). These alloys can be machined using practices appropriate for wrought alloys of similar composition and hardness. As shown in Table 6.1, Ni-base alloy MA 754 (277 BHN) has lower machinability than Inconel 718 and Fe-base alloy A-286. The Fe-base alloy MA 956 of approximately the same hardness (270 BHN) as MA 754, appears to be more machinable than MA 754. The machinability of the Ni-base alloy MA 6000 (450 BHN) appears to be similar to that of Udimet 700 [2]. Table 6.2 lists the nominal speeds and feeds for turning the mechanically alloyed versions, while Table 3.4 illustrates their MRs.

From the foregoing, it is depicted that the easiest-to-machine from super alloys are the Fe- base alloys such as A-286, Discaloy, N-155, and so on, as well as the alloy TD-Nickel, while the hardest-to machine are the Ni-base, and Co-base super alloys, such as IN-100, René 77, AiResist 13, 215, and NASA Co-W-Re (Table 6.1).

Similarly, the AISI, and others [1–4], tested many metals and alloys, and compared cutting speeds to those obtained when machining a reference alloy AISI-1212, a resulfurized, and rephosphorized plain carbon steel under the same cutting conditions, AISI-1212 has got a score 1. Materials score above 1 are easier-to-machine, on the other hand materials scoring less than 1 are more-difficult-to-machine (Table 6.3). In such cases, it is highly recommended to include the material hardness (e.g., in BHN), because the material hardness has a considerable effect on the MR. So, if a certain super alloy of a hardness value H1 has a rating MR1, has got a certain metal working, or heat treatment processing, which has raised its hardness number from H1 to H2; then it acquires a MR2, that can be evaluated as:

$$MR2 = \frac{H_1}{H_2} \times MR1$$

Table 6.1 Classification and grouping super alloys as based on recommended cutting speeds (for carbide tools) in descending order

No.	Category and group	BHN	Cutting speed (m/min) (carbides)	Super alloy
	Easy-to-machine			
1	Fe-C1 [Ann]	135–185	76	ASTM A 297 Grade HC
2	Fe-C2 [Ann or N]	135–185	60	ASTM A 351 Grades HK30, HK 40, HT30
3	Ni-W5 [R]	180–200	60	TD-Nickel
4	Fe-C3 [C]	160–200	53	Other types specified in Table 3.3
5	Fe-W [ST or ST-Ag]	180–320	30–58	A-286, Discaloy, [Incoloy 800, 801, 802], N-155, V-57, W-545, 16-25-6, 19-9DL, Incoloy MA 956 (PM)
	Medium-to-machine			
6	Ni-W4 [Ann or ST, Cd or Ag]	140–310	15–35	[Hastelloy B, B-2, (C-276), G, S, X], [Incoloy 804, 825], [Inconel 600, 601], refractory 26, Udimet 630
7	Ni-W1[Ann or ST, ST-Ag]	200–400	15–30	Haynes 263, Incoloy 901, [Inconel 617, 625, 702, 706, 718, 722, X-750, 751], M252, [Nimonic 75, 80], Waspaloy, Inconel MA 754 (PM)
8	Co-W [ST, ST-Ag]	180–320	15–27	AiResist 213, [Haynes 25 (L605), 188], J-1570, [MAR-M905, M918], S-816, V-36
9	Ni-C1 [C or C-Ag]	200–375	14–26	[Hastelloy B, C], ASTM A297 (Grades HW, HX), ASTM A608 (Grades XW50, HX50)
10	Ni-W2 [ST, ST-Ag]	225–400	15–24	Astroloy, IN-106, Inconel700, [Nimonic 90, 95], [René 41, 63], [Udimet 500, 700, 710]
	Hard-to-machine			
11	Co-C [C or C-Ag]	220–425	9–18	[AiResist 13, 215], [HS 6, 21, 25, 31 (X-40)], [MAR-M302, M322, M509, NASA Co-W-Re, W1-52, X-45
12	Ni-C2 [C or C-Ag]	250–425	9–18	B-1900, IN-100, [IN-738, 792], [Inconel 713 C, 718], M252, [MAR-M200, M246, M421]
13	Ni-W3 [ST, ST-Ag]	275–475	9–15	[René 77, 95], Unitemp 1753, Inconel MA 6000 (PM)

ST: solution treated, Ann: annealed, N: normalized, C: cast, W: wrought, C-Ag: cast and aged, ST-Ag: solution treated and aged, R: rolled, and Cd: cold drawn.
Adapted from: Metcut Associates [1].

Referring to Table 6.3, Inconel 901 has got such a treatment that increased the hardness from 200 to 300 BHN. Before the treatment, it had a rating of MR1 = 0.2, then the expected new rating MR2 will be:

$$MR2 = \frac{200}{300} \times 0.2 = 0.13$$

Based on Table 6.1 and 6.3, Figure 6.1 illustrates the MR of the most commonly used super alloys considering AISI-1212 as a reference material of MR = 1, that arranged in a descending

Table 6.2 Nominal speeds and feeds for turning mechanically alloyed super alloys

Alloy	BHN	DOC (mm)	HSS-tools			Indexable carbide tools		
			v (m/min)	f (mm/rev)	Tool material	v (ml min)	f (mm/rev)	Tool material
Fe-base								
Incoloy	270	Rough 6.5	12–15	0.75	T-15, M-36	50–75	0.5	C-6
M1956		Finish 1.3	18–21	0.25	T-15, M-36	75–90	0.2	C-8
Ni-base								
Incoloy	277	Rough 6.5	3–6	0.25	T-15, M-36	12–18	0.25	C-2
MA 754		Finish 1.3	5–6	0.20	T-15, M-36	15–30	0.20	C-2
Inconel	450	Rough 2.0	3–4	0.10	T-15, M-36	9–18	0.2	C-2
MA 6000		Finish 0.2	3–4	0.12	T-15, M-36	9–18	0.12	C-2

Adapted from Metcut [1].

Table 6.3 Machinability rating (MR) in turning of commonly used super alloys, as based on resulfurized and rephosphorized plain carbon steel AISI-1212 as a reference material, tool material HSS

Alloy grade	v (m/min)	MR	Alloy grade	BHN	MR
AISI-1212	60	1.0			1.0
Super alloys			*Super alloys*		
A-286	17	0.28	Discaloy	135	0.4
Haynes 25 (L-605)	5	0.09	Hastelloy B	200	0.12
Inconel 600	15	0.22	Hastelloy C	170	0.20
Inconel 625	6	0.12	IN-100	320	0.09
Inconel 718	6	0.12	Haynes 31-Cast	-	0.06
Inconel X-750	6	0.12	Inconel X	360	0.15
Waspaloy	14	0.2	Inconel 718-Cast	290	0.09
MP35N	14	0.2	Inconel 702	225	0.11
MP159	14	0.2	Inconel 901	200	0.20
Hastelloy C-276	12	0.18	Inconel 901	300	0.13[a]
Hastelloy X	14	0.2			
René 41	5	0.09			

[a] MR2 = (200/300) × 0.2 = 0.13.
Compiled from: High Performance Alloys, Inc. [3] and All Metals & Forge Group [4].

order. The literature provides similar data and recommendations for machining super alloys; however, they should be only considered as general guides. Machining of super alloys is so difficult that careful study should be undertaken for any alloy to develop a set of machining parameters that result in reasonable tool life and secure economical aspects.

Generally speaking, a material that is machinable by a certain process may not be machinable by another one. Moreover, a particular machining process found suitable under given conditions may not be equally efficient for machining the same material under other conditions. Also, for super alloys, besides the large variations in their machinability, the same alloy differs in response for different machining operations. This is illustrated by the comparison

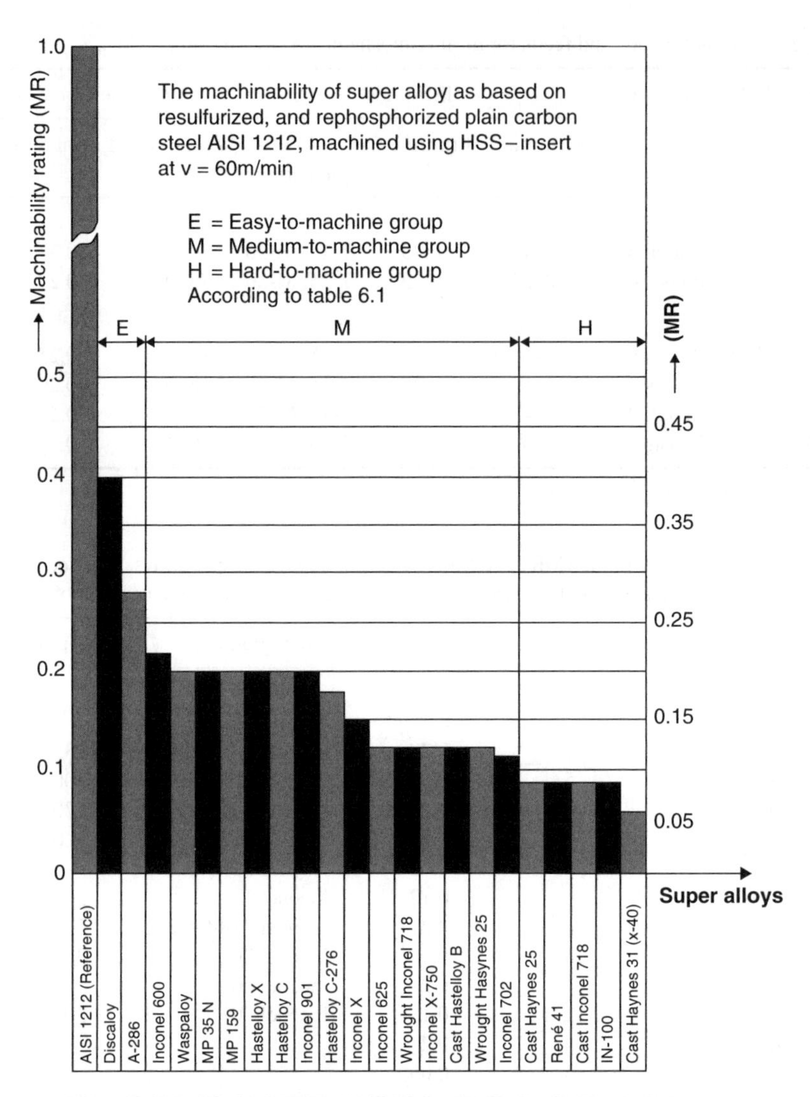

Figure 6.1 Machinability rating of the most commonly used super alloy considering AISI-1212 as a reference material.

presented between three widely used super alloys, representing three alloy classes, namely, A 286, Inconel X-750, and Haynes 25, when using different machining operations (Table 6.4) [2]. Although the average MR of the Inconel X-750 and Haynes 25 are closely similar (Figure 6.1), marked differences occur for several processes.

6.2.2 Machinability as Based on Specific Cutting Energy

One of the most important interpretations of the machinability is the specific cutting energy of the material at given working conditions. As previously described in Chapter 4, the specific cutting energy k_s can be determined either directly through force measurement

Table 6.4 Comparison of machinability rating of Inconel X-750, Haynes 25, and A-286 versus AISI 4130 steel (15 RC and UTS, 700 MPa)

Operation	Machinability rating		
	Ni-base	Co-base	Fe-base
	Inconel X-750	Haynes 25	A-286
	35 RC	24 RC	35 RC
Face milling	0.045	0.025	0.085
Turning	0.15	0.23	0.155
Drilling Φ6/Φ12 mm	0.1/0.09	0.12/0.10	0.035/0.07
Reaming Φ6/Φ12 mm	0.07/0.1	0.15/0.16	0.20/0.22
Threading	0.8	0.95	0.74

UTS, ultimate tensile strength.
Adapted from: MHB16 [2].

Table 6.5 Specific cutting energy of super alloys

Super alloy	Processing	Heat treatment	BHN	Specific cutting energy $k_{s.1}$ (N/mm^2)	Thickness exponent z
Fe-base alloy	Not specified	Annealed	200	2400	0.25
		Aged	280	2500	0.25
Ni-base alloy	Wrought	Annealed	250	2650	0.25
		Aged	350	2900	0.25
	Cast	Not specified	320	3000	0.25
Co-base	Wrought	Annealed	200	2700	0.25
		Aged	300	3000	0.25
	Cast	Not specified	320	3100	0.25

Adapted from: Sandvik [5].

using a machine tool dynamometer, or indirectly through input electrical power measurement using a wattmeter.

Sandvik [5] provided data regarding the specific cutting energy $k_{s.1}$, as well as the related chip thickness exponents, of basic types of super alloys (Table 6.5). Figure 6.2 illustrates the global values of k_s, for Fe-, Ni-, and Co-base alloys, along with those of some commonly used austenitic stainless steels for comparison.

6.3 Traditional Machining Processes (TMPs) of Super Alloys

TMPs find much use with super alloys, because they provide much higher metal removal rates than those achieved using nontraditional methods that will be dealt with in Chapters 7–9. The most traditional processes used to machine super alloys are turning drilling, reaming, milling broaching, and grinding. However, others may be used such as planing, shaping, threading,

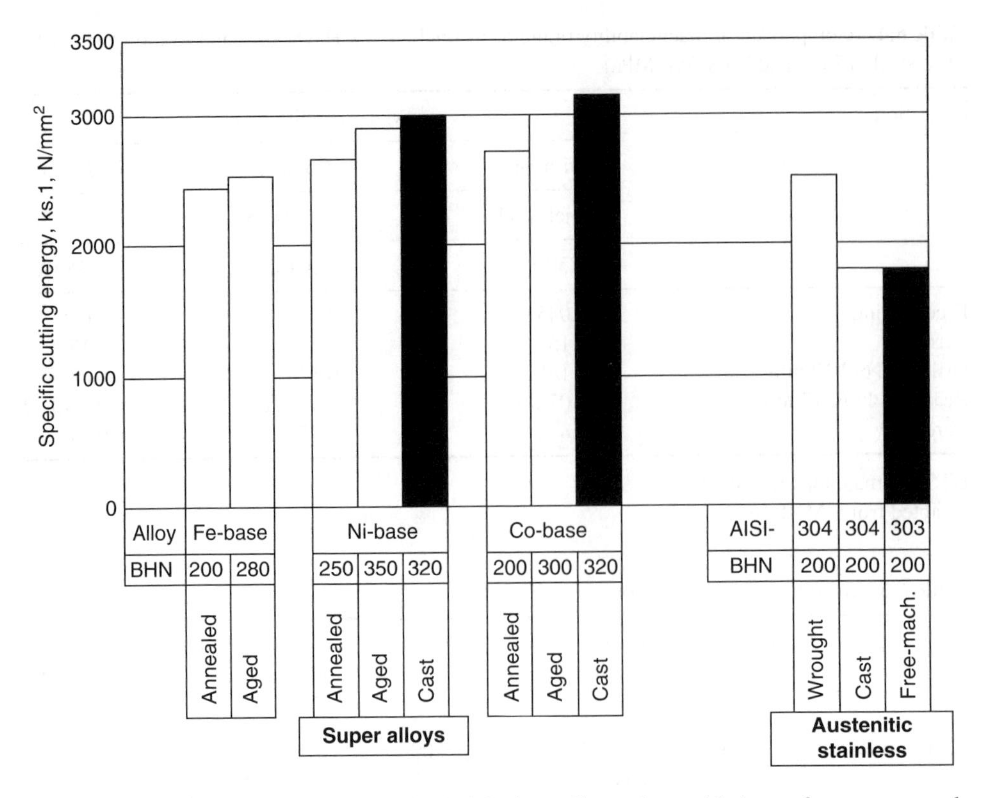

Figure 6.2 Global values of k_s, for Fe-, Ni-, and Co-base alloys, along with those of some commonly used austenitic stainless steels for comparison.

boring, gear cutting, and so on. Single-point tool turning is the most frequently used machining process for super alloys.

6.3.1 Challenges and Machining Guidelines for Super Alloys

Due to low thermal conductivity of super alloys, temperature during machining can be as high as 1000–1300 °C, thus causing crater wear and severe plastic deformation of the cutting tool. Crater wear can, in turn, weaken the cutting edge, leading to catastrophic failure. Therefore, resistance against crater wear is an important tool property requirement for machining super alloys. Plastic deformation, on the other hand, can blunt the cutting edge, thereby, increasing the cutting forces.

The chemical reactivity of these alloys facilitates BUE-formation and coating delamination, which severely degrades the cutting tool leading to short tool life. An ideal cutting tool should exhibit chemical inertness when machining such alloys. The hard abrasives inter-metallic compounds in the microstructure cause severe abrasive wear of the tool tip.

Heat generated during machining can potentially alter the microstructure. The chip produced when machining super alloys is tough and continuous, and requires superior chip breaker geometry.

As general guidelines, when machining super alloys, always high feed rate and depth of cut should be maintained to minimize work hardening. The tool is never allowed to dwell to avoid the possibility of work hardening and problems in subsequent processes. Except with ceramic tooling, generous quantity of coolant is recommended to reduce high temperature, and consequently rapid tool wear.

The following is the practical guide for machining high temperature super alloys (after Seco Technical guide, Turning Difficult-To-Machine Alloys [6]):

- Machine alloys in the softest state possible.
- Use a positive rake insert or groove.
- Use relatively sharp edges.
- Use strong geometry.
- Use a rigid set-up.
- Prevent part deflection.
- Use a high lead angle.
- When more than one pass is required, vary then depth of cut.

A positive rake cutting edge is recommended for semi-finishing and finishing operations whenever possible. Positive rake geometry minimizes work hardening of the machined surface by shearing the chip away from the work piece in an efficient way in addition to minimizing built-up-edge. Very light hones or even sharp insert edges are useful in preventing material build-up and improving surface finish during machining. Dull or improperly ground edges increases cutting forces during machining, causing metal build-up, tearing and deflection of the work material. It is however important to note that sharp insert edges are more fragile and susceptible to chipping during machining, thus honed edges are recommended for most roughing operations where concerns about surface roughness are at a minimum. Sharp edges are then used for finishing operations.

Using a large nose radius wherever part geometry does not demand otherwise can reinforce the cutting edge. This has the effect of subjecting more of the tool edge into the cut, decreasing the force at any one point, and preventing localized damage. Machining with a rigid set-up prevents vibration and subsequent chatter that deteriorates surface finish and can cause tool fracture. Tighter tolerance can be maintained with rigid set-ups. Deflection of the work material should be prevented, especially when machining thin walled components or parts. Special precautions like the use of filler metals, special fixtures, or back up may be used to prevent movement during machining.

6.3.2 Turning

As previously discussed in Chapter 4, all types of cutting tools are applicable in turning operations of super alloys. Carbide tools are usually used, although coated carbides, ceramics, cubic boron nitride (CBN), and even HSS are also used. A carbide tool of grade C-2 is frequently selected for roughing, while C-3 is frequently dedicated for finishing (Table 6.6). To realize high production rate when machining the harder Ni-base (wrought or cast), and Co-base, strengthened and aged (ST-Ag) alloys, HSS, and carbide tooling should be replaced by Borazon (CBN) or ceramics. Borazon realizes a cutting speed ranging from 120 to 185 m/min (Table 6.6), while ceramics may be used, without cooling, for higher cutting speeds. Table 6.7 summarizes

Table 6.6 Recommended speeds, feeds, and cutting tools for cylindrical turning of super alloys

Group of super alloy	Condition	BHN	DOC (mm)	HSS		Type	Indexable carbides			Borazon	
				v (m/min)	f (mm/rev)		v (m/min)	f (mm/rev)	Type	v (m/min)	f (mm/rev)
Fe-C1	Ann	135–185	1	37	0.18	M-2, M-3	115	0.18	C-7	—	—
			8	23	0.5	M-2, M-3	76	0.5	C-6	—	—
Fe-C2	Ann or N	135–185	1	24	0.18	M-2, M-3	84	0.18	C-7	—	—
			8	15	0.5	M-2, M-3	60	0.5	C-6	—	—
Ni-W5	R	180–200	0.8	30	0.13	T-15, M-42	90	0.13	C-3	—	—
			2.5	24	0.18	T-15, M-42	85	0.18	C-2	—	—
			5.0	18	0.40	T-15, M-42	60	0.40	C-2	—	—
Fe-C3	C	160–210	1	21	0.18	M-2, M-3	76	0.18	C-7	—	—
			8	12	0.5	M-2, M-3	53	0.5	C-6	—	—
Fe-W	ST	180–230	0.8	14	0.13	T-15, M-42	5.8	0.13	C-3	—	—
			2.5	11	0.18	T-15, M-42	49	0.18	C-2	—	—
			5.0	—	—	—	37	0.25	C-2	—	—
	ST-Ag	250–320	0.8	12	0.13	T-15, M-42	5.2	0.13	C-3	—	—
			2.5	9	0.18	T-15, M-42	44	0.18	C-2	—	—
			5.0	—	—	—	30	0.25	C-2	—	—
Ni-W4	Ann or ST	140–220	0.8	8	0.13	T-15, M-42	35	0.13	C-3	—	—
			2.5	6	0.18	T-15, M-42	30	0.18	C-2	—	—
			5.0	—	—	—	24	0.40	C-2	—	—
	Cd or Ag	240–310	0.8	6	0.13	T-15, M-42	27	0.13	C-3	—	—
			2.5	5	0.18	T-15, M-42	21	0.18	C-2	—	—
			5.0	—	—	—	15	0.40	C-2	—	—
Ni-W1	Ann or ST	200–300	0.8	8	0.13	T-15, M-42	30	0.13	C-3	—	—
			6	6	0.18	T-15, M-42	24	0.18	C-2	—	—
			—	—	—	T-15, M-42	18	0.40	C-2	—	—
	ST-Ag	300–400	0.8	8	0.13	T-15, M-42	29	0.13	C-3	185	0.08
			2.5	5	0.18	T-15, M-42	23	0.18	C-2	150	0.13
			5.0	—	—	—	15	0.40	C-2	135	0.13

Co-W	ST	180–230	0.8	8	0.13	T-15, M-42	27	0.13	C-3	—	—
			2.5	6	0.18	T-15, M-42	21	0.18	C-2	—	—
			5.0	—	—	—	17	0.25	C-2	—	—
	ST-Ag	270–320	0.8	6	0.13	T-15, M-42	24	0.13	C-3	—	—
			2.5	5	0.18	T-15, M-42	20	0.18	C-2	—	—
			5.0	—	—	—	15	0.25	C-2	—	—
Ni-C1	C or C-Ag	200–375	0.8	5	0.13	T-15, M-42	24	0.13	C-3	—	—
			2.5	3.6	0.18	T-15, M-42	21	0.18	C-2	—	—
			5.0	—	—	—	17	0.25	C-2	—	—
Ni-W2	ST	225–300	0.8	5	0.13	T-15, M-42	24	0.13	C-3	—	—
			2.5	3.6	0.18	T-15, M-42	21	0.18	C-2	—	—
			5.0	—	—	—	17	0.25	C-2	—	—
	ST-Ag	300–400	0.8	3.6	0.13	T-15, M-42	14	0.13	C-3	185	0.08
			2.5	3	0.18	T-15, M-42	12	0.18	C-2	135	0.13
			5.0	—	—	—	9	0.25	C-2	120	0.13
Co-C	C or C-Ag	220–290	0.8	6	0.13	T-15, M-42	18	0.13	C-3	—	—
			2.5	3.6	0.18	T-15, M-42	15	0.18	C-2	—	—
			5.0	—	—	—	—	—	—	—	—
	C or C-Ag	220–425	0.8	3.6	0.13	T-15, M-42	14	0.13	C-3	185	0.08
			2.5	3	0.18	T-15, M-42	9	0.18	C-2	135	0.13
			5.0	—	—	—	—	—	—	120	0.13
Ni-C2	C or C-Ag	250–300	0.8	5	0.13	T-15, M-42	18	0.13	C-3	—	—
			2.5	3.6	0.13	T-15, M-42	14	0.18	C-2	—	—
			5.0	—	—	—	11	0.25	C-2	—	—
	C or C-Ag	220–425	0.8	3.6	0.13	T-15, M-42	15	0.13	C-3	185	0.08
			2.5	3	0.13	T-15, M-42	11	0.18	C-2	135	0.13
			5.0	—	—	—	9	0.25	C-2	120	0.13
Ni-W3	ST	275–390	0.8	5.0	0.13	T-15, M-42	15	0.13	C-3	—	—
			2.5	3.6	0.18	T-15, M-42	14	0.18	C-2	—	—
			5.0	—	—	—	11	0.25	C-2	—	—
	ST-Ag	400–475	0.8	3.6	0.13	T-15, M-42	15	0.13	C-3	185	0.08
			2.5	3.0	0.18	T-15, M-42	12	0.18	C-2	135	0.13
			5.0	—	—	—	9	0.25	C-2	120	0.13

NB: The abbreviations as in Table 6.1.
Adapted from: Metcut Associates [1].

Table 6.7 Speed range for rough and finish turning of super alloys using CBN and coated carbide tools

Super alloy group as defined in Table 6.1	BHN	Roughing		Finishing	
		CBN	Coated	CBN	Coated
		v (m/min)		v (m/min)	
Easy-to-cut	135–210	—	70–100	—	100–145
Medium and hard-to-Cut	200–475	120–135	—	160–185	—
The cutting speed decreases as the hardness of the alloy increases					

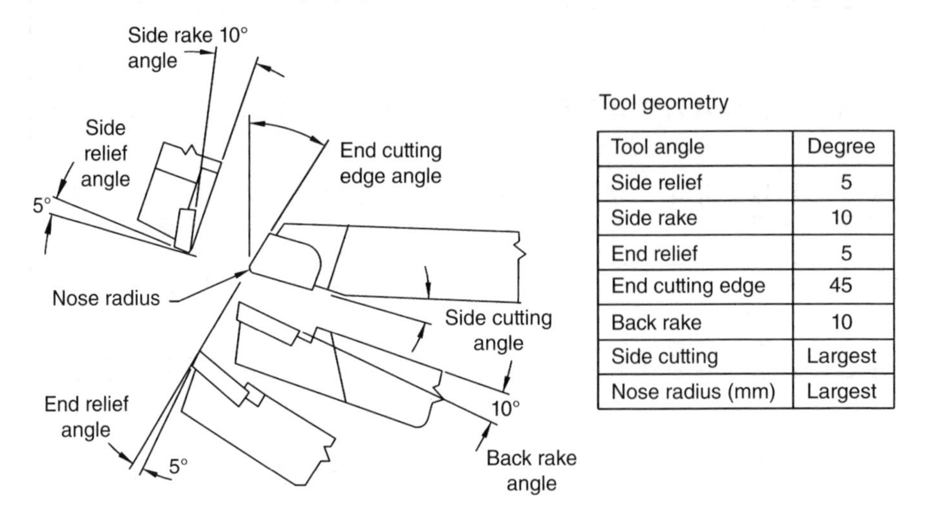

Figure 6.3 HSS-turning tool for super alloys (wrought and cast) of BHN ranging from 140 to 475.

the nominal speed ranges for rough and finish turning of super alloys using CBN and coated carbide tools. Figure 6.3 shows a HSS-turning tool for super alloys (wrought and cast) of BHN ranging from 140 to 475.

6.3.3 Drilling

The high cutting forces when drilling super alloys necessitate maximum rigidity of the tool, and workpiece. In terms of tool design or selection, the most important requirement is that the drills be as short and rigid as possible of heavy web thickness (Figure 6.4). Both HSS and carbide drills are used when drilling super alloys.

In applications involving small diameter twist drills (<6 mm diameter), a T-15 drill material has shown erratic performance. HSS-drill of grade T-15 is capable of developing very high hardness (67 RC or higher), and small drills of T-15 usually fail by microscopic chipping of the cutting edges. This is probably the result of inherent lack of rigidity of small drills and the slightly coarser grain size of T-15 HSS-twist drills [2].

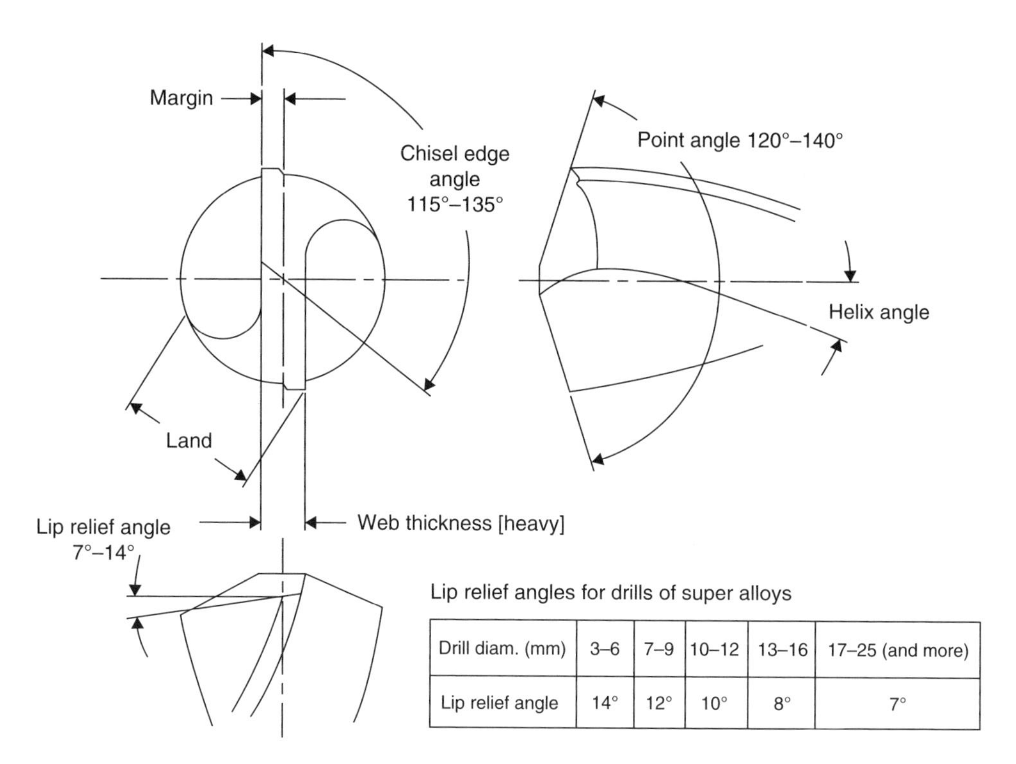

Drill diam. (mm)	3–6	7–9	10–12	13–16	17–25 (and more)
Lip relief angle	14°	12°	10°	8°	7°

Figure 6.4 Recommended tool geometry for a twist drill used for drilling super alloys.

Carbide-tipped-twist drills are more suitable for drilling the Ni-W3 category (Table 6.1), and the mechanically alloyed Ni-base alloy MA 6000 (Table 6.3). These alloys present high resistance if machined by conventional HSS-drills. A carbide drill is also preferred when drilling alloys in the Ni-C2, and Co-C categories (Table 6.1), if they exhibit hardness of 320–425 BHN, and 250–425 BHN, respectively [1]. Figure 6.4 shows the recommended tool geometry for a twist drill used for drilling super alloys. Nominal speeds and feeds along with the preferred drill material for drilling various types of super alloys are shown in Table 6.8.

When drilling deep holes, for example, of an aspect ratio of 8:1, the cutting speeds should be reduced by 40%, and the feed rates by 20% [2]. Depending on the type of super alloy, cutting speeds ranging from 2 to 18 m/min are used when drilling with HSS-drill. For the more difficult-to-cut super alloys such as René 41 and Haynes 25, speeds as low as 2 m/min may be required. Speeds lower than 3 m/min are seldom used, because in this case the shear action is poor and the drills are liable to fail by chipping. When carbide drills are used, speeds are usually two- to three-times as fast. A steady rate of feed is also important when drilling super alloys that work harden readily. Hand feeding is sometimes used, however, it is not recommended for drilling super alloys.

The high cutting forces when drilling super alloys necessitates maximum rigidity of the workpiece fixturing. When the walls of the workpiece are too thin to withstand clamping and

Table 6.8 Recommended speeds, feeds and tool materials for drilling super alloys using HSS and carbide drill

Super alloy machinability group	Condition	BHN	Speed v (m/ min)	Feed (mm/rev)						Tool material
				Nominal diameter (mm)						
				3	6	12	18	25	50	
Fe-C1	Ann~	135–185	18	0.05	0.1	0.18	0.25	0.35	0.4	M-1, M-7, M-10
Fe-C2	Ann or N	135–185	16	0.05	0.1	0.18	0.25	0.30	0.4	M-1, M-7, M-10
Ni-W5	R	180–200	16	0.05	0.1	0.18	0.25	0.4	0.45	M-1, M-7, M-10
Fe-C3	C	160–210	14	0.05	0.08	0.12	0.2	0.25	0.3	M-1, M-7, M-10
Fe-W	ST	180–230	8	0.05	0.1	0.15	0.2	0.25	—	T-15, M-42
	ST-Ag	250–320	6	0.05	0.1	0.15	0.2	0.2	—	T-15, M-42
Ni-W4	Ann or ST	140–220	6	0.05	0.075	0.075	0.1	0.1	—	M-1, M-7, M-10
	Cd or Ag	240–310	5	0.05	0.075	0.075	0.1	0.1	—	M-1, M-7, M-10
Ni-W1	Ann or ST	200–300	6	0.05	0.075	0.075	0.1	—	—	T-15, M-42
	ST-Ag	300–400	5	0.05	0.075	0.075	0.1	—	—	T-15, M-42
Co-W	ST	180–230	6	0.05	0.075	0.075	0.1	—	—	T-15, M-42
	ST-Ag	270–330	5	0.05	0.075	0.075	0.1	—	—	T-15, M-42
Ni-C1	C or C-Ag	200–375	2	0.05	0.075	0.075	0.1	—	—	T-15, M-42
Ni-W2	ST	225–300	5	0.05	0.075	0.075	0.1	—	—	T-15, M-42
	ST-Ag	300–400	3.6	0.05	0.075	0.075	0.1	—	—	T-15, M-42
Co-C	C or C-Ag	220–290	2.4	0.05	0.075	0.075	0.1	0.15	—	T-15, M-42
	C or C-Ag	290–425	5	0.025	0.05	0.075	0.1	—	—	C-2
Ni-C2	C or C-Ag	250–320	2.4	0.05	0.075	0.075	0.1	—	—	T-15, M-42
	C or C-Ag	320–425	5	0.025	0.05	0.075	0.1	—	—	C-2
Ni-W3	ST	275–390	6	0.025	0.05	0.075	0.1	—	—	C-2
	ST-Ag	400–475	5	0.025	0.05	0.075	0.1	—	—	C-2

ST: solution treated, Ann: annealed, N: normalized, C: cast, W: wrought, C-Ag: cast and aged, ST-Ag: solution treated and aged, R: rolled, and Cd: cold drawn.
Adapted from: Metcut Associates [1].

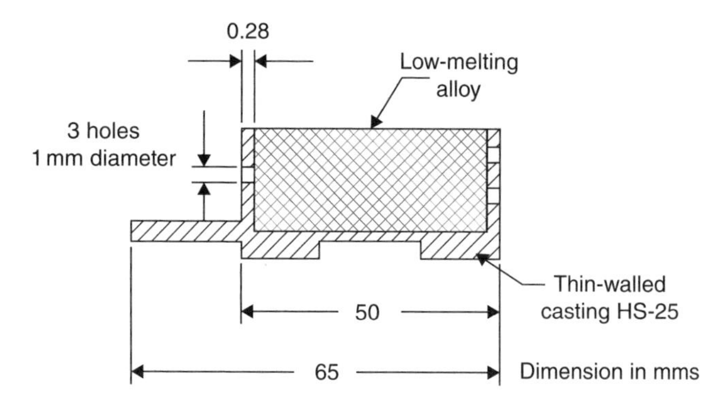

Figure 6.5 Filling the work cavity with a low-melting alloy.

cutting forces, some special techniques should be undertaken to secure the rigidity, such as filling the work cavity with a low-melting alloy (Figure 6.5).

Gun drills are recommended for deep hole drilling depths greater than three times of hole diameters (diameter up to 50 mm) in hard-to-cut super alloys, related to categories Ni-W3, Ni-C2, and Co-C (Table 6.1). This type of drills avoids the work hardening that occurs at the point extremes of standard twist drill. Depending on the hole diameter, the feed rates are ranging from 4 to 30 μm for a diameter range from 2 to 50 mm. These feeds are valid for all types of super alloys using carbide-inserted (C-2 grade), single-flute gun drills. Depending on the MR of super alloy categories defined in Table 6.1, the cutting speed, if carbide C-2 is used, ranges from 26 m/min for easy-to-cut super alloys to 15 m/min for hard-to-cut super alloys.

6.3.4 Reaming

For reaming super alloys, speeds of less than 3 m/min are seldom practical, because cutting edges are likely to chip. The feed must be great enough to maintain cutting action with a practical size of chip thickness. The reaming allowance is especially critical with super alloys. Removing an excessive amount of stock overloads the reamers, while insufficient stock will induce burnishing, causing the alloy to work harden. For super alloys, the optimum amount of reaming allowance varies with the hole size, but 0.13 mm on the radius should be a minimum, even for the smallest holes [2].

It is generally preferable to use carbide tipped reamers, although HSS-reamers are also used for easy-to-cut super alloys. Carbide-tipped reamers should be used when reaming hard-to-cut super alloys and mechanically alloyed MA 6000 products. The recommended speeds when using carbide-tipped reamers are ranging from 3 to 14 m/min depending on the machinability of super alloy. If HSS reamers are used, only two-third of these values are recommended [1].

6.3.5 Milling

Climb milling of super alloys is generally preferred than conventional. It requires milling machines, that equipped with backlash eliminator, however, cuts deeper than 1.5 mm are seldom attempted with climb milling of super alloys, because it is virtually impossible to attain the required rigidity [7]. Cutting edges of milling cutters must be with minimum edge rounding to prevent adherence of chip fragments when edge leaves the cut. It is a common practice to employ a fairly low cutting speed in combination with a moderately high feed/tooth (not less than 0.1 mm/tooth) to prevent work hardening of the material. The flank wear should not exceed 0.2–0.3 mm, otherwise the chances of catastrophic failure will increase rapidly.

For milling super alloys, two issues of cutter design must be specially considered. First, the tooth strength must be greater than that required for milling steel or cast iron, and second, relief angles must be large enough to prevent rubbing action and consequent work hardening of super alloy being cut. Regardless of the cutter material, and except small cutters, inserted blades are used on nearly all cutters, because even under the most favorable machining conditions, the tool life of the cutting edge is short. Mechanical methods of securing the blades on the cutter body are preferred, because replacement of chipped or broken blades (not brazed) is easier. Figure 6.6 illustrates a HSS face-milling cutter for milling wrought and cast super alloys of BHN ranging from 200 to 475 m/min.

Because of the interrupted cutting action of milling, HSS is used in most application for super alloys. However, carbide is more economical than HSS when milling the more difficult-to-machine super alloys, such as René 41 and MA-6000. Small solid carbide end-mills have been successfully used in some application. The more highly alloyed grades of HSS usually outperform the general purpose grades, but there is a less difference in performance than other efficient tools used for machining super alloys [2].

Nominal speeds and feeds for face and end (peripheral) milling of super alloys are given in Tables 6.9 and 6.10. The parameters listed in these tables assume that the milling operations are conducted under optimum conditions, regarding adequate rigidity of the setup, optimum tool geometry, and plentiful supply of cutting fluid. When one or more factors are less favorable (such as excessive super alloy hardness or inadequate setup rigidity), speed must be reduced. In some applications it may be necessary to mill Ni-base and Co-base alloys at speeds as low as 1.5 m/min to realize acceptable tool life [2].

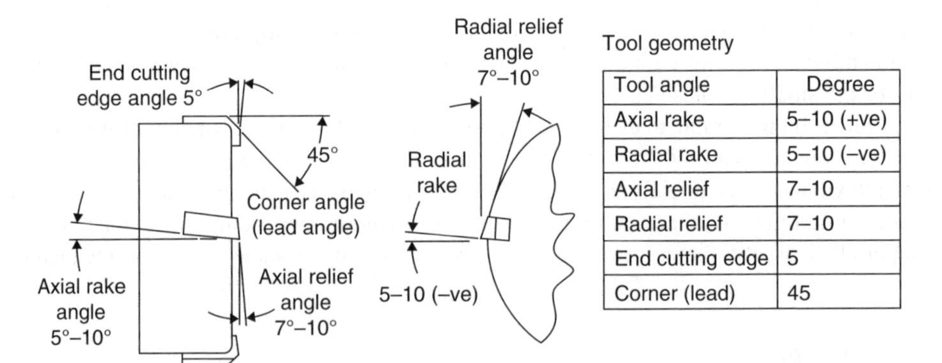

Tool angle	Degree
Axial rake	5–10 (+ve)
Radial rake	5–10 (–ve)
Axial relief	7–10
Radial relief	7–10
End cutting edge	5
Corner (lead)	45

Figure 6.6 HSS face-milling cutter for milling wrought and cast super alloys of BHN ranging from 200 to 475 m/min.

Table 6.9 Recommended speeds, feeds, and cutting tools for face milling

Group of super alloy	Condition	BHN	DOC (mm)	HSS			Uncoated indexable carbides		
				v (m/min)	f (mm/tooth)	Type	v (m/min)	f (mm/tooth)	Type
Ni-W5	R	180–200	1	35	0.10	T-15, M-42	84	0.15	C-2
			4	27	0.15		67	0.25	
			8	21	0.20		52	0.25	
Fe-W	ST	180–230	1	18	0.13	T-15, M-42	30	0.15	C-2
			4	9	0.20		24	0.20	
			8	5	0.25		18	0.25	
	ST-Ag	250–320	1	12	0.10	T-15, M-42	21	0.15	C-2
			4	6	0.15		20	0.20	
			8	5	0.20		12	0.25	
Ni-W4	Ann or St	140–200	1	9	0.07	T-15, M-42	24	0.13	C-2
			4	8	0.13		23	0.15	
			8	6	0.18		—	—	
	Cd or Ag	240–310	1	8	0.05	T-15, M-42	20	0.13	C-2
			4	6	0.10		18	0.15	
			8	5	0.15		—	—	
Ni-W1	Ann or ST	200–300	1	8	0.10	T-15, M-42	—	—	—
			4	6	0.15		—	—	—
			8	—	—		—	—	—
	ST-Ag	300–400	1	6	0.07	T-15, M-42	—	—	—
			4	5	0.13		—	—	—
			8	—	—		—	—	—
Co-W	ST	180–230	1	9	0.05	T-15, M-42	21	0.13	C-2
			4	8	0.07		20	0.13	C-2
			8	6	0.1		—	—	—
	ST-Ag	270–320	1	5	0.05	T-15, M-42	18	0.13	C-2
			4	3	0.07		17	0.15	C-2
			8	2.5	0.10		—	—	—
Ni-C1	C or C-A	200–375	1	8	0.10	T-15, M-42	23	0.13	C-2
			4	6	0.15		21	0.15	C-2
			8	5	0.20		—	—	—

(continued overleaf)

Table 6.9 (*Continued*)

Group of super alloy	Condition	BHN	DOC (mm)	HSS			Uncoated indexable carbides		
				v (m/min)	f (mm/tooth)	Type	v (m/min)	f (mm/tooth)	Type
Ni-W2	ST	225–300	1	6	0.1	T-15, M-42	—	—	—
			4	5	0.15		—	—	—
			8	—	—		—	—	—
	ST-Ag	300–400	1	5	0.07	T-15, M-42	—	—	—
			4	3	0.13		—	—	—
			8	—	—		—	—	—
Co-C	C or C-Ag	220–290	1	5	0.05	T-15, M-42	15	0.13	C-2
			4	3	0.07		14	0.15	C-2
			8	2.5	0.1		—	—	—
	C or C-Ag	290–425	1	3.5	0.05	T-15, M-42	11	0.13	C-2
			4	2.5	0.05		8	0.15	C-2
			8	2.0	0.07		—	—	—
Ni-C2	C or C-Ag	250–300	1	6	0.05	T-15, M-42	—	—	—
			4	3.5	0.05		—	—	—
			8	2.8	0.07		—	—	—
	C or C-Ag	300–425	1	5	0.05	T-15, M-42	—	—	—
			4	3	0.05		—	—	—
			8	2.5	0.07		—	—	—
Ni-W3	ST	275–390	1	5	0.05	T-15, M-42	—	—	—
			4	3	0.07		—	—	—
			8	—	—		—	—	—
	ST-Ag	290–425	1	3.5	0.05	T-15, M-42	—	—	—
			4	2.5	0.07		—	—	—
			8	—	—		—	—	—

The abbreviations as in Table 6.1.
Adapted from: Metcut [1].

Table 6.10 Recommended speeds, feeds, for slab milling of super alloys using HSS-milling cutters

Group of super alloy	Condition	BHN	DOC (mm)	Speed v (m/min)	Feed f (mm/tooth)	HSS-AISI
Ni-W5	R	180–200	1	32	0.10	M-2, M-7
			4	24	0.13	
			8	17	0.15	
Ni-W4	Ann or ST	140–200	1	8	0.10	T-15, M-42
			4	6	0.13	
			8	5	0.15	
	Cd or Ag	240–310	1	6	0.10	T-15, M-42
			4	5	0.13	
			8	—	—	
Ni-W1	Ann or ST	200–300	1	8	0.10	T-15, M-42
			4	6	0.13	
			8	5	0.15	
	ST or Ag	300–400	1	6	0.1	T-15, M-42
			4	5	0.1	
			8	—	—	
Ni-C1	C or C-Ag	200–375	1	6	0.10	T-15, M-42
			4	5	0.13	
			8	—	—	
Ni-W2	ST	225–300	1	6	0.10	T-15, M-42
			4	5	0.13	
			8	3	0.15	
	ST-Ag	300–400	1	6	0.10	T-15, M-42
			4	5	0.13	
			8	—	—	

ST: solution treated, Ann: annealed, N: normalized, C: cast, W: wrought, C-Ag: cast and aged, ST-Ag: solution treated and aged, R: rolled, and Cd: cold drawn.
Adapted from: Metcut Associates [1]

6.3.6 Broaching

Broaching dates back to the early 1850, when it was originally adopted for cutting keyways in pulleys and gears. However, its obvious advantages quickly led to its development for mass-production of various surfaces and shapes to tight tolerance. Today, almost every conceivable form and material can be broached. Therefore, broaching is extensively used, because it is the practical method of machining complex contours of blades, and related components of gas turbines.

Broaching is a fast process, in which both roughing and finishing operations are completed in one tool pass. The process is, however, impractical for blind holes and pockets. Moreover, the broaches are costly to produce and sharpen.

Successful broaching of super alloys requires the following important considerations:

- broach design that provides ample strength and clearance for swarf;
- rigid machine combined with adequate power;
- rigid tool and workpiece set up;

Table 6.11 Nominal cutting speed and super elevation (chip load) in broaching super alloys using HSS-broaches

Group of super alloy	Condition	BHN	Speed (m/min)	Super-elevations (mm/tooth)	HSS-grade
Ni-W5	R	180–200	6	0.075	M-2, M-7
Fe-W	ST	180–230	3.5	0.05	T-15, M-42
	ST-Ag	250–320	3.0	0.05	T-15, M-42
Ni-W4	Ann or ST	140–220	3.0	0.05	T-15, M-42
	Cd-Ag	240–310	2.5	0.05	T-15, M-42
Ni-W1	Ann or ST	200–300	2.5	0.05	T-15, M-42
	ST-Ag	300–400	2.0	0.05	T-15, M-42
Co-W	ST	180–230	2.5	0.05	T-15, M-42
	ST-Ag	270–320	2.0	0.05	T-15, M-42
Ni-C1	C or C-Ag	200–375	2.5	0.05	T-15, M-42
Ni-W2	ST	225–300	2.5	0.05	T-15, M-42
	ST-Ag	300–400	2.0	0.05	T-15, M-42
Co C	C or C-Ag	220–290	2.5	0.05	T-15, M-42
	C or C-Ag	290–425	2.0	0.05	T-15, M-42
Ni-C2	C or C-Ag	250–320	2.5	0.05	T-15, M-42
	C or C-Ag	320–425	2.0	0.05	T-15, M-42
Ni-W3	ST	275–390	2.0	0.05	T-15, M-42

NB: The abbreviations as in Table 6.1.
Adapted from: Metcut [1].

- avoidance of cutting edge rubbing against the workpiece;
- careful selection of cutting oil.

Tool angles and gullet shape are important design issues due to the behavior of super alloys in shearing and chip formation. The use of short, replaceable broach inserts can provide cost savings as well as better control of surface finish and accuracy. The pitch of the teeth should be approximately 25% more than that for broaching plain-carbon and low-alloy steels in order to provide the necessary greater chip clearance.

The large pitch also will decrease the total load by reducing the number of teeth in engagement. For the same desired stock removal, it is therefore necessary to use longer broaches or more broaches to a set. The rake angle can be increased by a maximum of 15°, promoting freer chip flow and minimizing the work-hardening effect. Rubbing contact should be avoided by providing as large relief angle as possible [7]. Nominal speeds and super-elevations (chip loads) for broaching using HSS-broaches are illustrated in Table 6.11.

The selection of solid broaches versus those with inserted cutting edges depends on the size and design of the broach as well as on cost. Cost is usually the deciding factor. In many applications, particularly in case of large broaches, the cost can be decreased using HSS inserts in an alloy steel body. Whether broaches are solid or having insert, it should not influence its performance [2].

Figure 6.7 illustrates the tool geometry of three different HSS-broach designs. The first is a standard broach for super alloys (Figure 6.7a). The second and third are specially designed broaches for A-286 (Figure 6.7b), while the last one is for broaching René 41 (Figure 6.7c), all of which are made of HSS.

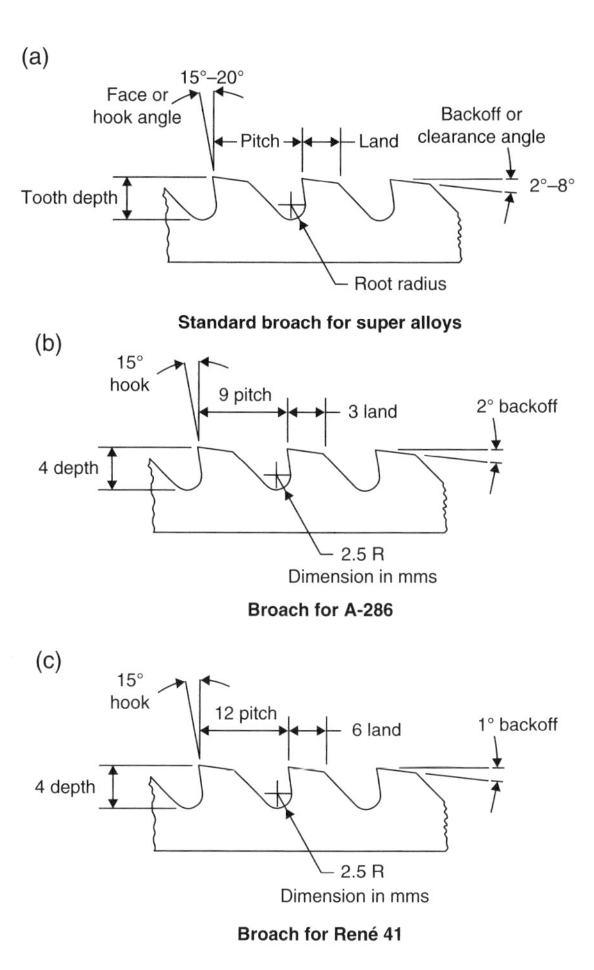

Figure 6.7 Tool geometry of three different HSS-broach designs for super alloys: (a) standard broach for super alloys; (b) broach for A-286; (c) broach for René 41. (Adapted and compiled from: Metcut [1], and MHB 16 [2].)

6.3.7 Grinding

Super alloys are generally more difficult and costly to grind than low-alloy steels. Specifically, Ni-base and Co-base super alloys are sensation to the grinding heat, metallurgical alterations, and micro cracking that occur within a considerable thickness results in a deleterious effect of the surface integrity of the component. Therefore, machining parameters during grinding of super alloys should be continue to achieve an optimum surface integrity.

6.3.7.1 Selection of Grinding Wheel Designation

Corundum wheels are selected for most super alloys, although, for some precision applications, CBN grinding wheels are used. Depending on the type of grinding operation, most super alloys are ground using medium-hard wheels (F to L). For surface grinding it is

Table 6.12 Recommended machining condition for surface grinding of super alloys

Work material	BHN	GW-designation	GW-speed vs (m/min)	Table speed vw (m/min)	Cross-feed: fraction of GW-width/pass
All types of super alloys	140–475	A-46-H-V	900–1200	15–30	1/12

Down feeds: 25 μm, and final finishing feed = 12 μm, Grinding fluid: water-base soluble-oil emulsion or sulfurized oil.
Adapted from: Metcut Associates [1]

Table 6.13 Recommended machining conditions for external cylindrical grinding of super alloys

Work material	BHN	GW-designation	GW-speed vs (m/min)	In-feed (μm)/pass for	
				Roughing	Finishing
All types of super alloys	140–475	A-60-J-V	900–1200	25	5

Work speed: 15–30 m/min, traverse feed rate: 0.2 GW-width/rev (roughing). 0.1 GW-width/rev (finishing), grinding fluid: water-base soluble-oil emulsion or sulfurized oil.
Adapted from: Metcut Associates [1].

recommended to use a grinding wheel (H), for internal grinding (F/G), while for external cylindrical grinding a grinding wheel of the grade (J K/l) is preferred. Medium wheel structure number from 7 to 10 is recommended for grinding super alloys, depending on the ductility of super alloy.

Wheel designations are provided here without the structure number. Vitrified bonded grinding wheels (designated by V) are most commonly used for grinding super alloys. However, resinoid-bonded wheels (B) are used for high speed grinding wheels (≥1400 m/min), and consequently frequently with resinoid-bonded CBN wheels.

Table 6.12 illustrates the recommended machining condition when surface grinding all types of super allays using vitrified bond wheels, while Table 6.13 illustrates the same, but for external cylindrical grinding.

References

[1] Machinability Data Center (1980) *Machining Data Handbook*, 3rd edn, Cincinnati, OH, Metcut Research Associates.
[2] ASM International (1989) *Metals Handbook: Machining*, 19th edn, Vol. **16**, ASM International, Materials Park, OH.
[3] High Performance Alloys, Inc. http:www.hpalloy.com/alloysdescriptions/Machinability Ratings.html (accessed November 30, 2011).
[4] All Metals & Forge Group http://steelforge/machinability ratings.html) (accessed November 30, 2011).
[5] Sandvik (2013) Workpiece Materials-ISO-S HRSA Titanium.
[6] Seco (2014) Technical Guide, Turning Difficult-To-Machine Alloys.
[7] M.J. Donachie, S.J. (2002) Donachie *A Technical Guide*, 2nd edn, pp. 189–194.

7

Nontraditional Machining Processes – an Overview

7.1 Nontraditional Machining Processes

Nontraditional machining processes (NTMPs) have evolved out of the increasingly needs of modern society. One invention that has created new manufacturing challenges is trans-atmospheric vehicles, and this has required continual improvements in materials to meet the demands for improved engine and aircraft-skin operating temperatures. Technological developments have prompted the creation of difficult-to-cut materials such as metal matrix composites (MMCs), ceramics, stainless steels, super alloys, and high performance polymers. The difficulty in machining these and other new materials results from their high hardness and brittleness, high refractoriness, poor thermal conductivities, chemical reactivity with the cutting tool, and inhomogeneous micro-structures. In many cases, the only effective way to machine such materials is by nontraditional processes. The traditional machining processes (TMPs) in use today for material removal primarily rely on hard tool materials to perform the chipping action. In contrast to TMPs, NTMPs, the material removal is accomplished with electrochemical reaction, high temperature plasma beams, and powerful electric sparks, and high-velocity jets of liquids and abrasives. Materials that in the past have been extremely difficult to machine, are now formed with magnetic fields. Machining capabilities have been expanded with the use of high-frequency sound waves and beams of electrons and coherent light of lasers.

NTMPs are generally classified according to the type of energy utilized in material removal. They are classified into three main groups (Table 7.1):

1. *Mechanical processes*: In these, the material removal depends on mechanical abrasion or shearing.

Machining of Stainless Steels and Super Alloys: Traditional and Nontraditional Techniques,
First Edition. Helmi A. Youssef.

Table 7.1 Classification of NTMPs according to the type of fundamental energy

Fundamental energy	Removal mechanism	NTMP
Mechanical	Abrasion, shearing	AJM, WJM, USM, MFM, AFM
Chemical	Ablative reaction (etching)	CH milling, PCM
Electrochemical	Anodic dissolution	ECM, ECT, ECD
Thermoelectric	Fusion and vaporization	EDM, LBM, EBM, IBM, PBM (PAC)

2. *Chemical and electrochemical processes*: In which, the material is removed in layers by ablative reaction, where acids or alkalis are used as etchants. Electrochemical machining (ECM) is characterized by a high removal rate. The machining action is due to anodic dissolution (AD) caused by the passage of high-density DC current in the machining cell.
3. *Thermoelectric processes*: In these, the metal removal rate depends on the thermal energy acting in the form of controlled and localized power pulses, leading to melting and evaporation of the work material.

An important and latest development has been realized by adopting what is called hybrid machining processes (HMPs). These are produced by integrating one NTMP with one or more TMPs and NTMPs to improve the performance and promote the removal rate of the hybrid process (HP). Examples of these processes are electrochemical grinding (ECG), and abrasive water jet machining (AWJM).

7.2 Mechanical Nontraditional Processes

7.2.1 Jet Machining

Jet machining (JM) comprises three processes based on using high-energy jets to cause machining due to mechanical abrasion. These are abrasive jet machining (AJM), water jet machining (WJM), and the HP AWJM.

7.2.1.1 Abrasive Jet Machining

In AJM a fine stream of abrasives is propelled through a special nozzle by a carrier gas (CO_2, N_2, or air) of a pressure ranging from 1 to 9 bar. Thus the abrasives attain a high speed ranging from 150 to 350 m/s, exerting impact force and causing mechanical abrasion of the workpiece (target material). The workpiece is positioned from the nozzle at a distance called the stand-off distance (SOD). In AJM, Al_2O_3, or SiC, abrasives of grain size ranging from 10 to 80 μm are used (Figure 7.1). The nozzles are generally made of tungsten carbides (WCs) or synthetic sapphires of diameters 0.2–2 mm. To limit the jet flaring, nozzles may have rectangular orifices ranging from 0.1 × 0.5 to 0.18 × 3 mm. The optimum jet angle is determined according to the ductility or brittleness of the workpiece material to be machined.

This process is not meant for bulk material removal. Its removal rate when machining the most brittle materials, such as glass, quartz, and ceramics, is about 30 mg/min, whereas only a

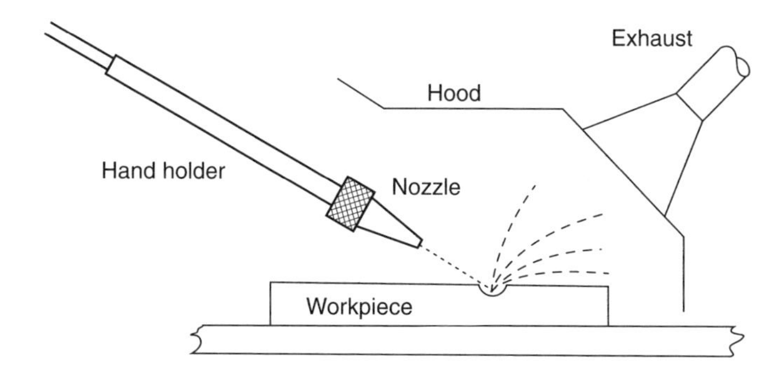

Figure 7.1 AJM – setup. (Adapted from: Machinability Data Center [1].)

fraction of that value is realized when machining soft and ductile materials. Owing to the limited removal rate, and also the significant taper, AJM is not suitable for machining deep holes and cavities. However, the process is capable of producing holes and profiles in sheets of small thicknesses. AJM is applicable for cutting, slitting, surface cleaning, frosting, and polishing [2].

The performance of AJM in terms of material removal rate (MRR) and accuracy is affected by the selected machining conditions. The MRR is mainly affected by the kinetic energy of the abrasives, which depends on gas pressure at nozzle, nozzle diameter, abrasive size and density, and SOD. The accuracy improves by selecting smaller SOD, which reduces the MRR. The grain size is a decisive factor for determining the surface finish.

The AJM station must be equipped by a vacuum dust collector to limit the pollution. Strict measures and precautions should be undertaken in case of machining toxic materials such as beryllium to collect produced dust and debris, and polishing.

Advantages of AJM include: The process is capable of producing intricate shapes in hard and brittle materials. It is used to cut fragile materials of thin walls. AJM can be used to clean surfaces, especially in inaccessible areas. The produced surfaces after cleaning by AJM are characterized by their high wear resistance. Finally, it is characterized by low capital investment and low power consumption.

Limitations of AJM include: AJM is not recommended to machine soft materials. Abrasives cannot be reused because they lose their sharpness. Nozzle clogging occurs if fine grains of $d_g < 10\,\mu m$ are used. The process accuracy is poor owing to the flaring effect of the abrasive jet. Excessive nozzle wear causes additional cost, and the process tends to pollute the environment.

7.2.1.2 Water Jet Machining

WJM has been utilized for a wide spectrum of materials, including Pb, Al, Cu, Ti, steels, and granite. It is hard to believe that a jet of water can cut steel and granite. However, in scientific terms, it is explainable, as illustrated in Figure 7.2.

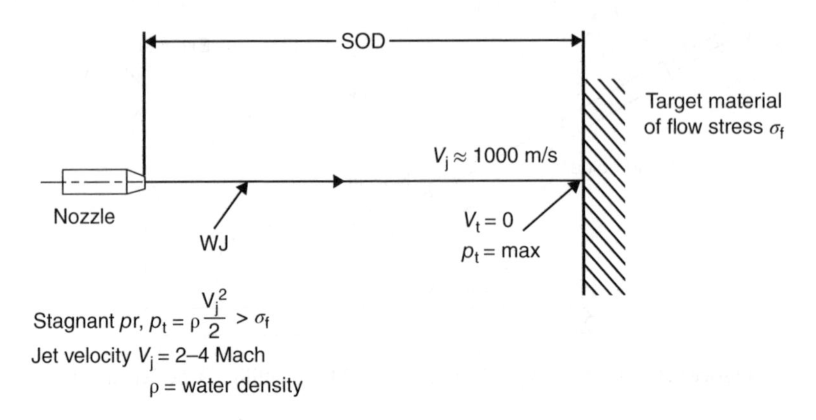

Figure 7.2 Cutting principle of WJM. (From Youssef *et al.* [3].)

When a stream of water is propelled at high pressure (2000–8000 bar) through a converging nozzle, it gives a coherent jet of water of high speed of 600–1400 m/s at the target. The kinetic energy (KE) of the jet is converted spontaneously to high-pressure energy, inducing high stresses exceeding the flow strength of target material, causing mechanical abrasion.

The WJM apparatus consists of three stations. These are the multistage filtering station, intensifier station, and cutting nozzle station. The nozzle provides a coherent water jet stream for optimum cutting. The jet coherency can be enhanced by adding long chain polymers such as polyethylene oxide (PEO). Such addition provides the water higher viscosity and hence increases the coherent length up to 600 d_n, where d_n is the nozzle orifice diameter that falls between 0.1 and 0.35 mm. For optimum cutting, the SOD is selected within this range. Even beyond this range, the stream is still capable of performing noncutting operations such as cleaning, polishing, or degreasing.

Nozzles are generally made from very hard materials such as WC, synthetic sapphire, or diamond. Diamond provides the longest nozzle life, whereas WC gives the lowest one. About 200 h of operation are expected from a nozzle of synthetic sapphire. The cutting station must be equipped by a catcher, which acts as a reservoir for collecting the machining debris entrained in the jet. Moreover, it absorbs the rest energy after cutting, which is estimated by 90% of the total jet energy. It reduces the noise levels (105 dB) associated with the reduction of the water jet from mach 3 to subsonic levels.

Process Capabilities

The MRR, accuracy, and surface quality are influenced by the WP material and the machining parameters. Brittle materials fracture, while ductile ones are cut well. The quality of cutting improves at higher pressures and lower traverse speeds. Under such conditions, greater thicknesses can be cut.

WJM is used in cutting of metals and composites applied in aerospace industries, and underwater cutting for ship-building industries. It is ideal in cutting both soft materials such as wood, paper, cloth, leather, rubber, and plastics, as well as hard materials such as rocks, granite, and marble. It is also used in slicing and processing of frozen foods, and meat. In such cases, alcohol, glycerin, and cooking oils are used instead of water. WJM is also used in cleaning, polishing, and degreasing of surfaces, and removal of nuclear contaminations.

Advantages of WJM over other cutting methods: Practically all types of material even with complicated contours can be cut with good quality without any diverse mechanical, thermal, or chemical influence on the workpiece.

Disadvantages of WJM include: WJM is unsafe in operation if safety precautions are not strictly followed. The process is characterized by high production cost due to the high capital cost of the machine and the need for highly qualified operators.

7.2.1.3 Abrasive Water Jet Machining

AWJM is a HP as it is an integration of AJM and WJM processes. The addition of abrasives to the water jet increases the range of materials that can be cut with a water jet and maximizes the MRR of this HP. The MRR is based therefore on using the kinetic energies of the abrasives and water in the jet. AWJM process is capable to machine both soft and hard materials at high speeds as compared with those realized by WJM. Moreover, the cuts performed by AWJM have better edge and surface qualities.

AWJM uses a comparatively lower water pressure than that of WJM (about 80%). The mixing ratio of abrasive to water in the jet is about 3/7 by volume. Abrasives (garnet, sand, Al_2O_3, and so on) of a grain size 10–180 μm are often used. As previously mentioned, apart from its capability to machine soft and hard materials at very high speeds, AWJM process has the same advantages of WJM. However, owing to the existence of the abrasives in the jet, there is an excessive wear in the machine and its elements. Moreover, the process is not environmentally safe as compared to WJM.

The AWJM process is mainly applicable in cutting of metallic materials such as Cu, Al, Pb, Mo, Ti, and W, carbides, ceramics, marble, granite, plastics, asbestos, composites, stainless steels, super alloys, acrylic, and glass (Figure 7.3). In the field of machining technology, the AWJM has two promising applications that include milling of flat surfaces and turning of cylindrical shapes, assisted by AWJ. The equipment of AWJM does not differ greatly from that of the basic WJM. However, the cutting station of the AWJM is provided with a jet former instead of a nozzle.

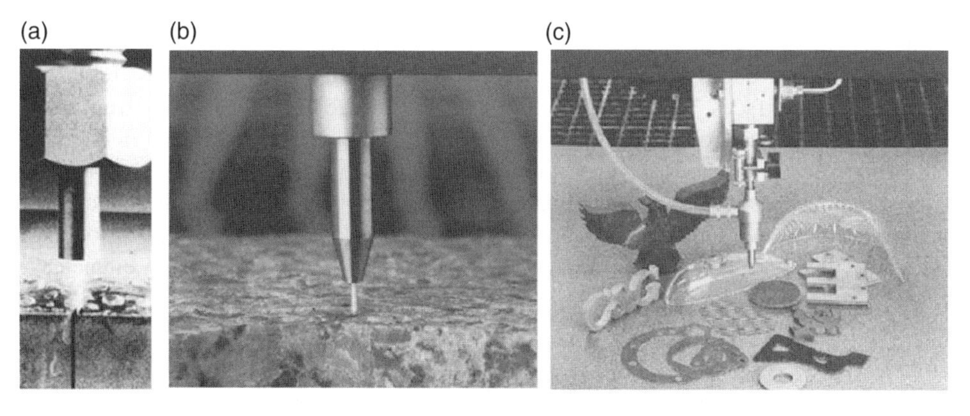

Figure 7.3 Cutting with AWJ: (a) 25-mm-thick carbon steel, (b) marble, and (c) plastic and asbestos. (Courtesy of: (a) ESAP Automation Ltd, Andover, and (b,c) Ingersoll-Rand [3].)

Table 7.2 Traverse velocity (mm/min) when machining different materials by AWJM

Materials	Material thickness				
	6 mm	15 mm	19 mm	25 mm	50 mm
Titanium	250	150	100	50	16
Aluminum	250	150	100	50	16
Fiber reinforced plastic	500	280	130	75	25
Stainless steel	200	90	60	40	15
Glass	2000	1000	700	500	150

Adapted from: Youssef [4].

Process Capabilities

The typical machining variables that affect the AWJM process include water pressure, nozzle diameter, geometry of focusing tube (length and diameter), SOD, size and type of abrasive grits, abrasive/water ratio, and hardness and strength of the workpiece material.

When machining glass by AWJ, a cutting rate of about 16–20 mm³/min is achieved. When cutting steel plates (or metallic materials), the surface roughness R_t ranges from 3.8 to 6.4 μm, while tolerances of ±130 μm are obtainable. The SOD has an important effect on the MRR and the achieved accuracy. It attains values between 0.5 and 5 mm. The smallest value realizes higher accuracy and smallest kerfs width, whereas the largest value realizes the maximum MRR. Beyond 5 mm, the jet loses gradually its cutting capability till it reaches 50–80 mm, at which the jet is used efficiently in surface cleaning and peening. Table 7.2 illustrates the traverse velocities when cutting different materials of different thicknesses using AWJ. It can be depicted that, pure metals (Ti, Al) have the same machinability, and glass is cut at 8–10 times faster than metals and alloys.

7.2.2 Abrasive Flow Machining

Abrasive Flow Machining (AFM) is purely mechanical process that finishes surfaces and edges by extruding a viscous abrasive media flowing, under pressure, through or across a workpiece. Abrasion occurs only when the flow of the media is restricted. AFM is used to burr, polish, radius edges, remove recast layers, reduce compressive residual stresses, and provide smooth surfaces. The process embraces a wide range of applications – from critical aerospace and medical components to high-production volume of parts. It can yield production rates of up to hundreds, or even thousands of parts per hour [1, 4]. It is not a mass material removal process, but it is particularly useful for polishing or deburring inaccessible internal passages and hard-to-reach-locations. Materials from soft aluminum to tough Ni-base super alloys are being processed with AFM. Holes smaller than 0.4 mm diameter are sometimes difficult to process with AFM. Blind hole polishing is impractical because AFM requires a flowing media [4].

Many types of abrasives are used in AFM. Mostly, B4C and SiC, however, Al_2O_3 is sometimes used. Media viscosity relates directly to the size of the restricting passages to be abraded. The most abrasive action occurs during the process if the hole changes size or direction. Fundamentally, viscosity must be high enough to maintain extrusion-type flow, and high enough to hold the abrasive grains at the outermost surface of the slug with sufficient force to allow abrasives to cut edges and surfaces of the restricting passage. At the same time the media must be soft enough to flow at reasonable rate to perform the operation in acceptable cycle time.

7.2.2.1 Parameters Affecting MRR of AFM

- Viscous media flow rate
- Media viscosity
- Size of abrasive particles
- Abrasive concentration
- Particle hardness
- Work material hardness.

7.2.2.2 Advantages of AFM

- AFM can be applied to any difficult-to-cut material.
- Material can be removed from targeted and hard-to-reach locations.
- Media can be engineered to match the application requirements.
- AFM improves air, gas, or liquid flow behavior, reduces cavitations tendencies, generating desirable laminar flow.

7.2.3 Ultrasonic Machining

Ultrasonic machining (USM) is an economically viable operation by which a hole or a cavity can be pierced in hard and brittle materials, whether electric conductive or not, using an axially oscillating tool. The tool oscillates with small amplitude of 10–50 μm at high frequencies of 18–40 kHz to avoid unnecessary noise (the audio threshold of human ear is 16 kHz). During tool oscillation, abrasive slurry (B_4C or SiC) is continuously fed into the working gap between the oscillating tool and the stationary WP. The abrasive particles are therefore hammered by the tool into the WP surface, and consequently they abrade the WP into a conjugate image of the tool form. Moreover, the tool imposes a static force ranging from 1 N to some kilograms depending on the size of the tool tip (Figure 7.4). The static pressure is necessary to sustain the tool feed during machining. The process productivity is realized by the large number of impacts per unit time (frequency), whereas the accuracy is achieved by the small oscillation amplitude employed. The tool tip, usually made of relatively soft material, is also subjected to an abrasion action caused by the abrasives; thus it suffers from wear, which does affect the accuracy of the machined holes and cavities. The process is characterized by the absence of any deleterious or thermal effects on the metallic structure of the WP. However, USM process is hampered by the following disadvantages:

- USM is not capable of machining holes and cavities with a lateral extension of more than 25–30 mm.
- The tool suffers excessive frontal and side wear when machining hard materials such as steels and carbides. The side wear deteriorates the accuracy of holes and cavities.

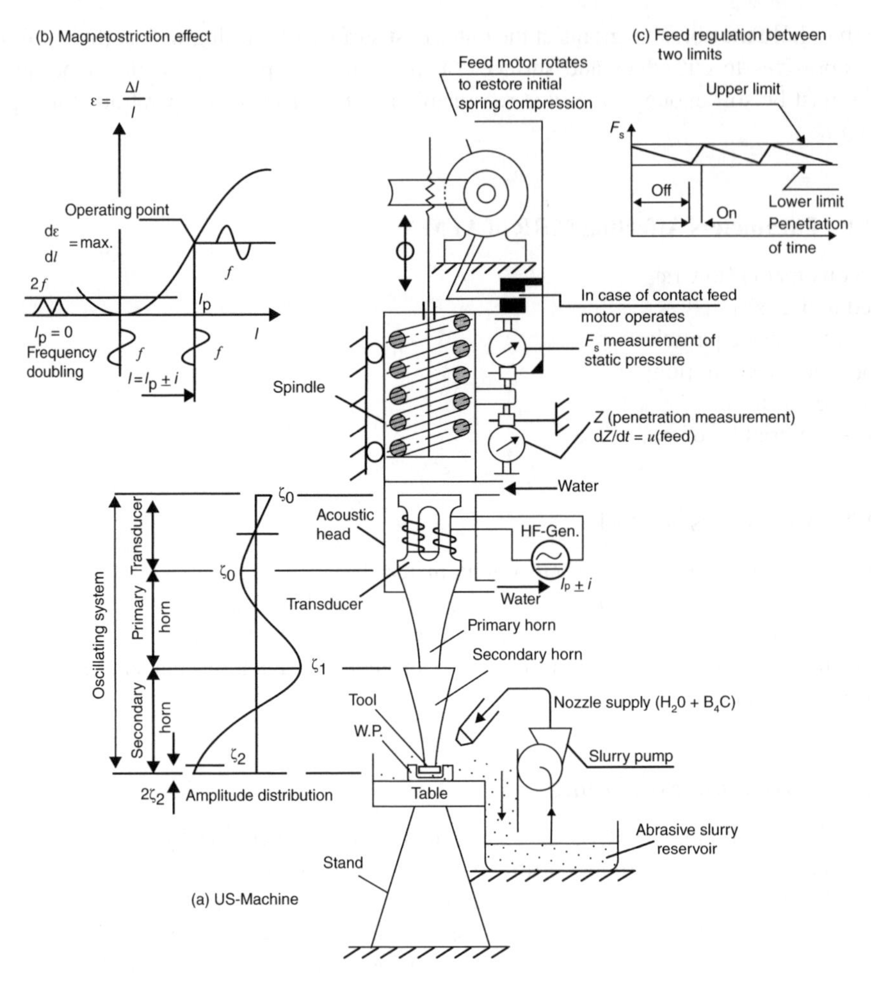

Figure 7.4 Schematic of vertical USM equipment: (a) US-Machine; (b) magnetostriction effect; (c) feed regulation between two limits. (Adapted from Youssef [4].)

- Every job needs a high-cost tool, which adds to the machining cost.
- In case of blind holes, the designer should not allow sharp corners, because these cannot be produced by USM.
- The abrasive slurry should be regularly changed to get rid of worn abrasives, which means additional cost.

The main elements of the USM-equipment comprise the oscillating system, the tool feed mechanism, and the slurry system (Figure 7.4). The oscillating system includes the transducer contained in the acoustic head, the primary acoustic horn, and the secondary acoustic horn.

7.2.3.1 Transducer and Magnetostriction Effect

The transducer transforms electrical energy to mechanical energy in the form of oscillations. Magnetostrictive transducers are generally employed in USM, but piezoelectric ones may also be used. According to this effect, in the presence of an applied magnetic field, ferromagnetic metals, and alloys change in length. An electric signal of US frequency f_r is fed to a coil that is wrapped around a stack made of magnetostrive material (iron–nickel alloy). This stack is made of laminates to minimize eddy current and hysteresis losses; moreover, it must be cooled to dissipate the generated heat (Figure 7.4a).

The alternating magnetic field produced by the HF-AC generator causes the stack to expand and contract at the same frequency. To achieve the maximum magnetostriction effect, the HF-AC current i must be super-imposed on an appropriate DC pre-magnetizing current I_p that must be exactly adjusted to attain an optimum working point. This point corresponds to the inflection point ($d^2\varepsilon/dI^2 = 0$) of the magnetostriction curve (Figure 7.4b). The pre-magnetizing direct current has the following functions:

- When precisely adjusted, it provides the maximum magnetostriction effect (maximum oscillating amplitudes).
- It prevents the frequency doubling phenomenon.

If the frequency of the AC signal, and hence that of the magnetic field, is tuned to be the same as the natural frequency of the transducer (and the whole oscillating system), so that it will be at mechanical resonance, then the resulting oscillation amplitude become quite large and the exciting power attains its minimum value. The required resonance condition is realized if the transducer length l equals half the wave length λ (or a positive integer number n of it).

$$l = \frac{n}{2}\lambda = \frac{\lambda}{2} \quad (if\, n = 1)$$

and

$$\lambda = \frac{c}{f_r} = \frac{1}{f_r}\sqrt{\frac{E}{\rho}} \quad \left(c = \sqrt{\frac{E}{\rho}}\right)$$

Where

c	=	acoustic speed in magnetostrictive material (m/s)
f_r	=	resonant frequency (1/s)
E,ρ	=	Young's modulus (MPa), and density (kg/m^3) of magnetostrictive material.

Hence,

$$l = \frac{1}{2f_r}\sqrt{\frac{E}{\rho}}$$

7.2.3.2 Acoustic Horns (Mechanical Amplifiers or Concentrators)

The oscillation amplitude ξ_o as obtained from the magnetostrictive transducer does not exceed 5 μm, which is too small for effective removal rates. The amplitude at the tool should therefore be increased to practical limits of 40–50 μm by fitting one or more amplifiers into the output end of the transducer (Figure 7.4a). To attain resonance, the acoustic horns, like transducers, should be half-wavelength resonators whose terminals oscillate axially in an opposite direction relative to each other. The nodal points (points of zero amplitude $\xi_n = 0$) are a little displaced toward the upper end in the case of tapered concentrators. Figure 7.4a illustrates the amplitude distribution of the cascaded oscillating system along its longitudinal axis. For detailed information about the calculations and design of acoustic horns, the reader may consult references [2, 4].

Figure 7.4c illustrates an automatic tool feeding mechanism, which operates precisely through the application of roller frictionless guides. A centrifugal pump is used to supplement the abrasive slurry into the working zone (Figure 7.4a).

7.2.3.3 Process Capabilities

Material Removal Rate

In USM, the stock removal rate (SRR) depends mainly on the work material, amplitude and frequency of tool oscillation, abrasive size and type, static pressure, and abrasive concentration (mixing ratio in the slurry). The efficiency of the slurry supplement in the working gap affects the material removal rate (MRR) considerably. The conventional method of supplying the abrasive slurry is the nozzle supply system, in which the slurry is directly supplied at the oscillating tool. Pumping in or suction from the working gap through a central hole in the horn is found to be more effective regime.

The highest machining rates are realized when machining brittle materials such as glass, quartz, ceramics, and germanium, whereas the lowest machining rates are expected when machining hard and tempered steels and carbides. USM is not applicable for soft and ductile materials, such as copper, lead, ductile steels, and plastics, which absorb energy by deformation.

In practice, the oscillation amplitude is mainly selected with reference to the size of abrasive grits used. It should be selected to be approximately the same as the grit size. MRR increases with increasing oscillation amplitude (or abrasive grit size). The maximum amplitude value is governed by the maximum allowable strength of the material of the acoustic horn.

Accuracy and Surface Quality

Factors affecting the accuracy and surface quality of holes and cavities produced ultrasonically are work material, tool material, tool geometry, oscillation amplitude, grain size of abrasives, hole depth and machining time, and the expected cavitations.

A main feature of the USM operation is that the abrasives start to cut for themselves a sideway between the tool and the WP (side gap) to move through it downward to the frontal gap, where the material removal takes place. Accordingly, the ultrasonically produced holes are somewhat larger than the tool used by a certain oversize (overcut), which approximately equals the size of the abrasive grains used. It should be emphasized that the hole accuracy does not mean the hole oversize. It means the repeatability of the oversize. Tolerances of ±25 μm can be easily obtained by USM. However, it is possible to obtain tolerances as close as ±2.5 μm if some measures are considered.

The wall roughness of the ultrasonically machined holes is mainly governed by the material to be machined, the oscillation amplitude, and the abrasives grain size. The surface quality deteriorates if cavitations conditions prevail [5].

7.3 Electrochemical and Chemical Machining Processes

7.3.1 Electrochemical Machining

ECM is well-suited for hard or tough conductive materials requiring to be machined in complex shapes, including those with holes. This process provides excellent and burr-free surface. In addition, fully heat-treated materials can be machined without distortion or heat affected zone (HAZ). Compared with electric discharge machining, ECM proceeds at a much faster rate without a recast layer or HAZ.

Electrolytes used in ECM could be corrosive. Fortunately, stainless steels when electrochemically machined will resist the corrosive effect of these electrolytes. Another characteristic of this process is that, once the tool cathode has been manufactured, it does not wear at all, which contributes to enhancing the accuracy of the process.

The theoretical removal rate as calculated from the basic Faraday's equation is given by ***Faraday's Equation***:

$$m = \frac{1}{F} . \frac{N}{n} . I . t$$

Where

m	=	mass of anodic dissolution of work material (g)
N	=	atomic weight of work material (g/mol)
n	=	valence of work material (–)
N/n	=	EC-equivalent of work material (g/mol)
I	=	machining current (A)
t	=	time of current passage (s)
$\frac{1}{F}$	=	constant (F = Faraday's constant = 96 487 A·s/mol)

The EC-equivalent of SS can be calculated according to the equation:

$$\left(\frac{N}{n}\right)_{SS} = \frac{1}{100}\left[X_1 \frac{N_1}{n_1} + X_2 \frac{N_2}{n_2} + ...\right]$$

where, X_1, X_2, … the percentages of alloying elements in stainless steel, or super alloy, N_1, N_2, … are the atomic weights, and n_1, n_2, … are the valances, respectively [4].

Figure 7.5 illustrates schematically a typical EC sinking machine. EC machines equipped with power generators of DC current capacities ranging from 50 to 40 000 A, which are available on the market. The power sources supply constant voltages ranging from 5 to 30 V. They are generally characterized by a high power factor, high efficiency, and should be equipped

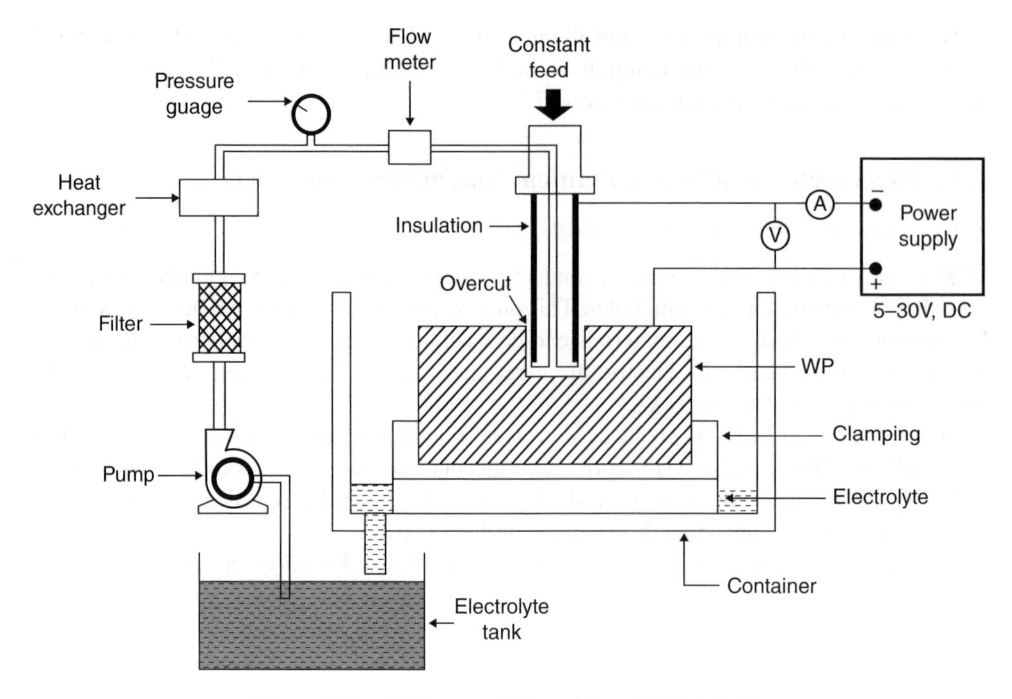

Figure 7.5 ECM setup. (Adapted from El-Hofy [6].)

with short-circuit protection within a small fraction of a second to prevent catastrophic short circuits across the machining cell.

The machine must be rigid enough to withstand the hydrodynamic pressure of the electrolyte in the machining gap, which tends to separate the tool from WP. A servomechanism is necessary to control the tool movement in such a way that the material dissolution is balanced by the constant feed rate of the tool. In contrast to conventional machine tools, EC machines are designed to resist corrosion attack by using nonmetallic materials. To eliminate the danger of corrosion on other machinery, EC machines should be perfectly isolated in separate rooms in the workshop. The electrolyte feeding unit supplies electrolyte at a given rate, pressure, and temperature. Facilities for electrolyte filtration, temperature control, and sludge removal are also included.

7.3.1.1 Process Capabilities

In ECM, the machining rate is solely a function of the ion exchange rate, irrespective of the hardness of the work material. The process provides metal removal rates for steels in the order of 1.5–2 cm^3/min/1000 A. Penetration rates up to 2.5 mm/min are routinely obtained when machining carbides, and steels, either hardened or not. Table 7.3 shows the EC removal rate of most common metals, assuming an electrolyzing current of 1000 A and a current efficiency of 100%.

A well-known and unique characteristic of ECM, among all traditional and nontraditional processes, is that both the accuracy and surface quality improve when applying higher removal rates (i.e., higher current densities). A major problem of ECM is the overcut (side gap), which affects accuracy. Roughly speaking, the side gap is governed by a complex set of parameters, of which the type of electrolyte and the electrolyte flow are most crucial. A typical dimensional tolerance of ECM is ±0.13 mm; however, through proper control of the

Table 7.3 Electrochemical removal and specific removal rates of commonly used metals

Metal	ρ (g/cm³)	N atomic weight (g/mol)	n valence	ε = N/n (g)	Removal rate i = 1000 A, η = 100%		Specific RR (cm³/A·min)
					g/min	cm³/min	
Aluminum (Al)	2.7	27	**3**	9.0	5.6	2.1	0.0021
Chromium (Cr)	7.2	52	**2**	26.0	16.2	2.3	0.0023
			3	17.3	10.8	1.5	0.0015
			6	8.7	5.4	0.8	0.0008
Cobalt	8.9	59	**2**	29.5	18.3	2.1	0.0021
			3	19.5	12.3	1.4	0.0014
Copper (Cu)	9.0	64	**1**	64.0	39.5	4.4	0.0044
			2	32.0	197	2.2	0.0022
Iron (Fe)	7.9	56	**2**	28	17.4	2.2	0.0022
			3	18.7	11.6	1.5	0.0015
Molybdenum (Mo)	10.2	96	**3**	32.0	20.0	2.0	0.0020
			4	24.0	14.9	1.5	0.0015
			6	16.0	10.0	1.0	0.0010
Nickel (Ni)	8.9	59	**2**	29.5	16.2	2.1	0.0021
			3	19.7	12.2	1.4	0.0014
Titanium (Ti)	4.5	48	**3**	16.0	10.0	2.2	0.0022
			4	12.0	7.5	1.6	0.0016
Tungsten(W)	19.3	184	**6**	30.7	19.0	1.0	0.0010
			8	23.0	14.3	0.7	0.0007
Zinc (Zn)	7.2	65	**2**	32.5	20.0	20.9	0.0029

Adapted from Youssef [4].

machining parameters, tight tolerance of ±0.025 mm can be achieved. It is difficult to machine internal radii smaller than 0.8 mm. A typical overcut of 0.5 mm and a taper of 1 μm/mm are possible. Typical surface roughness (R_a value) of ECM ranges from 0.2 to 1 μm is common; which deceases with increasing machining rate. There is no tool wear and the process produces stress free surfaces. The capability to cut the entire cavity in one stroke makes the process very productive, but the complicated tool shape increases the process cost.

Advantages of the ECM

These include:

- Three-dimensional surfaces with complicated profiles can be easily machined in a single operation, irrespective of the hardness and strength of the WP material.
- ECM offers a higher rate of metal removal as compared to traditional and nontraditional methods, when high machining currents are employed.
- There is no wear of the tool, which permits repeatable production.
- No thermal damage or HAZ.
- High surface quality and accuracy can be achieved at the highest MRR.
- The surfaces produced by ECM are burr-free and free from stresses.

Disadvantages of the ECM

These include:

- Nonconductive materials cannot be machined.
- The machine and its accessories are subjected to corrosion and rust, especially when NaCl electrolyte is used; expensive electrolytes like $NaNO_3$ are less corrosive.
- The endurance limit of parts produced by ECM is lowered by about 10–25%. In such a case, shot peening is recommended to restore the fatigue strength.
- Higher specific power consumption as compared to TMPs.
- Cavitations if present deteriorate the surface quality.

7.3.1.2 Pulsed Electrochemical Machining (PECM)

For the achievement of higher precision and better surface qualities during ECM, pulsed process modifications have been developed since about 2000. Pulsed (also termed "precise") electrochemical machining (PECM) is a vibration assisted development of ECM die-sinking by applying a low frequency oscillation of the tool electrode within the working gap [7, 8]. Using the combination of an additionally pulsed, high current density direct current, and an oscillating electrode enables precise machining at reduced working gaps of about 10–100 μm compared to typical values of 100–1000 μm during dc-ECM with surface roughness down to Ra = 20–30 nm [9]. Depending on the workpiece material, even polished surfaces can be achieved via PECM. The principle of PECM is shown in Figure 7.6.

Due to the off-times in PECM, a considerable amount of the process time is used for the replacement of the electrolyte in the pulse pauses. In these pauses, no metal removal takes place. Hence, the maximum removal rates of PECM compared to EC methods, in which continuous dc current is used, are significantly reduced (e.g., 20 times longer machining times [7]). However, due to refreshment of the fluid the electrochemical conditions within the working gap are kept much more constant. Therefore, for long fluid flow paths such as during

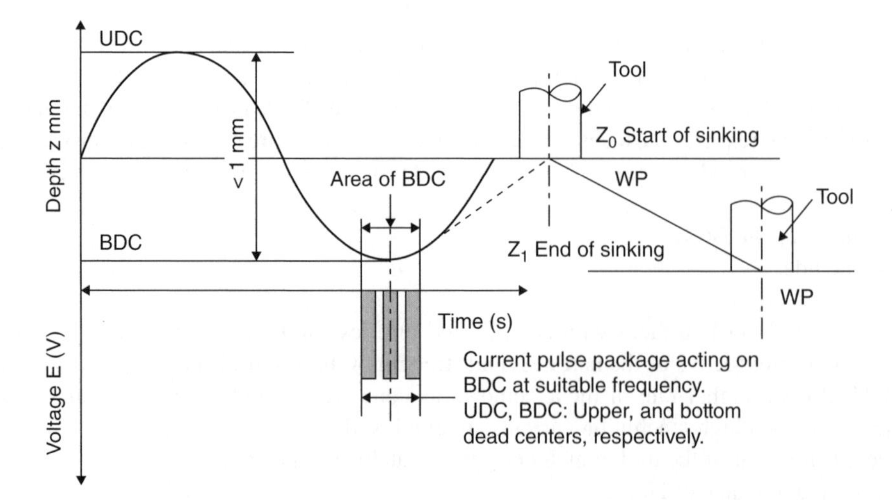

Figure 7.6 Principle of pulsed (precise) electrochemical machining (PECM) with oscillating cathode tool electrode. (From Fritz Klocke *et al.* [7].)

the machining of blades or other macro geometrical features of turbo-machinery components, gap widening effects over the flow channels length are significantly reduced. Figure 7.7 provides a detailed comparison between ECM and PECM.

Thus, a more uniform gap size distribution is achieved allowing more precise machining and also simplification of tool electrode development iterations. In addition, during off-times

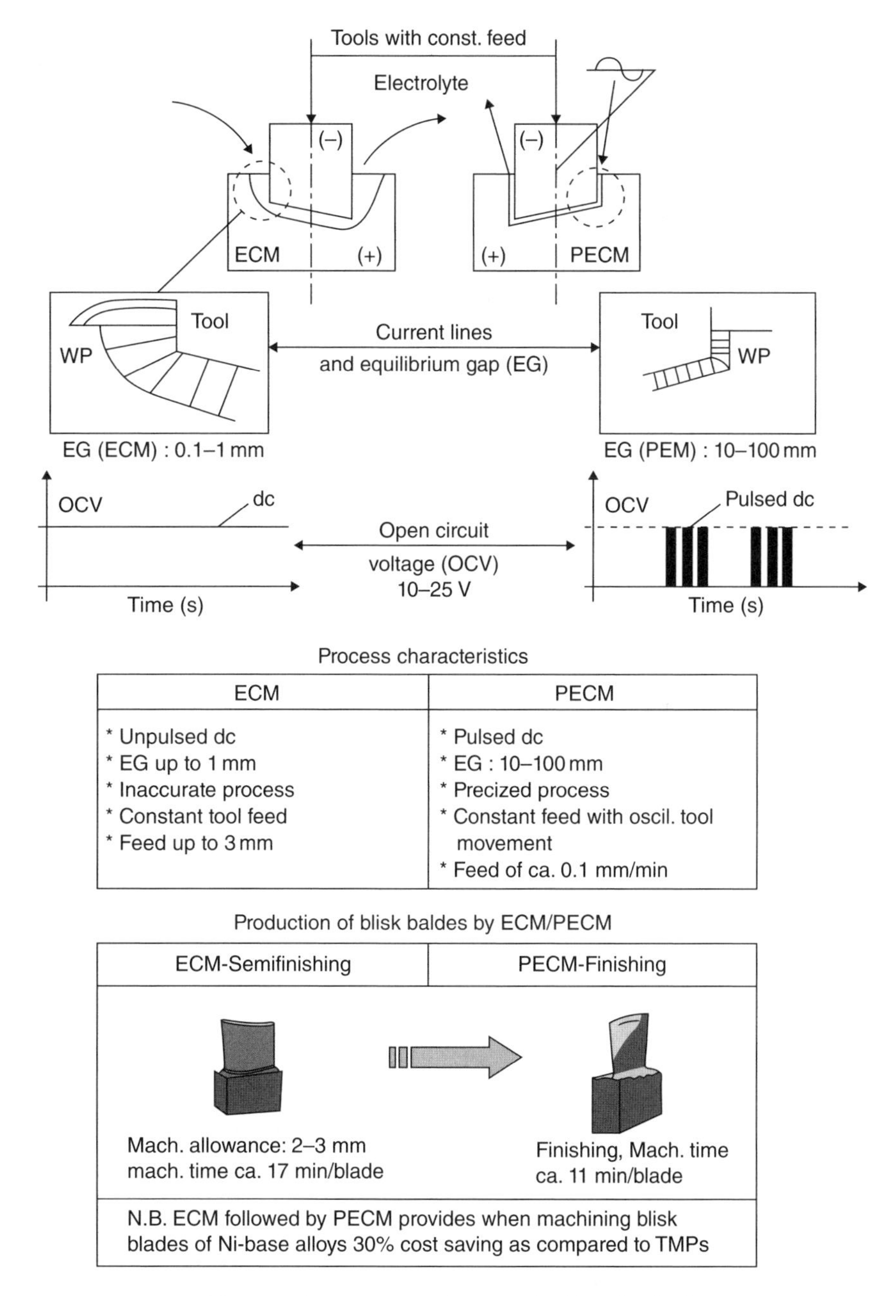

Process characteristics

ECM	PECM
* Unpulsed dc * EG up to 1 mm * Inaccurate process * Constant tool feed * Feed up to 3 mm	* Pulsed dc * EG : 10–100 mm * Precized process * Constant feed with oscil. tool movement * Feed of ca. 0.1 mm/min

Production of blisk baldes by ECM/PECM

ECM-Semifinishing	PECM-Finishing
Mach. allowance: 2–3 mm mach. time ca. 17 min/blade	Finishing, Mach. time ca. 11 min/blade
N.B. ECM followed by PECM provides when machining blisk blades of Ni-base alloys 30% cost saving as compared to TMPs	

Figure 7.7 Comparison between ECM and PECM.

the electrolyte saturation level with ions is reduced, allowing the application of higher current densities during on-times compared to dc applications. Another advantage is the fact that due to the high current densities in the small frontal gaps, reduced stray currents occur, and lower etching attack next to the machining areas takes place [7]. Currently, ECM followed by PECM is used to provide the best solution for machining turbine blisk blades from Ni-base alloys. This realizes 30% cost saving as compared to TMPs [9].

Investigations have shown that PECM improves fatigue life, and the process has been proposed as a possible method for eliminating the recast layer left on die and mold surfaces by electric discharge machining. Moreover, the purpose of pulsing action is to eliminate the need for high electrolyte flow rates, which limits the usefulness of ECM in die sinking [10].

7.3.1.3 Shaped Tube Electrolytic Machining (STEM)

It is a specialized ECM technique for drilling small deep holes utilizing acid electrolytes (Figure 7.8). Acid is used, so that the dissolved metal will go into solution rather than form a sludge as is the case with the salt-type electrolyte of ECM. The electrode is acid-resistant Ti- tube which is coated with a film of enamel-type insulation. The tubes are usually made from high quality Ti with acid resisting organic coating to provide insulation. The MRRs are also strictly governed by Faraday's law and are limited by the current carrying capacity of the tube electrode and the boiling point of the acid electrolyte.

This process is capable of producing small holes with diameters of 0.5–6.5 mm and a depth-to-diameter ratio of 180:1 in electrically conductive materials. It is difficult to machine such small holes using normal ECM as the insoluble precipitates produced obstruct the flow path of the electrolyte. The machining configuration is similar to that used in ECM. However, it has a periodically reverse polarity power supply. The cathodic tool electrode is made of titanium, its outer wall having an insulating coating to permit only frontal machining of the anodic workpiece.

The process also uses a 10% concentration sulfuric acid to prevent the sludge from clogging the tiny cathode and ensure an even flow of electrolyte through the tube. A periodic reversal

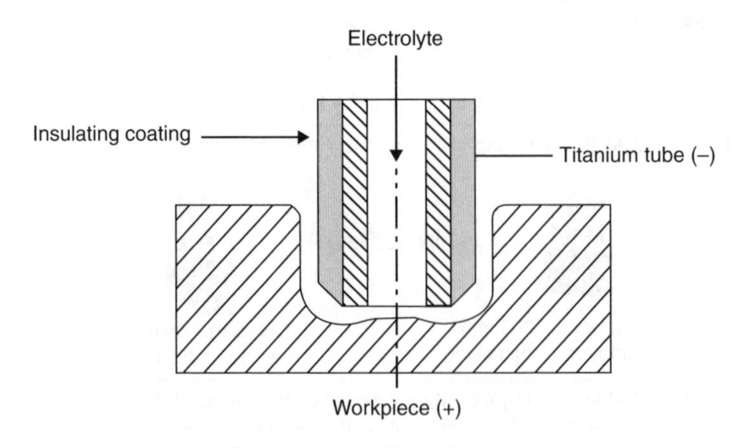

Figure 7.8 STEM schematic.

of polarity, typically at 3–9 s prevents the accumulation of the un-dissolved machining products on the cathode drill surface. The reverse voltage can be taken as 0.1–1 times the forward machining voltage.

The main operating parameters of shaped tube electrolytic machining (STEM) are given below:

- *Power supply*: 5–15 Vdc, forward (on time) 3–20 s, reverse time 0.2–0.4 s, curre t 1–40 A/hole.
- *Electrolyte*: 6–15 v/v H_2SO_4, Pressure: 70–300 kPa, Temp.: 32–43 °C, Flow rate: 20–150 cm^3/min, Contamination: 6 g/l max.
- *Feed rate*: 1–2 mm/min
- *Hole diameter*: 0.5–6.5 mm
- *Hole depth*: 600 mm
- *Depth/diameter*: 180:1
- *Hole/stroke*: 1–100
- *Hole tolerance*: ±5% of hole diam.
- *Roughness Ra*: 1–3 μm

Electrolyte concentration less than 10% reduces attack on tube insulation. The hole size and roughness increase with increasing electrolyte concentration levels. Reverse polarity is frequently at lower voltages (typically 4–5 V) than the forward voltages.

Critical Factors in STEM: There are three important factors that control the performance of STEM [11]

1. *Sludge formation*: Salt electrolytes are not used in STEM because they produce a large amount of sludge, that clogged the flow of electrolyte, and consequently limits minimum hole diameter, that can be drilled. Therefore, diluted acid electrolytes are used for STEM. However, acid electrolytes corrode the workpiece and produce poor surface finish, especially with hydrochloric acid.
2. *Overcut*: Minimizing overcut is an important consideration in hole making. Investigations have shown that machining with that show an abrupt passive to transpassive give better dimensional accuracy and finish than nonpassivating electrolytes.
3. *Surface finish*: The roughness of cooling hole may have an impact on the coefficient of heat transfer in turbine blade. The coefficient of friction and thereby the coefficient of heat transfer increase with the surface roughness. Smooth inner surfaces, however, are required to prevent clogging of holes in the event of precipitate formation. It appears that the surface finish with its contradicting effects has not been sufficiently investigated.

Advantages of STEM

- The depth-to-diameter ratio can be as high as 300.
- A large number of holes (up to 200) can be drilled in the same run.
- Nonparallel holes can be machined.
- Blind holes can be drilled.
- No recast layer or metallurgical defects are produced.
- Shaped and curved holes as well as slots can be produced.

Limitations of STEM

- The process is used for corrosion-resistant metals.
- STEM is slow if single holes are to be drilled.
- Special workplace and environment are required when handling acid.
- Hazardous waste is generated.
- Complex machining and tooling systems are required.

7.3.1.4 Electro-stream (ES) or Capillary Drilling

This is another version of ECM adapted for drilling very small holes using high voltages. The voltages are more than 10 times those employed in ECM or STEM, so special provisions for protection are required. Electrolyte temperature, pressure, concentration, and flow rate are needed for acid electrolyte, which is chosen to be chemically compatible with the workpiece metallurgical state. Holes ranging from 0.1 to 1 mm diameter, with depth to diameter ratios up to 50:1 can be made in any conductive material. The tool is a drawn-glass nozzle, 25–50 μm, smaller than the desired hole size. To conduct the machining current through the acid electrolyte, a platinum electrode is fitted inside the glass nozzle (Figure 7.9). Multiple hole drilling predominates [1]. The process can also be used in drilling side holes in inaccessible positions (Figure 7.10). Wire electrical discharge machining (EDM) starting holes of less than 0.5 mm can be drilled using ES. Up to 100 holes per stroke have been accomplished on multiple parts. Automatic indexing is important.

The operating parameters of ES are as follows:

- *Power supply*: 150–850 Vdc, 20–200 mA
- *Electrolyte*: 5–15 v/v H_2SO_4, HCl, or mixture, pressure: 140–700 kPa, temperature: 15–50 °C, Contamination: 6 g/l max.
- *Feed rate*: 1–5 mm/min
- *Hole diameter/depth*: 0.2–1 mm/20 mm
- *Depth/diameter*: 50:1

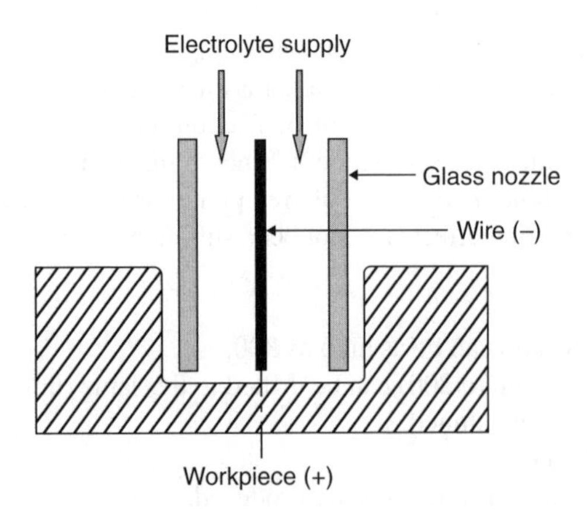

Figure 7.9 ES drilling schematic.

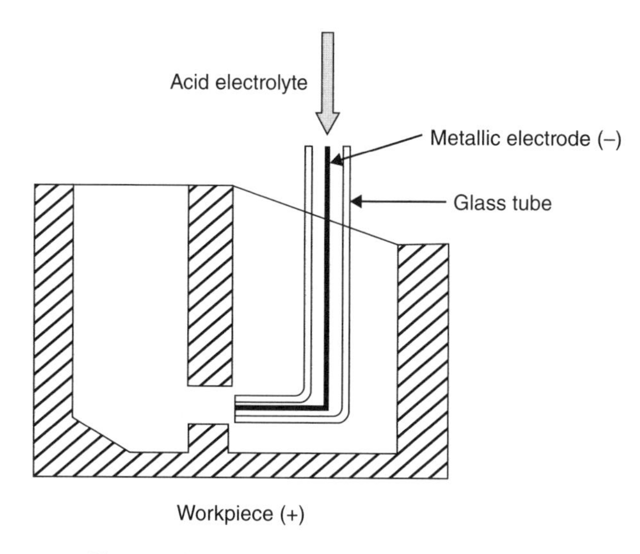

Figure 7.10 ES drilling in inaccessible positions.

- *Hole/stroke*: up to 100
- *Hole tolerance*: ±5% of hole diam.
- *Roughness Ra*: 0.5–1.5 am.

Advantages of ES
- High depth-to-diameter ratios are possible.
- Many holes can be drilled simultaneously.
- Blind and intersecting holes can be machined.
- There is an absence of recast and metallurgical defects.
- Powder metallurgy hard-to-cut materials can be tackled.
- Burr-free holes are produced.

Limitations of ES
- Can only be used with corrosion-resistant metals.
- Hazardous waste is generated.
- The process is slow when drilling a single hole.
- The handling of acid requires a special environment and precautions.

7.3.2 Electrochemical Grinding

ECG is another variant of ECM. It is a hybrid process in which the metal is removed by a combination of EC dissolution and mechanical abrasion. MRRs are higher than traditional grinding, particularly for hard alloys, which can be ground without distortion or metallurgical changes. Burrs are eliminated and smooth surfaces are obtainable.

The equipment used in ECG (Figure 7.11) is similar to a traditional grinding machine, except that the GW is metal bonded with diamond or borazon (CBN) abrasives. The wheel is the negative

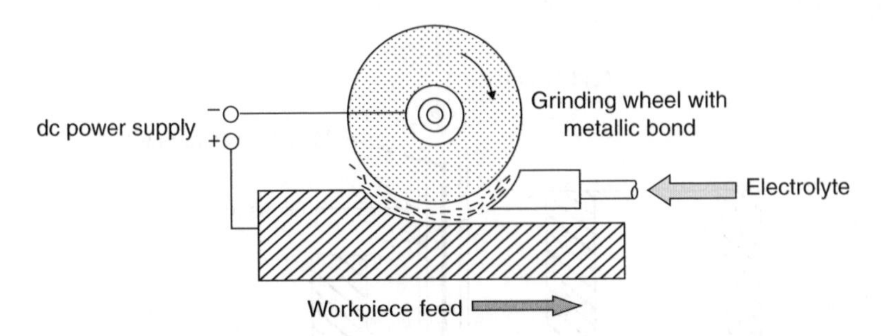

Figure 7.11 Electrochemical grinding.

electrode, whereas the work is connected to the positive terminal of the DC power supply. A flow of electrolyte, usually the noncorrosive $NaNO_3$ is provided in the direction of wheel rotation to achieve the ECM phase of the operation. The abrasives in the wheel are nonconductive, and thus they act as an insulating spacer, maintaining a separation (electrolytic gap) of 12–80 μm between electrodes. Therefore the removal of WP material occurs through electrochemically decomposition owing to the flow of current (100–300 A/cm^2) enhanced by the grinding action of the abrasive grains. With proper operation, typically, 95% of material removal is due to electrolytic dissolution, and only about 5% is due to the abrasion effect of the grinding wheel. Consequently, the wear of the wheel is very low, thus eliminating or considerably reducing the need for grinding wheel redressing, which reduces the sharpening costs approximately by 60%.

The lack of heat damage, distortions, burrs, and residual stresses in ECG is very advantageous, particularly when coupled with high MRR, in addition to far less wear of the grinding wheel. That is why the process has been applied most successfully in sharpening cutting tools, die inserts, and punches made of hardened high-strength steel alloys. However, ECG has the following disadvantages:

- higher capital cost of the equipment;
- limited to electrically conductive materials can be machined
- hazard due to corrosive nature of electrolyte, for that reason $NaNO_3$ of limited corrosive nature is used;
- necessity of electrolyte filtering and disposal.

7.3.3 Chemical Machining

Chemical machining (CHM) depends on controlled chemical dissolution (CD) of the work material by contact with an etchant. Today, the process is mainly used to produce shallow cavities of intricate shapes in materials independent of their hardness or strength. CHM includes two main applications, namely chemical milling (CH-Milling) and photochemical machining (PCM).

7.3.3.1 Chemical Milling (CH-milling)

CH-milling is an important variant of CHM. A strippable maskant covers areas which are not to be removed. This process has a special importance in aerospace industries, where it is used to reduce the thickness of plates enveloping walls of rockets and airplanes, striving at improving stiffness to weight ratio. It is best suited for machining large shallow areas such as aircraft wings and fuselage panel for weight reduction. Cut surfaces are burr-free, and free from residual stresses and heat-affected-zones. Disadvantages include low cutting rates, and the fact that masked areas will be undercut (etch factor, EF) by the corroding solutions (Figure 7.12).

Tooling for CH milling is relatively inexpensive and simple. Four types of tools are required: maskants, etchants, scribing templates, and accessories. Etchants are highly concentrated acidic or alkaline solutions maintained within a controlled range of chemical composition and temperature. They are capable of reacting with the WP material to produce a metallic salt that dissolves in the solution. Table 7.4 shows the machined material, the recommended etchant, its concentration and temperature, EF, and the etch rate.

Using fresh solutions, the etch rate ranges from 20 to 40 μm/min. Generally, the high etch rate is accompanied by a low surface roughness and hence closer machining tolerances.

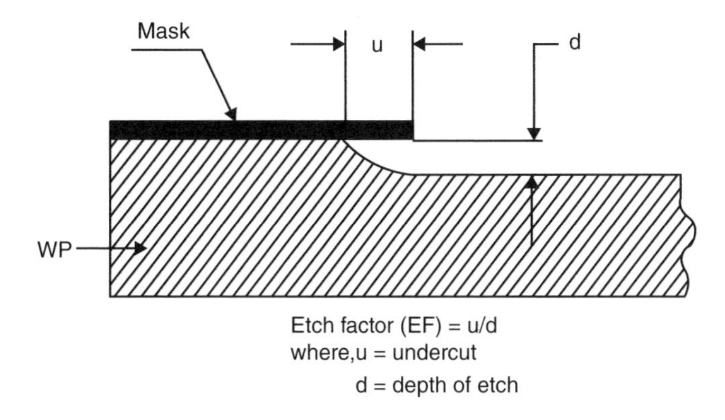

Figure 7.12 Etch factor in CH-milling.

Table 7.4 Machined materials, recommended etchant, etch factor (EF), and etch rate for CH-milling

Etchant	Concentration	Temperature (°C)	Etch rate (μm/min)	Etch factor (EF = u/d)	Metal to be machined
$FeCl_3$	12–18° Be[a]	50	20	1.5 : 1	Al alloys
$HCl:HNO_3:H_2O$	10 : 1 : 9	50	20–40	2 : 1	
$FeCl_3$	42° Be[a]	50	20	2 : 1	Cold rolled steel
HNO_3	10–15% (vol.)	50	40	1.5 : 1	
$FeCl_3$	42° Be[a]	50	40	2.5 : 1	Cu and Cu alloys
$CuCl_2$	35° Be[a]	55	10	2 : 1	
HNO_3	12–15% (vol.)	50–70	20–40	—	Magnesium
$FeCl_3$	42° Be[a]	50	10–20	(1–3) : 1	Nickel
$FeCl_3$	42° Be[a]	55	20	2 : 1	Stainless steel, tin
HNO_3	10–15% (vol.)	50–70	20	—	Zinc

[a] Baume specific gravity scale.
Adapted from Youssef [4].

Typically, surface roughness of 0.1–0.3 μm (R_a value), depending on the initial roughness, can be obtained. However, under special conditions, surface roughness of 0.023–0.05 μm are possible.

Advantages of CH-milling

- Weight reduction is possible on complex contours.
- Several parts can be machined simultaneously.
- No burr formation.
- No induced stresses, thus minimizing distortion of delicate parts.
- Low capital cost of equipment and minor tooling cost.

Disadvantages of CH-milling

- Only shallow cuts are practical. Deep narrow cuts are difficult to produce.
- Handling and disposal of etchants can be troublesome and hazardous.
- Masking, scribing, and stripping are repetitive, time-consuming, and tedious.

7.3.3.2 Photochemical Machining (Spray Etching)

PCM or spray etching is another important variant of CHM, where the resistant mask is applied to the workpiece by photographic techniques. The two processes are quite similar, because both use etchant to remove material by chemical dissolution. CH-milling is usually used on the 3-D parts originally formed by other manufacturing processes such as forging and casting of irregular shapes. However, PCM is more applicable for machining of foils and sheets of thicknesses ranging from 0.013 to 1.5 mm to produce accurate and micro-shapes. So, PCM becomes a realistic alternative to shearing and punching operations performed by mechanical presses.

Visser *et al.* [12] claimed that the etch rate of PCM is 10 times and more that achieved by CH-milling. Also, the accuracy is considerably greater than that realized by CH-milling. Of course, in PCM, highly developed expensive equipment are needed to provide high pressure/high temperature (HP/HT). The PCM equipment is generally provided with a system of upper and lower nozzles, a unit for cleaning the worksheet by water then drying by hot air, a unit for measuring and controlling the etchant, and finally a unit for product inspection (Figure 7.13). Aluminum, copper, zinc, steels, stainless steels, lead, nickel, titanium, molybdenum, glass, germanium, carbides, ceramics, and some plastics are photo-chemically machined. The process also works well on springy materials, which are difficult to punch. The materials must be flat so that they can later be bent to shape and assembled into other components. Products made by PCM are generally found in the electronic, automotive, aerospace, telecommunication, computer, and other industries.

Typical products such as PCB (Figure 7.14), fine screens, flat springs, and so on, machined from foils are produced by PCM.

In addition to the previously mentioned advantages of CH milling, PCM is characterized by the following:

- The accuracy and etch rate are considerably greater than those realized by EC-milling.
- Because tooling is made by photographic techniques, they can be easily stored, and patterns can be reproduced easily. Lead times are correspondingly smaller.

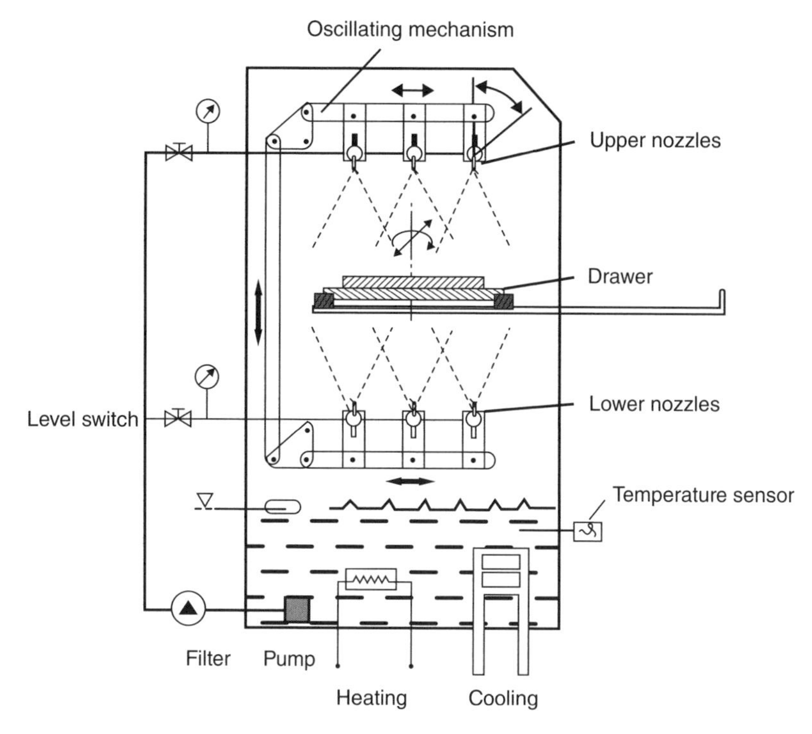

Figure 7.13 Schematic of PCM equipment. (From Visser *et al.* [12], with permission.)

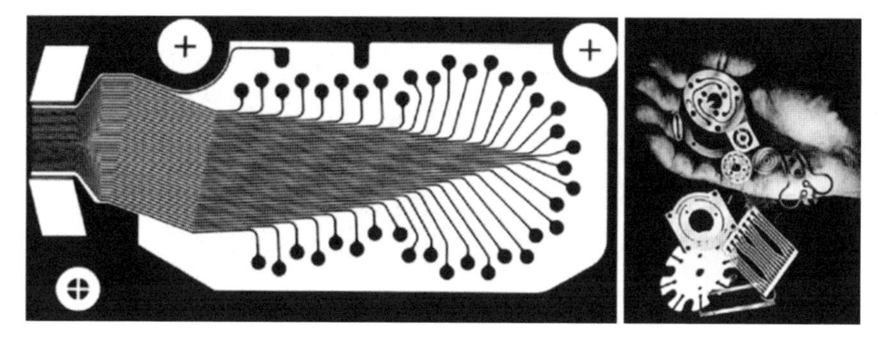

Figure 7.14 Typical products by PCM. (From Visser *et al.* [12], with permission and courtesy of Chemical Corporation.)

Apart from the disadvantages of CH milling, PCM also has the following limitations:

- Requires highly skilled operator.
- Requires more expensive equipment.
- The costly machine should be protected from the corrosive action of etchants.

7.4 Thermoelectric Processes

7.4.1 Electrical Discharge Machining

Although the erosive effect of electrical discharge has been detected by Priestly since 1790, it was not until the 1940s that a machining process, EDM, based on this principle was developed by Lazarenkos in Russia. EDM is a thermo-electrical NTMP, in which material is removed by sparks generated by a pulsating current flow between the workpiece and a shaped electrode. The workpiece and tool electrode are separated by a dielectric fluid, such as oil or kerosene. The thermal energy due to sparking is mainly consumed in generating high temperature plasma, eroding the workpiece (and the electrode) material. When a potential difference between the tool and the WP is sufficiently high, the dielectric in the gap is partially ionized, so that a transient spark discharge ignites through the fluid, at the closest points between the electrodes. Each spark of thermal power concentration, typically 10^8 W/mm^2, is capable of melting or vaporizing very small amounts of metal from the WP and the tool (Figure 7.15). A part of the total energy is absorbed by the tool electrode, yielding some tool wear, which can be reduced to 1% or less if adequate machining conditions are carefully selected.

The instantaneous vaporization of the dielectric produces a high-pressure bubble that expands radially. The discharge ceases with the interruption of the current, and the metal is ejected, leaving tiny pits or craters in the WP and metal globules suspended in the dielectric (Figure 7.15). A sludge of black carbon particles is formed from hydrocarbons of the dielectric produced in the gap and expelled by the explosive energy of the discharge; it remains in suspension until removed by filtering. Immediately following the discharge, the dielectric surrounding the channel de-ionizes and, once again, becomes effective as an insulator.

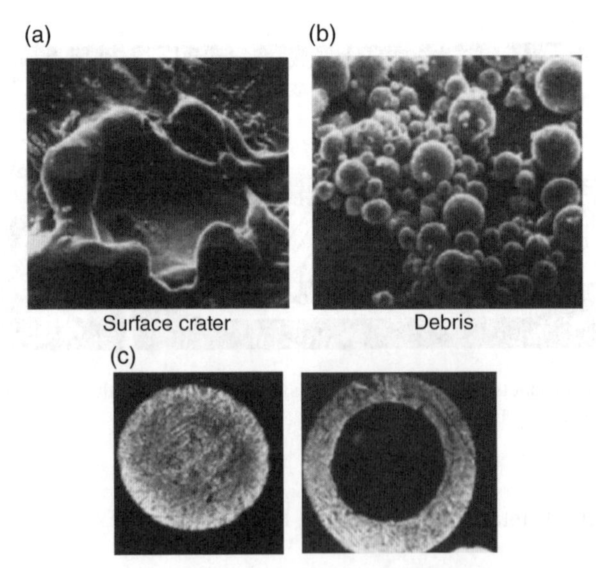

Figure 7.15 SEM surface crater and debris produced by EDM: (a) surface crater; (b) debris; (c) section in solid and hollow debris with trapped gas. (From König [13], with permission.)

The capacitor discharge is repeated at rates between 0.5 and 500 kHz, at a voltage between 50 and 380 V and currents from 0.1 to 500 A. In EDM, the tool and the WP are separated by a small gap of 10–500 μm. The gap between the electrode and the WP is critical; hence the down feed should be controlled by a servomechanism that automatically maintains a constant gap. A short circuit across the gap causes the servo to reverse the motion until proper control is restored. At the beginning, the fresh supply of dielectric is clean and has a higher insulation strength than a contaminated dielectric. When spark discharges commence, debris are created and the dielectric insulation strength is diminished. If too many particles are allowed to remain due to bad flushing, a bridge is formed, resulting in arcing (not sparking) across the gap, which causes damage to the tool and WP. Therefore the contamination in the gap must be controlled by efficient flushing, and filtering.

7.4.1.1 Types of Generators, Applicable for ED-Machines

1. *RC circuit (relaxation)*: It is also called the Lazarenko circuit, which is basically a relaxation oscillator. It is simple, reliable, rigid, and low-cost power source that is ordinarily used with copper or brass electrodes. It provides a fine surface texture of $R_a = 0.25$ μm. However, the machining rate is slow, because the time required to charge the capacitors prevents the use of high frequencies. The reversed polarity encountered in the relaxation circuit leads to an additional tool wear
2. *Transistorized pulse generator circuits*: The adoption of the transistorized pulse generators in the 1960s allowed the process parameters (frequency and energy of discharges) to be varied with a greater degree of control, in which charging takes only a small portion of the cycle. Furthermore, the voltage of these machines is reduced to 60–80 V range, permitting discharge that is characterized by lower current pulses of a square profile. This results in shallower and wider craters, which means better surface texture. Alternatively, when required, they provide high MRRs at the expense of surface quality by permitting high discharge currents. Moreover, this type of generators provides considerably lower electrode wear as compared to simpler *RC* circuits.

7.4.1.2 Process Capabilities

EDM is a slow process compared to conventional methods. It produces matte and pitted surfaces composed of small craters, which are characterized by a nondirectional, randomly distributed nature due to succession of individual sparks of the process. The metal removal rates usually range from 0.1 to 600 mm^3/min. High removal rates produce a rough finish, having a molten and recast structure with poor surface integrity (SI) and low fatigue strength. The finish cuts are made at low removal rates, and the recast layer formed during rough cuts is removed later by finishing EDM operations. The MRR depends not only on the WP material but also on the machining variables, such as pulse conditions (voltage, current, and duration), electrode material and polarity, and the dielectric. EDM is not affected by the hardness or toughness of the material and is well suited for machining a variety of complex or irregular shapes, including holes.

In EDM, the surface finish and MRR vary widely as a function of the spark frequency, voltage, current, and other parameters. New techniques use oscillating electrodes to provide very fine surface quality. Alternatively, bad surfaces and surface defects characterize EDM using graphite electrodes.

Typically, oversize values vary from 10 to 300 μm, depending on the breakdown voltage, and the size of debris flowing in the side gap. In this respect, suction flushing is preferred, because the debris are not drawn past the side gap, and thus lateral sparking is minimized, leading to a smaller overcut and side taper. Typical taper varies from 1 to 5 μm/mm per side, depending on the machining conditions, and especially on the flushing technique used.

Features of EDM

- Because there is no contact between tool and WP, very delicate work can be machined.
- The process is widely used to produce accurate cavities of intricate shapes in extra-hard materials.
- EDM electrodes can be accurately produced from machinable materials.

Limitations of EDM

- It cannot be used if the WP material is a bad electric conductor.
- On machining materials, the process produces HAZ, which is characterized by hairline cracks and thin, hard recast layer.
- EDM cannot produce sharp corners and edges.

7.4.2 Electron Beam Machining

The pioneer work of electron processing is related to Steigerwald [14], who designed a prototype of electron beam equipment that has been built by Messer-Griessheim in Germany for welding applications. This new technology has quickly spread in industry to embrace other fields of applications such as machining and surface hardening. Electron Beam Machining (EBM) is a thermal NTMP that uses a beam of high-energy electrons focused to a very high power density on the WP surface, causing rapid melting and vaporization of its material. A high voltage, typically 120 kV, is utilized to accelerate the electrons to speeds of 50–80% of light speed in a vacuum (10^{-4}–10^{-5} Torr, 1 Torr = 1 mm Hg). The interaction of the electron beam with the WP produces hazardous X-rays, consequently, shielding is necessary, and the equipment should be used only by highly trained personnel. EBM can be used to machine conductive and nonconductive materials. The material properties, such as density, electric and thermal conductivity, reflectivity, and melting point, are generally not limiting factors. The greatest industrial use of EBM is the precision drilling of small holes ranging from 0.05 to 1 mm diameter to a high degree of automation and productivity. As with EDM, a thin recast layer and HAZ will be present and may have to be removed for critical applications.

The parameters that affect the performance of EBM are

- Density and thermal properties of WP (such as specific heat, thermal conductivity, and melting point)
- Accelerating voltage (50–150 kV)
- Electron beam current (0.1–40 mA)
- Power (1–150 kW)
- Pulse duration (4–60 000 μs)
- Pulse frequency (0.1–16 000 Hz)
- Minimum spot diameter (12–25 μm)

- Beam intensity 10^6–10^9 W/cm^2
- Traverse speed of the WP.

The following equation expresses the traverse speed v_f (m/s), in terms of the machining parameters of the process.

$$v_f = \frac{C_d}{d_f}\left[0.1\frac{P_e}{\theta_m t_1 k_t}\right]^2$$

Where

t_1	=	plate thickness (m)
P_e	=	power of electron beam (N m/s) = beam current × acceleration voltage = $i_b \times V_b$
d_f	=	beam focusing diameter (m)
k_t	=	thermal conductivity (N/s °C)
C_d	=	coefficient of thermal diffusivity = $k_t/\rho \cdot c_1$ (m^2/s)
c_1	=	specific heat of WP material (N m/kg °C)
θ_m	=	melting point of WP material (°C).

Figure 7.16 shows the effect of number of pulses (Vb = 130 kV, ti = 10 μs, fi = 500 Hz, Li = 60 × 10^{-9} A·s) on the hole depth. It is similar to laser percussion drilling (PD) (see Section 7.3.3); however, the maximum hole depth in EBM is attained after about 50 pulses instead of after 9 pulses for laser PD. For this reason it is not recommended to increase the number of pulses above these prescribed limits.

EBM is characterized by its minor volumetric removal rate, which reaches a maximum value of 0.1 cm^3/min. The volumetric removal rate increases by increasing pulse energy (power intensity), provided that the same number of pulses is used. The machinability depends on the melting point of the material to be machined. In this regard, tin and cadmium have the highest machinability, whereas

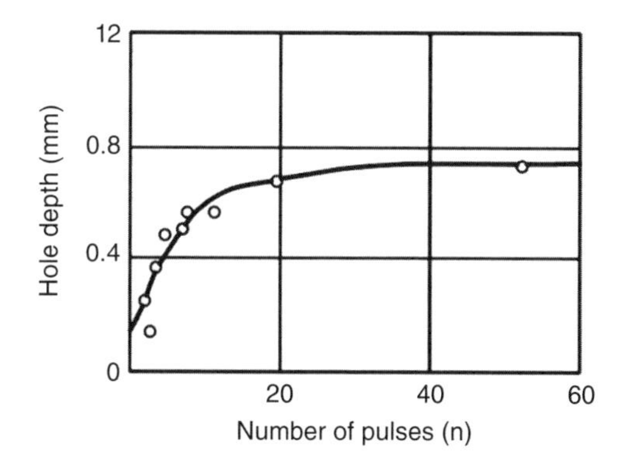

Figure 7.16 Effect of number of pulses on the hole depth in EBM. (Adapted from Visser [15].)

W and Mo are of low machinability. In EB hole drilling, the achieved tolerance depends on the hole diameter and WP thickness, whereas the surface roughness depends mainly on the pulse energy. Depending on the working conditions, the tolerance of the drilled holes (and slits) may attain values between ±5 and ±125 μm, while the surface roughness R_a ranges from 0.2 to 6.3 μm.

Advantages of EBM:

- Drilling of fine holes is possible at high rates (up to 4000 holes/s).
- Machining any material irrespective of its properties.
- Providing a high degree of automation and productivity.

Limitations of EBM:

- High capital cost of equipment.
- Time loss for evacuating the machining chamber.
- Presence of a thin recast layer and HAZ.
- Need for qualified personnel to deal with computerized numerical control (CNC) programming and the X-ray hazard.

Applications of EBM:

EBM is almost exclusively used in drilling and slitting operations. Drilling is preferred when many small holes are to be made or when holes are difficult to drill because of the hole geometry or material hardness. Textile and chemical industries use EB drilling as a perforating process to produce a multitude of holes for filters and screens.

7.4.3 Laser Beam Machining

Laser is an acronym for light amplification by stimulated emission of radiation. It is a highly collimated monochromatic and coherent light beam in the visible or invisible range. Laser beam machining (LBM) is a promising NTMP for machining any material, irrespective of its physical and mechanical properties. It is used to cut and machine both hard and soft materials, such as steels, cast alloys, refractory materials, ceramics, tungsten, titanium, nickel, CBN, diamond, plastics, cloth, alumina, leather, woods, paper, rubber, and even glass when its surface is coated with radiation-absorbing material such as carbon. However, machining of Al, Cu, Ag, and Au is being especially problematic, as these metals are of high thermal conductivity and have the tendency to reflect the applied light. But recently, yttrium aluminum garnet (YAG) with enhanced laser focusing has been used to cut such metals after treating their surfaces by oxidizing them or increasing their surface roughness. YAG is superior more than a CO_2 laser because it emits shorter wave lengths [1].

Laser is a versatile tool, useful in many areas ranging from precision watch-making to heavy metal-working industrial applications. The key of laser's effectiveness lies in its ability to deliver, in some cases, a tremendous quantity of highly concentrated power, as high as 10^{10} W/mm^2. One of the laser beams main advantages is that it does not take up time for the evacuation of the machining area, as does EB. A laser can operate in transparent environments like air, gas, vacuum, and in some cases even in liquids. However, LBM is quite inefficient and cannot be considered as a mass metal removal process. A significant limitation of laser drilling is that the process does not produce round and straight holes. This can, however, be overcome by rotating the WP as the hole is being drilled. A taper of about 1/20 is encountered. HAZ is produced in LBM, and heat-treated surfaces are also affected.

High capital and operating cost, and low machining efficiency, which could be as low as 1%, prevent LBM from being competitive with other NTM techniques. Protective measures are absolutely necessary when working around laser equipment. Extreme caution should be exercised with lasers; even a low power of 1 W can cause damage to the retina of the eye. In all cases, safety goggles should be used, and unauthorized personnel should not be allowed to approach the laser working zone [1].

Industrial lasers comprise in most cases the solid-state lasers such as neodymium yttrium aluminum garnet (Nd:YAG), neodymium glass (Nd:glass), and the ruby and gas lasers (CO_2, excimer, and He/Ne). Basically, four types prevail in metal working processes, namely, the CO_2, Nd:YAG, Nd:glass, and excimer lasers. Out of these, the CO_2 and YAG are considered the most dependable workhorses.

1. *CO_2 lasers*: In these lasers, the active lasing material is the CO_2 gas. However, a mixture of gases is used (CO_2:N_2:He = 0.8:1:7). Helium acts a coolant of the gas cavity. CO_2 lasers are characterized by their long wavelength of 10600 nm; thus, the material removal depends only on the thermal interaction with the WP. However, these lasers are bulky but economical.
2. *Nd:YAG lasers*: This laser is a single crystal of YAG doped by 1% neodymium as an active lasing material. It is compact and economical, its wavelength is 1060 nm, and it can operate in either pulsed (P) or continuous wave (CW) mode. It is characterized by relatively high efficiency and high pulsating frequency. Its pulsating frequency ranges from 1 to 10000 p/s and pulse energy 5–8 J/p. It has an average power output close to 1 kW.
3. *Nd:glass lasers*: This laser uses a glass rod doped by 2–6% neodymium as the acting lasing material. This laser is often uneconomical has the same wavelength as the Nd:YAG, and operates only in the (P) mode. Owing to the low thermal conductivity of glass, the pulse rate should be limited. Consequently, it is only used in drilling and welding where higher-energy output and low pulse frequency (1–2 p/s) are necessary.
4. *Excimer lasers*: Excimer lasers present a family of pulsed lasers (P) operating in the UV region of the spectrum. Excimer is an abbreviation of "excited dimmer." The beam is generated due to fast electrical discharges in a mixture of high-pressure dual gas, composed of one of halogen gas group (F, H, Cl) and another from the rare gas group (Kr, Ar, Xe). The wavelength of the excimer laser attains a value from 157 to 351 nm, depending on the dual gas combination. Excimer lasers have low power output, which removes the material photolithically, and has a remarkable application in the machining of plastics and micromachining and marking applications [4].

In some instances, the same laser can perform cutting, welding, marking, and heat-treating functions by varying the energy density, spot size (focus), and pulse duration. Other lasers with wave lengths outside of the infrared spectrum have also been used in material processing.

Gas-Assisted LBM:

Gas streams are often used to assist laser drilling or cutting. Oxygen or air can be used to enhance the cutting rate (Figure 7.17). The gas stream also serves to eject molten material, particularly from deep cuts. Selection of the gas (N_2, Ar, or He) depends on the type of work material, its thickness, and type of cut. Oxygen is the most commonly used assisting gas for steels and most metals. Inert gas is used also to prevent plastics and other organic materials from charring.

For machining deep holes, the laser should be adjusted to act in successive pulses (multiple shot drilling MSD or Percussion Drilling PD). It should be emphasized that the drilling depth is not linearly correlated with the number of pulses, but increases with a decreasing rate till it attains a maximum value after the ninth pulse as visualized in Figure 7.18. Percussion drilling is intended to produce micro holes ranging from 25 to 1000 μm, while trepanning is used to drill larger holes (>1 mm) (Figure 7.19). The holes can be produced through metal sections up to 25 mm, however the maximum metal thickness for small holes is limited [16]. In case of PD of small deep holes of larger aspect ratios, it is important to move the focus of the beam in the feed direction between consecutive pulses.

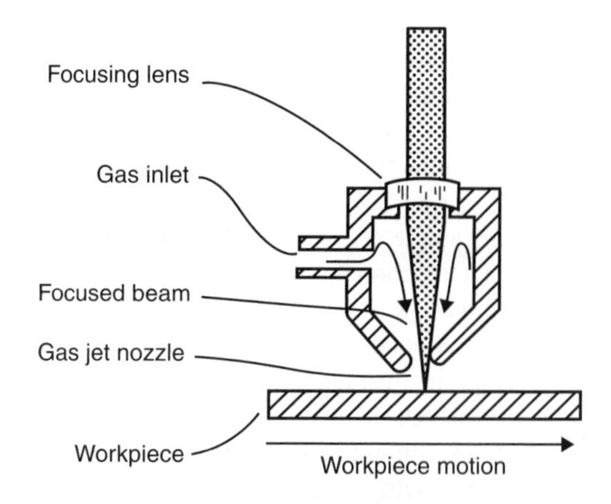

Figure 7.17 Schematic of gas-assisted laser slotting and drilling head with a gas jet.

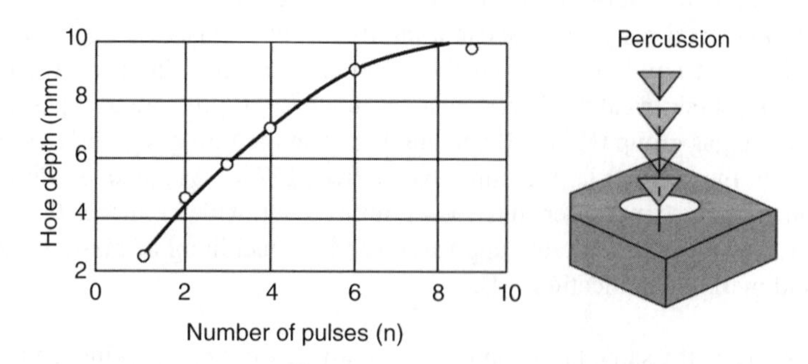

Figure 7.18 Percussion drilling of micro holes and the effect of number of pulses on the produced hole depth.

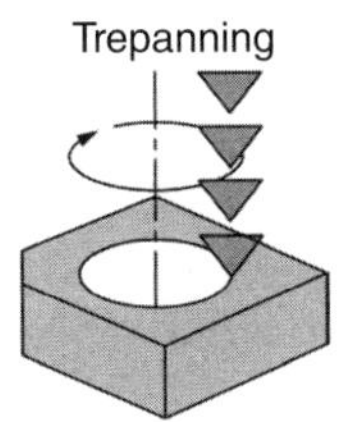

Figure 7.19 LBM of large holes by trepanning.

PD is most often performed with Nd:YAG lasers because of their higher energy pulse. The laser is performed in a pulsed mode with each pulse removing a certain volume of metal. As plate thickness increases, the range of hole diameter decreases [16]. The operating parameters of concern in PD are:

- *Power*: The average value ranges from less than 100–250 W.
- *Pulse duration*: This is selected to optimize the quality of the hole. Shorter pulse durations limit the maximum energy achievable in a single pulse. Typical pulse duration ranges from 0.5 to 2 ms.
- *Pulse frequency*: This ranges from 5 to 20 Hz with Nd:YAG, and up to 100 Hz with CO_2 lasers.
- *Pulse energy*: This is determined based on the material thickness, composition, and hole diameter. Higher pulse energy also provides faster drilling rates, however, can be detrimental to hole quality.
- *Focusing lens*: This determines the spot size, which corresponds to the desired hole diameter. The spot size is equal to the hole size in thinner plates <6 mm thick. Thicker plates (longer holes) dictate the use of lenses of longer focal lengths. Focal lengths usually range from 100 to 250 mm.
- *Focal position*: This is optimized above, below, or on the surface of the plate, depending on the desired results. Most frequently, the focus lies below the surface at a depth varying between 5 and 15% of the plate thickness. Optimum focus is found empirically for the best hole quality as evaluated by the roundness, taper, recast, and microcracks [16].

Trepanning is commonly performed with either CO_2 or Nd:YAG laser. It requires a percussion drilled pilot hole. Metal thickness range is the same as PD. Trepanning can be performed by operating the laser in the CW or pulsed mode. The operating parameters in trepanning are:

- *Power:* This is established at the level needed to maximize the output.
- *Pulse duration*: This is selected to optimize the quality of the hole. Typical pulse duration with both CO_2 and Nd:YAG lasers are generally less than 2 ms.
- *Pulse frequency*: Lower pulse frequencies are used as plate thickness increases.
- *Focusing lens*: Lens focal lengths are similar to those of PD, although CO_2 lasers require a focal length of 125 mm or shorter for trepanning.

To obtain best surface quality in PD, it is recommended to use a beam of low pulse energy and low pulse duration [4]. The surface quality is also affected by the carbon content of the

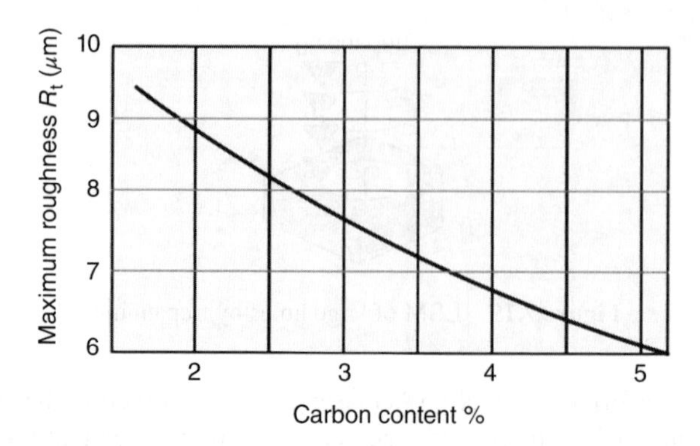

Figure 7.20 Effect of carbon content on the surface roughness of gas-assisted laser drilled holes. (Adapted from Rykalin *et al.* [17].)

work material. It increases considerably with increasing carbon content. This may be justified by the decreased molten metal viscosity, which can be effectively ejected by assisting gas leaving better surface quality (Figure 7.20).

7.4.4 Plasma Arc Cutting

In plasma arc cutting (PAC), a stream of gas is ionized and heated to a high temperature (about 20 000 °C) constricted in arc between a tungsten electrode and a workpiece anode. The material is removed by high velocity, high temperature gas stream (plasma). The plasma flows through a water-cooled nozzle that directs gas stream to the desired location. PAC can be used to cut electrically conductive metals and alloys such as carbon steel, aluminum, and stainless steel, with straight or profile cuts at high speeds. It has the ability to cut thick materials, up to about 150 mm. Recently, cutting of conductive and nonconductive materials by PAC has become much more attractive. The main attraction is that PAC is the only method that cuts faster in stainless steel than it does in mild steel.

However, PAC has many disadvantages which include the possibility of dross attached to the bottom of the cut and an appreciable recast layer (RL) and HAZ. It has also reduced accuracy and surface quality. Moreover, the process requires high power, and produces toxic fumes, IR, and UV radiations, that may cause eye injuries (cataracts) and loss of sleep. An external shielding gas around the primary arc may be used.

PAC systems operate either in a transferred arc mode or a nontransferred jet mode. In the transferred arc mode (Figure 7.21a) the arc is struck from the rear negative electrode of the plasma torch to the conductive WP (+ve electrode) causing a temperature as high as 30 000 °C. Owing to the greater efficiency of the transferred systems, they are often used in the cutting of any electrically conductive material, including those of high electrical and thermal conductivity that are resistant to oxy-fuel cutting as aluminum.

In the nontransferred jet mode (Figure 7.21b) the arc is struck with the torch itself. The plasma is emitted as a jet through the nozzle orifice, causing a temperature rise of about 16 000 °C. Because the torch itself is switched as the anode, a large part of the anode heat is extracted by cooling water and therefore is not effectively used in the material removal

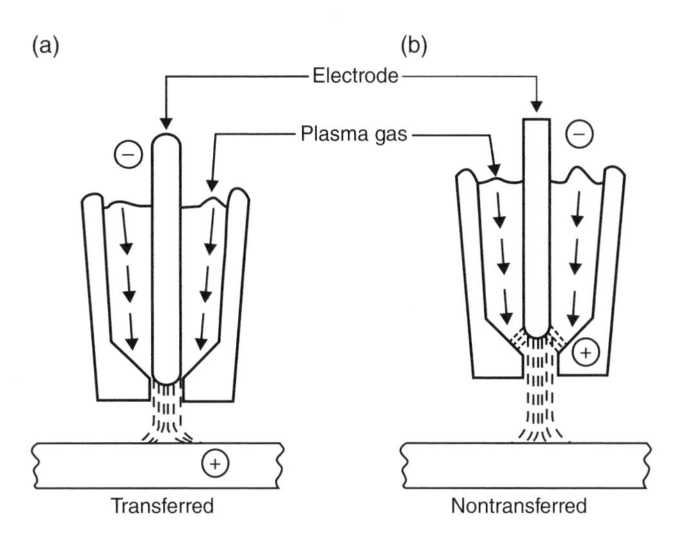

Figure 7.21 (a) Transferred and (b) nontransferred plasma torches. (Adapted from Kalpakjian [10].)

processes. Nonconductive materials that are difficult to cut by other methods are often successfully cut by the nontransferred plasma systems. The nozzle diameter depends on the arc current and the flow rate of the working gas. The commonly used working gases are He, H_2, N_2, or a mixture of them. The gas flow rate ranges from 0.5 to 6 m^3/h depending on the arc power and the plate thickness. The nonconsumable electrodes are made of 2% thoriated tungsten to resist wear.

7.5 Nontraditional Machining Processes – an Outlook

Most NTMPs typically have low MRRs and high specific removal rates compared to TMPs, which have considerably lower specific removal rates (0.1 $kW/cm^3 \cdot min$ for turning, 0.2–0.4 $kW/cm^3 \cdot min$ for milling, and 0.4–1 $kW/cm^3 \cdot min$ for grinding) as compared to specific removal rates of NTMPs, ranging from 1 to 5000 $kW/cm^3 \cdot min$, which means that the NTMPs, generally necessitate much higher power requirement (Table 7.5). The table also shows the tooling, working media, the typical operating parameters along with the specific removal rate for each NTMP. Figure 7.22 shows the range of surface roughness and dimensional tolerances of the commonest NTMPs, as based on nominal diameter of 25 mm, whereas Table 7.6 summarizes the main features and applications of NTMPs dealt with in this chapter.

The future of NTMPs will be characterized by steady growth. Although nontraditional processes will probably never replace the currently used conventional tools. Most NTMPs are computer controlled with respect to process parameters, which ensures process reliability and repeatability. Compared to TMPs, the NTMPs possess almost unlimited capabilities, with one exception, that they have limited volumetric MRRs. Every year more attention is directed toward nontraditional processes, as evidenced by the increasing number of technical papers, conferences, books, and technical symposia [16].

Table 7.5 Tooling, working media, typical operating parameters, and the specific removal rate for NTMP

NTM-process			Operating parameters	Working medium	Tool	Specific RR ($kW/cm^3 \cdot min$)
MECHANICAL	Jet machining	AJM	p_n = 1–9 atm, v_j = 0.5–1 Mach, SOD = 0.5–20 mm, d_g = 10–80 μm	Abrasives Al_2O_3, SiC, and so on	Nozzle	100–1000
		WJM	p_n = 1300–4000 atm, v_j = 1–3 Mach, SOD = 2.5–5 mm	Water, oil, alcohol	Nozzle	>5000
		AWJM	p_n = 1300–3500 atm, v_j = 1–2 Mach, SOD = 2.5–5 mm, d_g = 90–200 μm	Water + abrasives	Mixing tube	1000–4000
	USM		f = 18–25 kHz, ξ = 10–50 μm, F_s = 0.1–50 N, d_g = 10–100 μm	Slurry	Horn, tool, abrasives	10–100
CH	CH-milling		T_e = 20–95 °C, suitable etchant	Etchant	Mask	1–10
	PCM (SE)	Only for electric conductive materials	T_e = 35–100 °C, p_{et} = 0.2–1.5 MPa, suitable etchant	Etchant	Photo-resist	1–10
EC	ECM		I = 50–60 000 A, J = 1.5–8 A/mm^2, E = 5–15 V, h = 0.1–1 mm, p_e = 1–10 atm	Electrolyte	Conductive tool	3–10
	ECG		I = 50–3000 A, J = 1–3 A/mm^2, E = 5–15 V, v_g = 1200–2400 m/min	Electrolyte	Metallic bonded GW	2–8
THERMAL	EDM		i_d = 0.1–500 A, V_o = 50–300 V, h = 10–500 μm, t_d = 2–2000 μs, f_s = 1–500 kHz, τ = 0.1–0.95	Dielectric	Conductive tool	2–5
	EBM		V_b = 50–150 kV, i_b = 100–1000 μA, P = 2–60 kW, f = 0.1 Hz to 16 kHz, t_d = 4 μs to 60 ms	Vacuum 10^{-5} Torr	EB	450
	LBM		V = 4.5 kV, SOD = 1.5 mm, λ = 0.6–10.6 μm, E° = 20 kW	Air	LB	2700
	PBM (PAC)		V = 30–250 V, P = 200 kW, I_p = 50–1000 A, SOD = 6–10 mm	Plasma Ar, H_2, N_2	Nozzle	1–4

Operating parameters: p_n = nozzle pressure, v_j = jet velocity, SOD = stand-off-distance, d_g = abrasive grit diameter, f = frequency, ξ = oscillation amplitude, F_s = static force, Q = flow rate, T_e = etchant temperature, p_{et} = etchant pr., I = machining current, J = current density, E = cell voltage, h = gap thickness, p_e = electrolyte pr., v_g = tangential, speed of GW, i_d = discharging current, V_o = supply voltage, f_s = sparking frequency, t_d = discharge duration, τ = cycle duty, V_b = break-down voltage, λ = wave length, E° = laser peak power, i_b = machining current of EB, V = voltage, P = plasma power, and I_p = plasma current.
Adapted from: Youssef [4]. With permission.

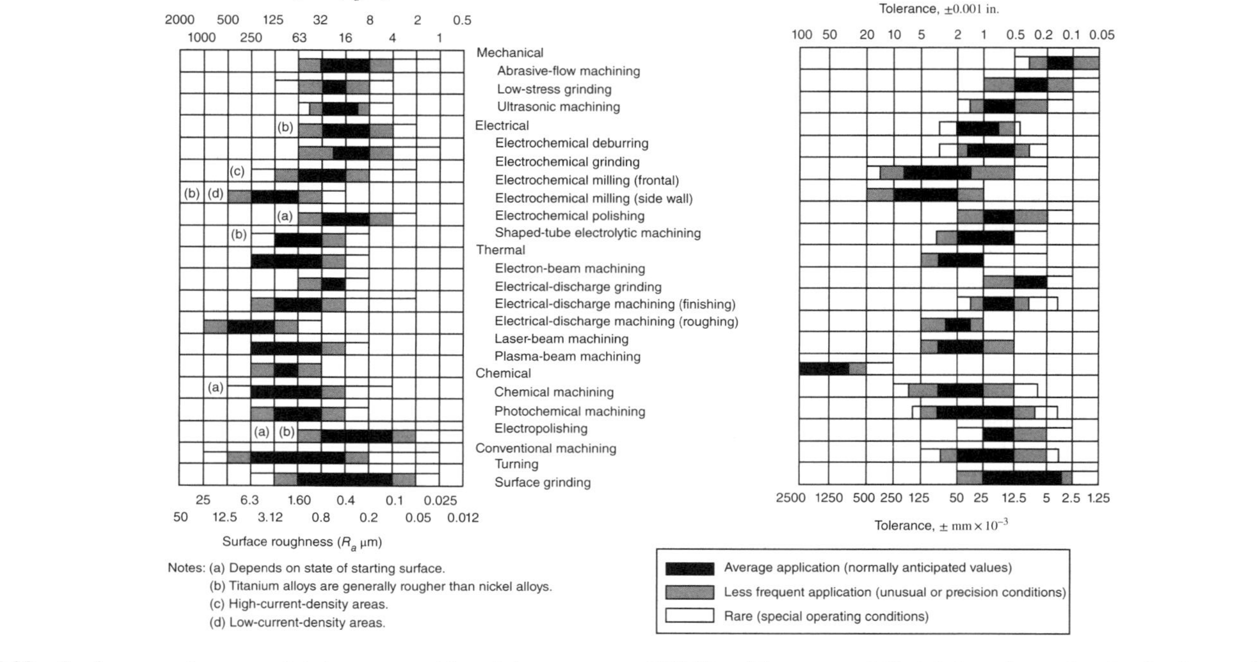

Figure 7.22 Surface roughness and tolerances achieved by common NTMPs. (Courtesy of TechSolve, formerly the Institute of Advanced Manufacturing Sciences.)

Table 7.6 Main features and applications of NTMPs

NTM-process			Characteristics and applications
MECHANICAL	AJM		Cutting, slotting, deburring, deflashing of brittle and hard materials/surface cleaning, glass defrosting, deburring of crossed holes in hydraulic valves, medical applications, deburring of Teflon, nylon, derlin/ environmentally pollutant
	WJM		Cutting rocks and granite, metals, machining of FRP and PCB, cutting soft materials like wood, leather, paper, meat and frozen foods/no thermal damage/noisy (catchers needed), expensive equipment/environmentally safe
	AWJM		Hybrid process for cutting all materials, soft and hard with high traverse speeds, granite, metals, glass, FRP, Ti-alloys with large thicknesses, no thermal effect/noisy (catchers needed), expensive equipment
	USM		Effective in machining hard and brittle materials like glass, ceramics, germanium, and carbides/not recommended for soft materials/accuracy and tool wear material dependent/not applicable for large size cavities (up to 40 mm)
CH	CH-milling		Almost all materials even glass and germanium are machinable/depth limited to 12 mm/suitable for low production runs/no surface stress/burr free/inexpensive tool and equipment cost/applicable in aerospace and metal working, industries
	PCM (SE)	Only applicable for electric conductive materials	Limited to metallic materials/thin sheets 2 mm maximum/burr free/low tool cost/high apparatus cost/higher penetration rate than CH-milling/applicable for machining micro electronics and PCB
EC	ECM		Applicable for difficult-to-cut electric conductive materials, production of complex shapes and turbine blades/stress and burr free/expensive tool/zero tool wear/RR is current and material dependant/problematic electrolyte disposal
	ECG		Hybrid process applicable for difficult-to-cut metals and alloys/burr free, less residual stresses and better accuracy than conventional grinding, better surface integration, reduced wear of metallic bonded GW/tool sharpening/expensive GW, expensive equipment, corrosion hazard, problematic electrolyte disposal
THERMAL	EDM		Most common NTMP, machining of complex cavities in electric conductive materials regardless its hardness/die-making in hardened steels and carbides/HAZ, RL/micromachining by ED milling preferred than conventional EDM due to reduced tool cost/expensive equipment
	EBM		Hole making and cutting thin sheets of any material, micro-holes of high aspect ratios/minor material removal/HAZ, RL/vacuum is needed/expensive equipment/safety precautions
	LBM		All materials can be machined, however, reflective materials difficult to machine/applicable for micro holes in thin sheets/minor material removal/HAZ, RL/necessitates no vacuum/expensive equipment/hazardous (extreme caution is required)
	PBM (PAC)		Rapid and effective process/cutting profiles in plates up to 200 mm/rough turning of difficult-to-cut materials (PTM), applicable for cutting stainless steels, super alloys, Al, and so on/severely HAZ, and RL

From Youssef [4]. With permission.

References

[1] Machinability Data Center (1980) *Machining Data Handbook*, 3rd edn, Metcut Research Associates, Cincinnati, OH.
[2] Youssef, H.A., El-Hofy, H. (2008) *Machining Technology, Machine Tools and Operations*. Boca Raton, FL: CRC Press.
[3] Youssef, H.A., El-Hofy, H.A., Ahmed, M.H. *Manufacturing Technology: Materials, Processes, and Equipment*, 1st edn 2011, CRC Press, Taylor & Francis, Boca Raton, FL.
[4] Youssef, H.A. (2005) *Non-Traditional Machining Processes: Theory and Practice*, Alexandria: El-Fath Press (in Arabic).
[5] Youssef, H.A. (1967) *Herstellgenauigkeit beim Stoßläppen mit Ultraschall frequenz*. Dr.-Ing. Dissertation. TH Braunschweig.
[6] El-Hofy, H. (2005) *Advanced Machining Processes – Nontraditional and Hybrid Machining Processes*, McGraw-Hill, New York.
[7] Lauwers, B., Klocke, F., Klink, A., *et al.* (2014) Hybrid processes in manufacturing. CIRP Ann. Manuf. Technol. **63**(2): 561–583.
[8] Schubert, A., Hackert-Oschätzchen, M., Meichsner, G. *et al.* (2011) Precision and micro ECM with localized anodic dissolution, in J.M. Slabe (ed.), *TECOS Slovenian Tool and Die Development Centre, Celje: Proceedings of the 8th International Conference on Industrial Tools and Material Processing Technologies*, 193–196.
[9] Burger, M., Platz, A., Werner, E. (2007) Herstellung/Nachbearbeitung von Turbi- nenblisks durch Präzises Elektrochemisches Bearbeiten, TUM, http://www.wkm.mw.tum.de/forschung/postergalerie/ (accessed April 19, 2015).
[10] Kalpakjian, S. (1985) *Manufacturing Processes for Engineering Materials*, Addison-Wesley, Reading, MA.
[11] Sharma, S., Jain, V.K., Shekhar, R. (2002) Electrochemical drilling of inconel super alloy with acidified sodium chloride electrolyte *Int. J. Adv. Manuf. Technol.* (2002) **19**:492–500.
[12] Visser, A., Junker, M., and Weissinger, D. *Sprühätzen mittallischer Werkstoffe*, 1st edn, 1994, Eugen G. Leuze Verlag, Bad Saulgau.
[13] König, W., *Fertigungsverfahren*, Band 3: Abtragen, VDI Verlag, Duesseldorf, 1990.
[14] Steigerwald, K.H., *Materialbearbeitung mit Elektronenstrahlen*. 4. International Kongress für Elecktronenmikroskopie, Springer, Berlin, 1958.
[15] Visser, A. (1966) Werkstoffabtrag mittels Elektrononstrahl. Dr.-Ing. Dissertation. TH Braunschweig.
[16] Davis, J.R. (1989) *Metals Handbook: Machining*, Vol. **16**, ASM International Materials Park, OH.
[17] Rykalin, N., Uglov, A., Zuev, I., Kokora, A. (1988) *Laser and Electron Beam Material Processing Handbook*, MIR-Publishers, Moscow.

References

8

Nontraditional Machining of Stainless Steels and Super Alloys

While the majority of stainless steels (SS) and super alloys are machined traditionally, nontraditional techniques are used when justifiable. Such justification involves cost savings when machining alloys at the extremes of toughness and hardness, or when machining intricate shapes. This chapter briefly describes, as based on the available data in literature, specialized company information, and handbooks some of the nontraditional machining processes which have been applied successfully to stainless steels and super alloys. Also, some hybrid processes are considered important and promising for machining these materials, especially the thermally-assisted machining (TAM) processes.

8.1 Mechanical Nontraditional Machining Processes of Stainless Steels and Super Alloys

8.1.1 Jet Machining

Although abrasive jet machining (AJM) is best suited for hard materials, it has been used to de-burr and to clean stainless steel alloys. One advantage of using jet machining processes is the fact that they are not thermal processes. Table 8.1 shows the machined surface finish of a soft austenitic SS AISI-316 processed by AJM. The starting surface had been ground to Ra = 0.47 μm.

Water jet machining (WJM) is also successfully used to cut SS-alloys of series 300 and 400 of stocks ranging from 0.25 to 100 mm. Abrasive water jet (AWJ) not only possesses the versatility of water jet (WJ) but also extends the applications to harder and denser workpieces. The addition of abrasives allows the cutting of difficult-to-cut materials such as super alloys, stainless steels, composites, and ceramics. Abrasive water jet machining (AWJM) is highly suitable for automation and cutting of complex shapes from stainless steel and super alloys.

Machining of Stainless Steels and Super Alloys: Traditional and Nontraditional Techniques,
First Edition. Helmi A. Youssef.

Table 8.1 Surface roughness for annealed SS AISI-316 in AJM

Abrasive	Grit size (μm)	Ra (μm)
Al_2O_3	25	0.25–0.53
	50	0.38–0.96
SiC	20	0.3–0.5
	50	0.43–0.86
Glass beads	50	0.30–0.96

Starting surface had been ground to Ra = 0.47 μm.
Adapted from Machinability Data Center [1].

Table 8.2 Traverse cutting rates for some SS-alloys using AWJM

Stainless steel alloy	Machining condition	Plate thickness (mm)	Cutting rate (mm/min)
PH-S 15500 (15 Cr-5Ni)	Pressure: 310 MPa, with 60 mesh garnet	3	230–380
		64	13–25
Austenitic-S31600 (bar stock)		76 (diameter)	13–50
Martensitic-S17400 (630)	Pressure: 200 MPa, with 60 mesh garnet	25	50

Adapted from Schwartz [3].

It is slower than plasma arc machining, and compared favorably with electric discharge machining (EDM), but it does not introduce residual stress or a heat affected zone (HAZ) [2]. Table 8.2 lists traverse cutting rates of some SSs processed by AWJM [3]. Figure 8.1 predicts the effect of SS-plate thickness on the cutting rate (traverse speed) when using AWJM [4].

Figure 8.2 illustrates a 760 mm diameter turbine wheel was machined with an AWJ from a solid, 45 mm thick disk of Inconel. The objective was to remove the material from between the turbine blades. Final shaping has been performed by electrochemical machining (ECM), where the total machining time was 48 hours [2].

8.1.2 Ultrasonic Machining (USM) of Stainless Steels and Super Alloys

Generally, USM is not recommended to machine all types of stainless steels and super alloys, since such a process is intended to machine only hard and brittle materials (see Chapter 7). However, it has been reviewed in literature that Neppiras and Fosket [5] tried in 1957 to cut SS 304 ultrasonically, where they used boron carbide B4C (mesh 100).They reported very low material removal rate (MRR) of 3 mm^3/min. Stainless steels and super alloys, when machined ultrasonically, they acquire an index 2, as based on soda glass of index 100, provided the same machining condition. Since that time, no encouraging technical data were found in literature related to USM of stainless steels and super alloys. However, recently *(2003), Park, Myung Ho Samcheok National Univ., Korea* developed an advanced USM technology for Inconel, using 60 and 75 kHz high frequency transducers, of amplitudes of about 8 and 4 μm, respectively. Such high frequencies and low amplitudes secure high efficient, and precise USM of Inconel.

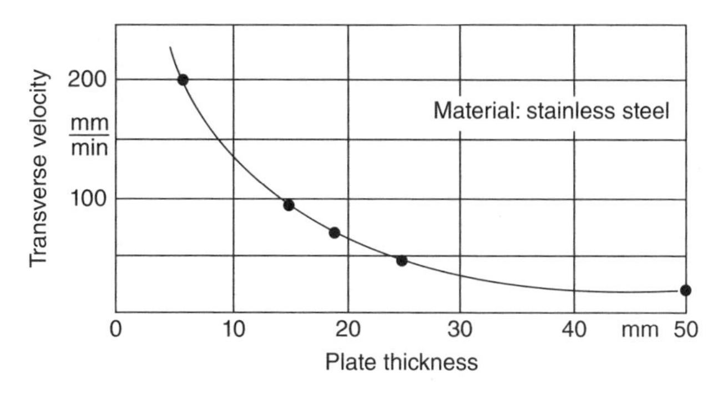

Figure 8.1 Effect of SS-plate thickness on the traverse speed in AWJM. (From Youssef [4] with permission.)

Figure 8.2 AWJM of an Inconel disk to produce integral turbine wheel blade. (Adapted from Metals Handbook [2].)

8.1.3 Abrasive Flow Machining of Stainless Steels and Super Alloys

Abrasive flow machining (AFM) was initially developed for the critical de-burring of aircraft valve bodies and components. Other applications include finishing impellers, integrally bladed rotors (IBRs), compressor wheels, turbine disks, and gears. Milled surfaces of an IBR, as finished by AFM, need only 15–30 minute operation time. Their surface finishes are considerably improved while eliminating hours of hand finishing (Figure 8.3).

Other important applications are:

- Cross-drilled and intersecting holes that present a major problem for conventional de-burring methods are easily handled by AFM media. De-burring of hole intersections in stainless steel part (Figure 8.4a).

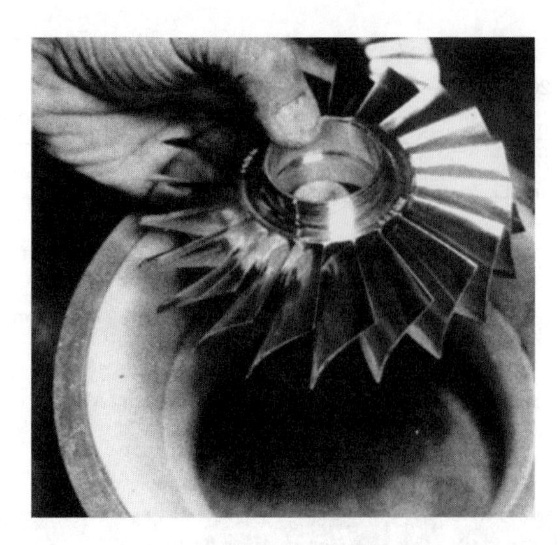

Figure 8.3 Milled surfaces of an aircraft IBR finished by AFM. (From Metals Handbook [2].)

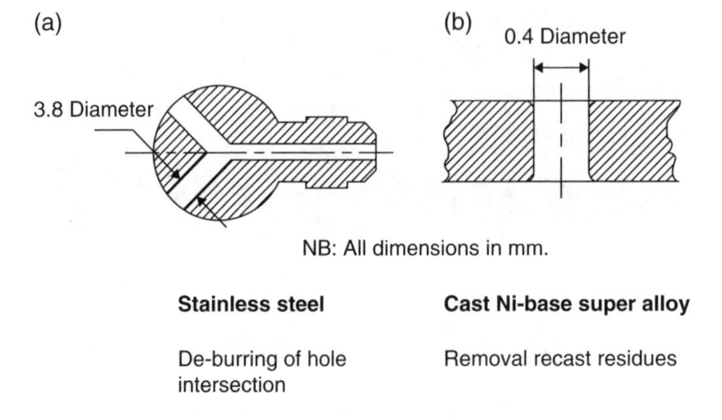

Figure 8.4 Processing of stainless steel and Ni-base super alloy using AFM: (a) stainless steel; (b) cast Ni-base super alloy. (Adapted from Machinability Data Center [1].)

Operating Conditions for AFM of Cross-drilled and Intersecting Holes:

Grit type SiC, 700 grit
Extrusion pressure 30 bar
Process time 1.5 minutes
No. of strokes 6
Pieces/fixture 1
Surface roughness Ra = 0.4 μm

- Removal recast residues from small holes of cast Ni-base super alloy (Figure 8.4b).

Operating Conditions for Removal of Recast Residues Using AFM:

Grit type SiC, 220 grit
Extrusion pressure 32 bar
Process time, controlled by volume, not strokes, and time
Pieces/fixture 1
Surface roughness Ra = 0.8 μm.

Viscous medium used to finish both parts was the Dynetics D080. The surface finish depends mainly on the abrasive grit size, which is ranging from 20 to 700.

8.2 Electrochemical and Chemical Machining Processes of Stainless Steels and Super Alloys

8.2.1 Electrochemical Machining

Turbo-machinery systems are characterized by the use of dedicated high temperature, high specific strength, and wear-resistant materials. Machining such difficult-to-cut materials using conventional means is very challenging, often resulting in low MRRs, reduced precision due to high cutting forces, high tooling costs due to increased wear and consequently low process efficiency. In addition, the resulting surface integrity is often characterized by thermo-mechanically altered or even damaged rim zones [6]. Thus, the utilization of technological as well as economically suitable manufacturing technologies is of great interest.

The major advantages of ECM are its process specific characteristics of high MRR in combination with almost no tool wear. ECM is specifically used in large batch size production and represents a viable alternative manufacturing technology for turbo-machinery components. In addition, high MRRs can be realized while achieving good workpiece surface quality without the occurrence of white layers, HAZs, or strain hardening [7].

Taking the aerospace sector as an example, it is found that Ti-based and Ni-based alloys are mostly preferred as constructional materials in aero-engines. The temperature capability of such materials is constantly increasing through the development of new materials with different primary manufacturing technologies [8]. Due to the absence of grain boundaries, single crystal materials exhibit far better creep properties than polycrystalline materials and can therefore be utilized at higher temperatures [9]. The use of such new materials and polymer matrix composites – PMCs (for fan blading components), require amongst others, the development of appropriate manufacturing technologies [10].

Table 8.3 illustrates the recommended electrolytes (along with their concentration and inlet temperatures), used for ECM of some SS-alloys. Table 8.4 shows theoretical removal rates for to SS-alloys, as calculated from the basic Faraday's equation.

ECM is not a highly accurate process. The distribution of the current lines leads to rounding the corners and edges. Thus sharp corners cannot be produced by ECM. Tolerances of about 0.12 are generally held to be typical for ECM of super alloys and stainless steels, while dimensional accuracies of 15 μm or less have been claimed under special shielding and masking of the tool cathode to direct the current flow only to the areas to be machined [11].

Table 8.5 shows the theoretical removed rates for ECM of some selected types of super alloys and stainless steels, as calculated according to Faraday's law using a current of 1000 A,

Table 8.3 Recommended electrolytes used for ECM of SSs

Stainless steel alloy	Electrolyte	Concentration (g/l)	Inlet temperature (°C)
Type 410 (martensitic)	NaCl or	36	27
	NaCl + $NaNO_3$	192–216	46
Type 302 (austenitic)	NaCl + NaF	30–32	38
Type 303 (austenitic)	NaCl + $NaNO_3$	120–140	21
Type 316 (austenitic)	NaCl	120	38
517400 (Custom 630) (enh., PH, Mart.)	NaCl or	96–120	27
	$NaNO_3$	240	38
Pyromet A-286	$NaNO_3$	240	38

Source: Compiled from Machinability Data Center [1].

Table 8.4 Theoretical removal rates for ECM of Custom 630 and Pyromet A-286 SS-alloys

Stainless steel alloy	Theoretical RR (cm^3/min) for 150 A/cm^2
S17400 (Custom 630)	2.0
Pyromet A-286	1.9

Source: Compiled from Machinability Data Center [1].

Table 8.5 Theoretical metal removal rates for ECM of super alloys and stainless steels as calculated by Faraday's low assuming current efficiency $\eta = 100\%$, using a current of 1000 A

Work material	MRR, based on I = 1000 A (cm^3/min), $\eta = 100\%$	Assumed valences (n)
Super alloys		Al = 3, Nb = 3, Co = 2, Cr = 3, Cu = 2,
A-286[a]	1.92	Fe = 2, Mn = 3, Mo = 4, Ni = 2, Ti = 4,
M 252	1.80	W = 6, V = 5, C = 0, and Si = 0
René41	1.77	
Udimet 500	1.80	
Udimet 700	1.77	
Haynes 25 (L605)	1.75	
Stainless steels		$MRR = (600/F)\ (N/n)_{eq.}\ I.\ 1/\rho\ cm^3/min$
17-4PH (UNS 17400)	2.02	F = 96 487 A·s/mol, $(N/n)_{eq}$ = chemical equivalent of anode, and ρ = anode density

[a] A-286 is the Pyromet Stainless Steel.

and assuming 100% current efficiency which is realized if NaCl electrolyte is used. The valences of alloying elements illustrated in the table are assumed to control the chemical dissolution of the machined alloys.

Table 8.6 provides the recommended electrolytes along with their recommended concentrations and inlet temperatures, used for ECM of some typical super alloys and stainless steels. In the majority of applications, try simple NaCl first, then try more complex electrolytes only

Table 8.6 Electrolyte selection guide (composition, concentration, and inlet temperature) for ECM of super alloys and stainless steels

Work material	Electrolyte			Remarks
	Composition	Concentration (g/l)	Intel temperature (°C)	
Super alloys				
Fe-base				
A-286(W)	$NaNO_3$	240	38	$NaClO_3$ for specialized applications
Ni-base				
M252(W)	NaCl	240	41	ST (not Ag) gives better results
Waspaloy (W)	NaCl	120	35	
Astroloy (W)	NaCl	240	24	
	$NaNO_3$ or	120–240	38	Cuts easily and rapidly
	NaCl + NaF	90 + 30	38	
Inconel 700(W)	NaCl	120	38	
Inconel 706 (W)	NaCl	96	24	
Inconel 718 (W)	NaCl or	108–120	35–38	ST and Ag gives better results
	$NaNO_3$	216–240	35–38	
René 41 (W)	NaCl	258 (or 120)	24 (or 35)	
Udimet 500 (W)	NaCl	120	38	
Udimet 700 (W)	NaCl	120	38	
Inconel X (W)	NaCl + $NaNO_3$	240 + 60	38	
IN-100(C)	NaCl	120	38	
René 80 (C)	NaCl	120	38	
René 125 (C)	NaCl	120	38	
René 95 (C)	$NaNO_3$ or	216–240	38	
	NaCl	120	38	
Co-base				
MAR-M509	NaCl	240	32–52	—
HS-21	NaCl	120	38	
Haynes 25(L605)	NaCl + $NaNO_3$	240 + 30	38	
HS-31(X-40)	NaCl + NaF	30 + 2.4	38	
Stainless steels				
302 (austenitic)	NaCl + NaF	30 + 2.4	38	—
303 (austenitic)	NaCl + $NaNO_3$	120 + 20	21	
316 (austenitic)	NaCl	120	38	
410 (martensitic)	NaCl or	96	27	
	NaCl + $NaNO_3$	192 + 120	115	
17-4 PH (Custom	NaCl or	192–120	26	
630, UNS 17400)	$NaNO_3$	240	38	

(W): wrought, (C): cast, ST: strengthened, and Ag: aged.
Adapted from Metcut.

if necessary. NaCl is highly corrosive and nonexpensive (costs 30% of $NaNO_3$). Electrolyte concentrations should be kept as low as compatible with productivity so as to reduce accumulations of salt deposits on equipment and tooling. The temperature control should be within +1°C of the recommended value, in order to attain consistent electrolyte conductivity at the inlet side of the electrolytic cell. The cell open circuit voltage is set to attain the desired current density that matches the feed rate. The typical voltage setting for ECM of super alloys and stainless steel ranges from 10 to 25 V [1].

It should be emphasized that electrolytes play an important role in the dimensional control of the produced holes and cavities. NaCl, for example, yields much less accurate components than nitrates ($NaNO_3$), the latter having far better dimensional control due its current efficiency/current density characteristic. As shown in Figure 8.5, with $NaNO_3$, the current efficiency is greatest at the highest current densities. In hole drilling these high current densities occur between the leading edge of the tool and workpiece surface. In the side gap, where there is no side movement between the tool and workpiece surface, so that the gap widens and current densities are lower. The current efficiencies are consequently lower there; much less metal than that predicted from Faraday's law is removed. Thus overcut in the side gap is reduced with $NaNO_3$. If NaCl were used instead, then the overcut could be much greater. Its current efficiency remain, steady at almost 100% for the wide range of current densities. Accordingly, even in the side gap, meted removal proceeds at a rate which is mainly determined in accordance of Faraday's law in which a high current efficiency of 100% is substituted. A wider overcut then ensues [11].

Sodium chlorate solution $NaClO_3$ has been investigated. Specialists are reluctant to employ it, because of its ready combustibility; however, this electrolyte is claimed to give even better throwing power and closer dimensional control than sodium nitrate solution [11]. Special safety precautions should be exercised when using a chlorate ($NaClO_3$) or a nitrite ($NaNO_2$) electrolyte.

Actual MRRs are less deviated in most cases from the Faraday's theoretically calculated values listed in Table 8.5. Such deviations are related to the not correctly assumed valences of

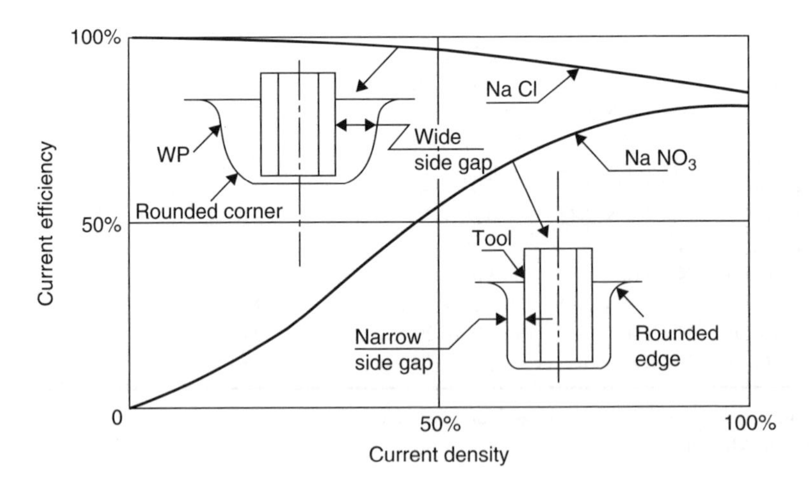

Figure 8.5 Current efficiency/current density characteristic of NaCl and $NaNO_3$ electrolytes and their associated gap overcuts.

the work material and the current efficiency which attains values not necessarily 100%, as considered in the Faraday's law. Table 8.7 provides typical MRR and the relevant operating conditions for ECM of stainless steels and super alloys as compared to Ti-base alloy Ti-6Al-4V.

Figure 8.6 shows the relation between the open circuit voltage and the current density for two selected frontal equilibrium gaps 0.5 and 0.25 mm respectively, when ECM of a Ni-base super alloy René 41 for minimum starting voltage $\Delta E = 2.6$ V.

The equilibrium frontal gap is not independently adjustable. However, when machining René 41, smaller gap (0.25 mm) which necessitates lower voltage (24 V) than larger gap (0.5 mm) which necessitates higher voltage (30 V) to attain the same current density and consequently the same penetration rate. The lower gap is recommended, since it provides less power requirement.

Figure 8.7 shows an airfoil, directly machined on a turned Inconel 718 compressor disk to eliminate the mechanical joining of the airfoils on the disk which realizes the highest rigidity.

Table 8.7 Typical MRRs and relevant operating conditions for ECM of stainless steels and super alloys

Work material	Electrolyte		Minimum starting voltage (ΔE)	MRR, based on I = 1000 A (cm^3/min)
	Type	Concentration (g/l)		
Super alloys				
Inconel 718 (ST-Ag)	NaCl	120	3.3	1.44
René 95 (ST-Ag)	$NaNO_3$	270	4.3	1.69
300 M (tempered)	NaCl	120	1.1	2.16
MAR-M509 (C)	NaCl	120	1.2	1.57
Astroloy	$NaNO_3$	240	4.0	2.05
Stainless steel				
17-4 PH (ST-Ag)	$NaNO_3$	270	3.6	1.41
Ti-alloy				
Ti-6Al-4V (Ann)	NaCl	120	3.8	1.64

(C): cast, ST-Ag: solution treated and aged, and Ann: annealed.
Adapted from Metcut.

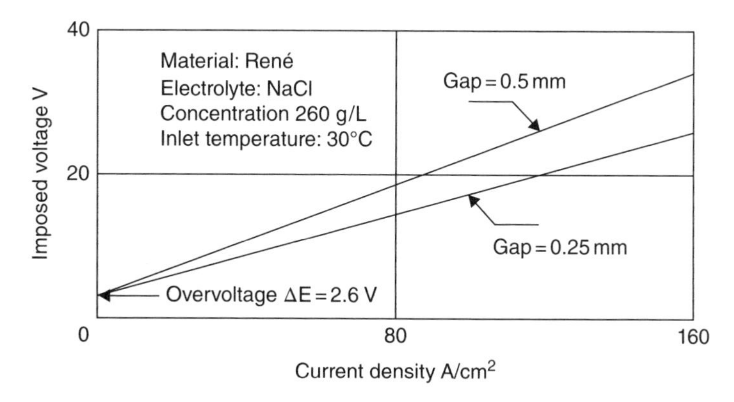

Figure 8.6 Relation between the open circuit voltage and the current density for two selected frontal equilibrium gaps when ECM of René 41. (Adapted from Bellows [12].)

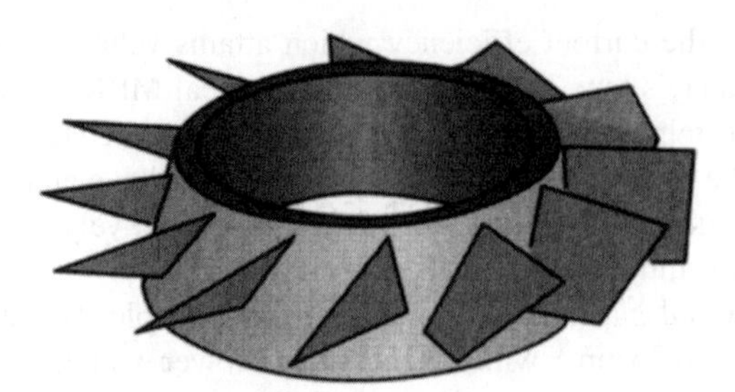

Figure 8.7 Airfoil, directly ECM- and PECM-machined on a turned Inconel 718 compressor disk. (Adapted from [13].)

A multi-axis numerical control (NC)-ECM machine is used to rough and finish the airfoils out of a turned hub. After the finish run, ECM yields a consistent surface finish and satisfactory part tolerances; accordingly, no further processing (machining or polishing) of the airfoils is required [13].

Some other examples are provided to illustrate the application of ECM in machining components made from stainless steels and super alloys (Figure 8.8 and Table 8.8).

Figure 8.8a: ECM of turbine blade made from either an Fe-base super alloy A-286, or a Ni-base Waspaloy, using a machine that provides 2000 A machining current, and an open circuit voltage (OCV) of up to 20 V. The tool electrode used was made of Cu-W-alloy. For both materials, the current starts at 100 A and ends at 150 A. The electrolyte was $NaNO_3$ (of concentration 260 g/l) for the first blade, whereas it was NaCl (of concentration 200 g/l) for the second blade. Other working conditions are shown in Table 8.8.

Figure 8.8b: Jet engine made from Ni-based alloy Udimet 700 was provided by small size deep holes using shaped tube electrolytic machining (STEM). Acid electrolyte H_2SO_4 (10% concentration) was used. Periodic voltage reversal as described in CH_7 was performed.

Figure 8.8c: This illustrates an Inconel 718 rotor in which multiple small cavities are to be machined electrochemically. Milling of such small cavities conventionally is costly because small milling cutters must be used. Seven hours are needed to machine 54 cavities. With ECM, however, these cavities can be machined either single or in groups. In the latter case, an assembly of brass electrode composed of 54 separate replaceable units, is mounted on vertical-ram. EC-machine equipped with 18 000 A, 11 Vdc-generator.

The machining proceeds in one stroke at a feed rate of 1 mm/min accordingly, the machining cycle necessitates g minutes. NaCl electrolyte is recommended with a concentration of 120 g/l and inlet temperature of 38 °C the data given in Table 8.6 are, however, valid for single cavity/stroke, where smaller generator of 1000A 18Vdc is needed. A tolerance of ±75 μm surface roughness of 1 μm (Ra) is expected [1].

Figure 8.8d: This shows the cross-section of a nozzle made from AISI-316 stainless tube. The EC-machine can provide a maximum current of 500 A at 24 Vdc. A Cu-electrode is fed into one end of the tube to machine the hole to a conical shape, with a current beginning at 20 and ending at 310 A. the open-circuit-voltage was 17 V. The electrolyte was NaCl (concentration 120 g/l, inlet temperature 27°C). A feed rate of 6 mm/min has been realized. A cutting time of 8 minutes produced an excellent surface quality of Ra = 0.12–0.25 μm.

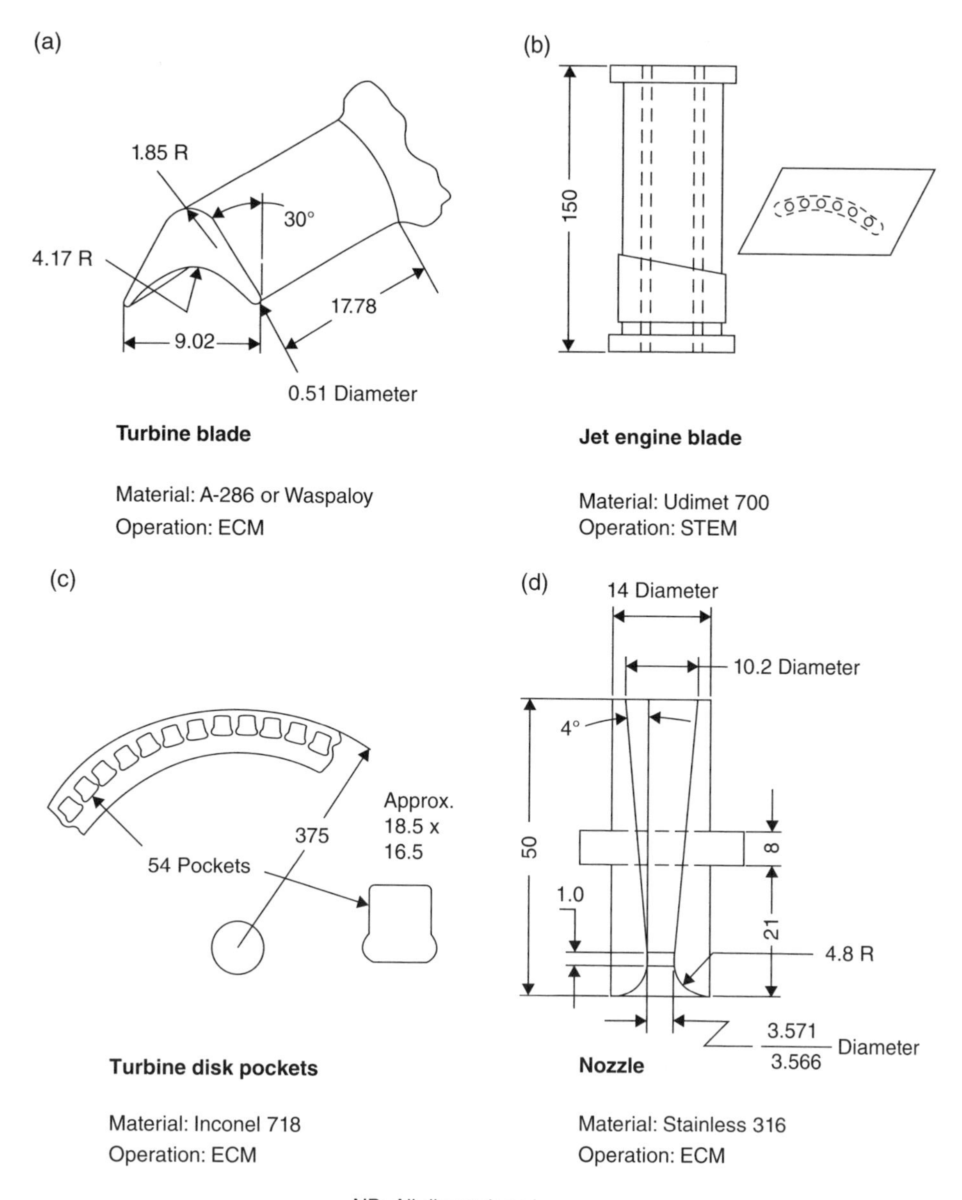

Figure 8.8 Application of ECM and STEM in machining aero-engine components made from stainless steels and super alloys: (a) turbine blade; (b) jet engine blade; (c) turbine disk pockets; (d) nozzle. (Compiled from Metcut.)

After this operation, the part is inverted, and another Cu-electrode was used to machine the radius at the other end of the nozzle. The current began at 20 A and ended at 220 A. the electrolyte and its circulation were the same as before. The tool feed was 2.5 mm/min, the cutting time was 2 minutes, producing a similar surface finish as before [1].

Table 8.8 Working conditions for ECM and STEM of stainless steel and super alloy components shown in Figure 8.8

Working conditions	Component				
	a		b	c	d
Operation	ECM	ECM	STEM	ECM	ECM
Material of component	A 286	Waspaloy	Udimet 700	Inconel 718	AISI-316
Electrolyte					
Type	$NaNO_3$	NaCl	H_2SO_4	NaCl	NaCl
Concentration (g/l)	260	200	10%	120	120
Inlet temperature (°C)	42	32	35	38	27
Inlet pressure (bar)	9–14	9–14	1	15–17	6
Flow rate (l/min)	8	10	2	10	8
Current (A)					
Start	100	100	3	130	20
End	150	150	10	630	310
OCV (V)	11	12	9[a]	18	17
Feed rate (mm/min)	7.5	8	1.25	1.5	6
Tolerance (μm)	±100	±100	±50	±75	4° taper
Surface roughness Ra (μm)	0.4–0.7	0.2–0.5	1.5–3	0.7–1	0.12–0.25

[a] Periodic reversal in STEM, see Chapter 7.

Filtration of electrolytes to the 50 μm level or less is most desirable. Placement of filters immediately ahead of the electrode is a good practice. The volume of metal hydroxide or metal hydrate is considerable. In case of salt electrolytes it can be 100–500 times the volume of the metal removal. For super alloys and stainless steels, the volumetric ECM sludge ratio is about 200 times, which is considerably less than that for Ti-base alloys which is about 500 times the volume of metal removal [1]. Settling and filtration to remove these chips (sludge) is highly recommended to avoid short circuiting (arcing) which may destroy both the work and the tool electrode. Sludge filtration, however, necessitates environmentally satisfactory measures.

In ECM and related processes, the micro structure of metals and alloys (specially the grain size) does generally affect the integrity of produced surfaces. Figure 8.9 illustrates an overview on surface integrities of different materials after specific electrochemical treatments. For Inconel 718 and Inconel 718 DA which has a more fine-grained microstructure, a flat and smooth surface finish without any rim zone can be detected in the cross-sections when employing optimized ECM parameters.

Figure 8.10 shows the relative machinability index of stainless steel AISI-316 and super alloy Inconel 901, when PECMed, as based on Al-2017 alloy of 100% machinability rating. Inconel 901 has the relative machinability index of 77.8%, whereas the stainless steel AISI-316 has lower machinability, compared to Inconel 901, of 58.4% [15].

Modern pulsed electrochemical machining (PECM) machine tools allow roughing, finishing, and polishing on one platform [16]. The machine tool consists of a compact and closed system with autonomous electrolyte management, with typically 7–8 axes (3–4 axes for the manipulation of the workpiece and 4 axes for tool movement, 2 of which are oscillating cathodes) [17].

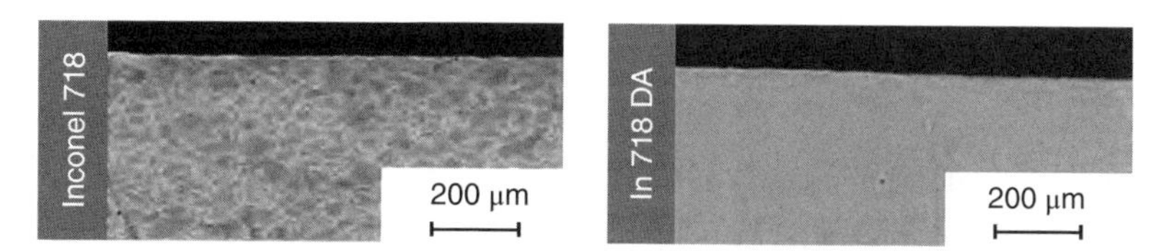

Figure 8.9 Achievable surface integrity of nickel-based alloys after ECM machining. (Adapted from Baumgärtner [14].)

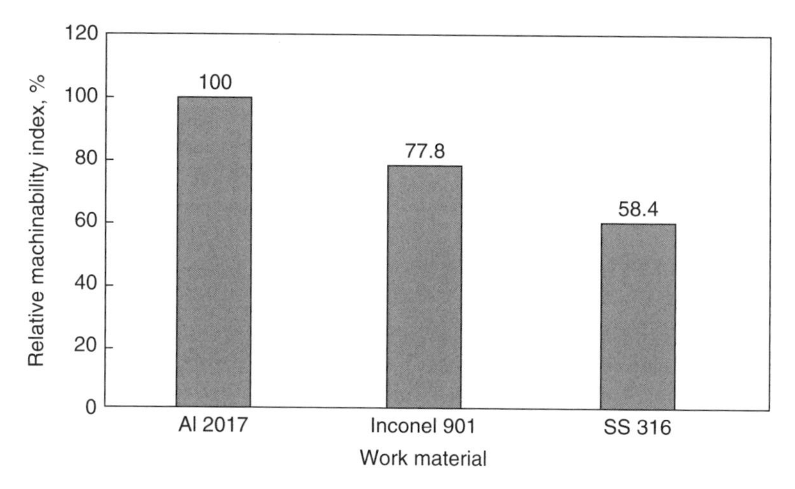

Figure 8.10 Machinability index of stainless steel AISI-316 and super alloy Inconel 901, when PECMed, as based on Al-2017 alloy of 100% machinability rating. Adapted from El-Hofy [15].

For the machining of titanium-based alloys, the electrolyte is typically aqueous solution of NaCl (avoiding passivation phenomena) while for nickel-based super alloys and titanium-aluminum alloys, $NaNO_3$ is used. Fluid flow rates up to 1000 l/min and pressures up to 40 bar are realized in a closed-loop system with control of temperature and pH-value (via acid-/base-dosage) [18]. In addition, a monitoring and chemical treatment of chromium VI (reduction to chromium III) is nowadays implemented [19]. For filtering of the hydroxide precipitation usually back-flushable slurry filter membranes and chamber filter presses are used [18]. Finally, recycling and/or disposal have to be managed by authorized specialists. Resulting costs for the electrolyte handling and the recycling/disposal of the slurry have therefore to be taken into account.

Figure 8.11 shows machine setups for blade and blisk manufacture (dc-ECM). On the left, the simultaneous machining of four blades is presented. The feed rate is independent of the area to be machined enabling a significant increase in productivity [21]. On the right the blisk fluid flow box which is completely flooded during machining is shown together with the tool electrodes. The mechanical tool oscillators of PECM have typical amplitudes of 100 mm at frequencies of up to 100 Hz. During mechanically pulsed machining, typical feed rates are only 0.1 mm/min. Therefore, the PECM process is mainly used for finish machining and it becomes economic if it is combined with preliminary steps of dc-ECM [17,22]. In addition to providing high MRRs and low surface roughness, both ECM and PECM produce burr-free geometries independent of shape complexity [23].

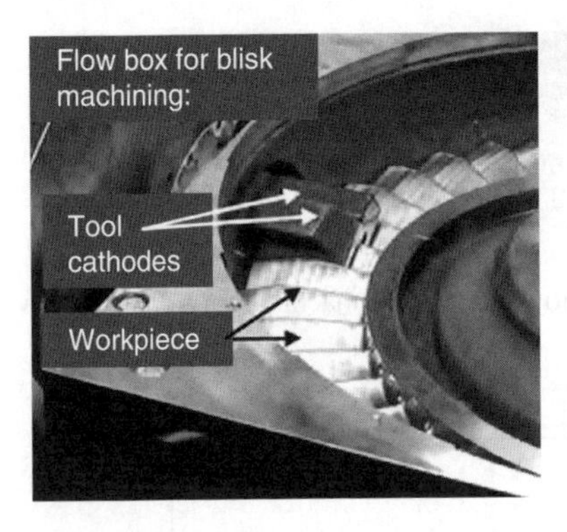

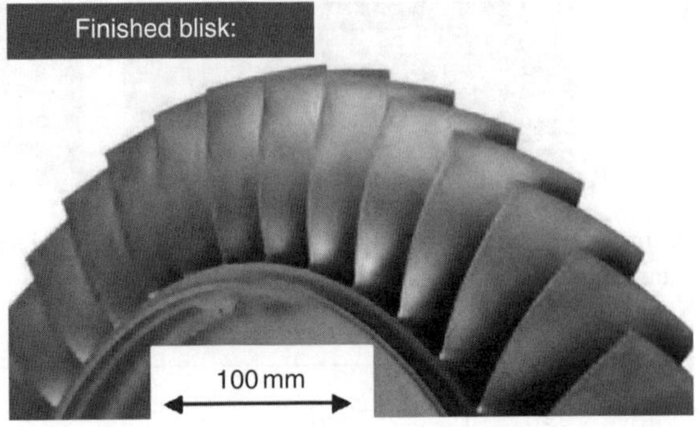

Figure 8.11 Blisk (left) manufacture (direct current electrochemical machining (DCECM)), Leistritz Turbomaschinen Technik LTT [7], and (right) finished low pressure compressor blisk out of Ti6Al4V machined via ECM [20] (compiled).

Once a process is configured and all electrical and fluid process parameters are maintained, excellent repeatability can be achieved. During machining of a blisk with 75 blades, 43 c_p values (defined as the number of times the spread of the process fits into the tolerance band) describing different geometrical features for blade position and thickness, chord length, and line angles, thickness as well as geometry of leading and trailing edges, were analyzed. All values were better than 1.3 and only one was lower than 1.33 (4σ). Maximum cp index values were around 6.65 indicating a highly stable process [17].

Operating voltages of 13–18 V, currents of 12–15 kA and a feed rate of 1 mm/min resulted in total machining time of 5 min per blade [7]. Continuous developments in cutting technology has outpaced the application of ECM for the machining of Ti-based blisks but it is still very competitive for machining of Ni-based alloys. In order to obtain the best surface qualities of aerofoils, a combination of ECM roughing and PECM finishing is suggested; such approach developed from machining a nickel-based high pressure (HP) compressor blisk made from Inconel 718. The machining sequence comprises dc-ECM pre-machining of basic slots between the blades starting from a turned raw part using $NaNO_3$ as electrolyte. The oversize around each blade is unequal and differs from about 1–3 mm. Sufficient gap is necessary between two blades in order to accommodate the finishing electrodes. This process is followed by an unpulsed ECM roughing step to an equidistant oversize. After this, the leading and trailing edges are prepared. In the last step, PECM finishing with oscillating tool electrodes takes place to finalize the aerofoils and annulus [17].

The machining results for both ECM technologies are presented on the left and right parts of Figure 8.12. In order to get a shiny, hyper-polished surface, an additional smoothing operation was performed to remove oxide particles via vibratory polishing with chemical support. Final roughness values were Ra < 0.1 μm and Rz < 1 μm. Besides complex blisk geometries, complete turbine wheels for turbo charger applications can also be successfully machined via ECM from solid or from near net shape (Figure 8.13) [24].

Successful application examples of ECM and PECM for the manufacture of turbo-machinery components include information on relevant process conditions and achievable machining performance. ECM is capable of producing single blade and vane geometries of different shapes both for aircraft engine and stationary steam and gas turbine applications. During dc-ECM

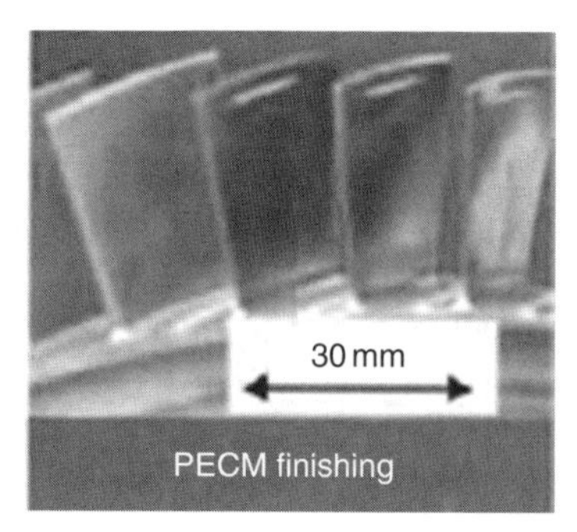

Figure 8.12 Production of a nickel-based HP compressor blisk through ECM: roughing, and PECM: finishing. (Adapted from Platz and Feiling [17].)

Figure 8.13 ECM machining of complex geometries like complete turbine wheels for turbo-charger applications. (Adapted from Giese [24].)

machining typical material removal can be several cm^3/min for different steel-based, titanium-based (e.g., Ti6242: 3.9 cm^3/min) and nickel-based (e.g., Inconel 718: 2.1 cm/min) alloys [24]. Form accuracy and surface roughness values of 0.1 mm and Ra = 0.8 μm respectively are possible. Calculated savings during manufacture of single blades amounts to 30% in comparison to traditional cutting operations [22]. This is especially so in continuous production due to reduced tooling costs. Further increase of productivity is possible during subsequent process optimization [25]. The introduction of forged or cast blades of TiAl-alloys and Ni- based super alloys in future engines will add even more significance to ECM [26].

The broad capabilities of ECM/PECM, when combined in one machine setup, to achieve both high MRR and good surface integrity together with high productivity is anticipated to see wider use in the future, especially for the machining of new advanced difficult-to-cut alloys.

8.2.2 *Shaped Tube Electrolytic Machining (STEM) of Stainless Steel and Super Alloys*

Because the process uses acid electrolytes, its use is limited to drilling holes in stainless steel or other corrosion-resistant materials in jet engines and gas turbine parts such as:

- turbine blade cooling holes;
- fuel nozzles;
- any holes where EDM recast is not desirable;
- starting holes for wire EDM;
- drilling holes for corrosion-resistant metals of low machinability;
- drilling oil passages in bearings where EDM causes cracks.

STEM is used to drill round and shaped holes in difficult-to-cut materials such as stainless steels and super alloys. Holes ranging from 0.5 to 6 mm diameter and as deep as 600 mm, can be produced with length-to-diameter ratios up to 300. Over 100 holes/machine stroke are practical.

Holes in the following stainless steels and super alloys have been drilled using STEM with H_2SO_4–acid electrolyte of 10% concentration.

Stainless steels:	AISI-304, AISI-321, AISI-414
Super alloys:	Udimet 500, 700, 710
	Stellite
	IN-100, –102, –738
	Inconel 625, 718, X-750, 825
	René 41, 80, 95, 100
	Haynes 25, 181
	HS – 31(X-40)
	Hastelloy C, X
	Greek Ascoloy

Table 8.9 provides the operating parameters for STEM of stainless steel AISI-304, and super alloys Udimet 700, Inconel 718, Rene 95, and Haynes 25, along with Ti-base alloy Ti-8Al-1Mo-1V for comparison, which indicates that for Ti, an HCl acid electrolyte of relatively lower concentration of 5% is preferred.

STEM is mainly applied in turbo-machinery component manufacture for the production of cooling holes. This includes holes both for blades/vanes as well as disks. The major advantages of applying ECM are the production of smooth, stress- and crack-free surfaces. In addition, low contour drilling angles can easily be realized [27].

Figure 8.14 shows the STEM of curved elliptical cooling holes in nickel-based HP turbine disks. The long axis of the ellipse has a length of 6.5 mm while the disc has a diameter of 500 mm. The curved design is necessary for an ideal stress distribution during load and cannot be machined via conventional cutting. Both inlet and outlet contours are simultaneously chamfered. The overall machining time for 74 cooling holes amounts to 20 hours [7].

Table 8.9 Specific operating parameters of STEM processing of stainless steels and super alloys as compared to Ti-8Al-1Mo-1V

Operating parameters	Work material					
	AISI-304	Udimet700	Inconel718	René95	Haynes25	Ti-8A1-1Mo-1V
Electrolyte (acid)	H_2SO_4	H_2SO_4	H_2SO_4	H_2SO_4	H_2SO_4	HCl
Concentration percentage by volume	10	10	9.5	10	10	4–5
Temperature (°C)	34	40	40	40	40	40–50
Pressure (bar)	0.7	3	1.8	1.6	0.7	3.5
Voltage Vdc	10	6	7.5	6.5	6	13.5
Forward (s)	8	7	10	10	3	10
Reverse (s)	0.3	0.1	0.3	0.3	0.3	0.3
MRR (cm^3/min)	1	0.8	0.8	0.8	0.6	1.1
Hole diameter (mm)	0.5	1.2	1.2	3	0.6	1.5
Ampere/hole (A)	1	1	2 (max)	7.5	0.5	4.3
Hole/stroke	1	16	6	1	30	3
Hole depth (mm)	15	22	10	25	1	20
(Depth/diameter) up to	300	180	80	20	1–2	130
Tolerance (μm)	±5	±6	—	±5	—	±2.5
Roughness Ra (μm)	—	—	—	2–3	—	3–4

Adapted from Metcut.

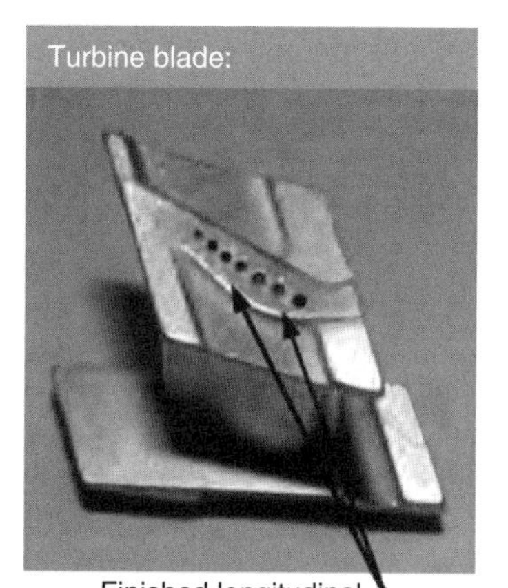

Figure 8.14 Machining of circular and curved elliptical cooling holes in nickel-based high pressure turbine disks via STEM. (Compiled from *Electrochemical Machining* [28] and *Elektrochemisches Abtragen* [29].)

8.2.3 Electro-stream (ES) Machining of Stainless Steel and Super Alloys

A principal application of electro-stream (ES) is the drilling of cooling holes in gas turbine components which are usually fabricated from Ni-, and Co-base super alloys. One example is the drilling of many holes simultaneously in the leading edge of a super alloy gas turbine vane.

As in STEM, an H_2SO_4–acid electrolyte is mainly used for drilling of Stainless steels and super alloys. Holes in the following stainless steels and super alloys have been drilled

Stainless steels:	AISI-304, AISI-316, AISI-321
Super alloys:	Udimet 700
	René 41, 77, 80, 100, 120, 125
	IN-102, –738
	Inconel 625, 718, X-750, 825
	Haynes 25, 181
	HS-31(X-40)
	Hastelloy C, X

8.2.4 Electrochemical Grinding (ECG) of Stainless Steels and Super Alloys

In electrochemical grinding (ECG), the wheels must have insulating grits (SiC and some forms of Borazon cannot be used because they are electrically conductive). This is why the wheels in Table 8.10 are corundum. Also, NaCl-electrolytes are rarely used in ECG because they are strongly corrosive to machine components.

Table 8.10 lists the recommended parameters of ECG of super alloys and stainless steels in general such as AISI-304, 316, 321, and 414. Pyromet SS A-286 (Fe-base super alloy) has also been considered. The maximum current density must be low enough to prevent overheating the low conductive materials. For comparison, the same parameters are listed for Cu-alloys, Al-alloys, and carbon steels. It is depicted from the table that greater current densities are allowed for the latter alloys than stainless steels and super alloys. Corundum grinding wheels, and $NaNO_3$-electrolytes are recommended for most cases of ECG.

8.2.5 Chemical Milling (CH-Milling)

As previously mentioned, the disadvantages of chemical (CH)-milling include low cutting rates, and the fact that masked areas will be under-cut (etch factor EF) by the corroding solutions. The corrosive effect is less serious in case of stainless steels; however, hydrogen embitterment may be a problem with hardened martensitic stainless alloys and intergranular corrosion may occur, depending on the chemical composition of that type.

Table 8.11 illustrates the main parameters for CH-Milling of austenitic and martensitic stainless alloys. The etching rate of martensitic alloys is considerably lower than that of austenitic alloys. Table 8.12 shows the same parameters for CH-Milling of some selected super alloys.

Table 8.10 Recommended parameters for ECG of super alloys and stainless steels

Work material	Abrasives of GW	Electrolyte/concentration		Maximum current density (A/cm^2)
			g/l H_2O	
Super alloy				
A-286 (Pyromet)		$NaNO_3$	120–140	116
Hastelloy X		$NaNO_3$	120–140	116
M 252		$NaNO_3$	120–140	116
Udimet 500,700		$NaNO_3$ or NaCl	110–120	116
Waspaloy		$NaNO_3$	120–140	116
Inconels		$NaNO_3$	120–140	116
René 41		$NaNO_3$	180–230	78
René 80		$NaNO_3$	120–140	78
HS-31(X-40)	Al_2O_3	$NaNO_3$ + NaCl	60–80	78
Stellite		$NaNO_3$	210–240	78
Stainless steels		$NaNO_3$	180–200	78
Other alloys				
Cu-alloys		$NaNO_3$ or KNO_3	180–200	233
Al-alloys		$NaNO_3$	120–140	233
Steels				
Low carbon		KNO_3:KNO_2 (9 : 1)	60–120	155
High carbon		$NaNO_3$	120–180	155

N.B: SiC and CBN (cubic boron nitride) not used for ECG since they are electric conductive.
Adapted from Machinability Data Center [1].

Table 8.11 Parameters of CHM of austenitic and martensitic SSs. (Machinability Data Center [1])

Stainless alloy	Etchant	Concentration	Temperature (°C)	Etch rate (μm/min)	Maskant	Etch factor (–)	Depth tolerance (±μm)	Ra (μm)
Austenitic	$FeCl_3$ or HCl:HNO_3	42°Be*	54	20–130	Polyvinyl chloride	1.5–2.0	100	1.6
Martensitic	$FeCl_3$ or HCl:HNO3	52°Be*	54	6	Polyvinyl chloride	—	100	3.2

Be*: Baumé specific gravity scale.
CHM, chemical machining.

8.2.5.1 MRR and Depth Tolerance

MRR is determined by etchant type, temperature, and concentration, all of which are selected to be compatible with the particular metallurgical state of the work material. The depth is controlled by the immersion time. Figure 8.15 shows the effect of nitric acid and sulfuric acid concentrations at 70 °C, respectively on the etch rate [1].

When machining Fe- and Ni-base super alloys under best conditions of control of time, temperature, and solution concentration, accuracies of ±25 μm can be achieved for shallow

Table 8.12 Parameters of CHM of super alloys

Super alloy	Etchant	Concentration	Temperature (°C)	Etch rate (μm/min)	Maskant	Etch factor (–)	Depth tolerance (±μm)	Ra (μm)
Co-base alloy	HCl:HNO3: $FeCl_3$	—	60	10–38	—	—	—	1–3.8
Inconel	$FeCl_3$ or	42°Be*	54	13–38	Polyethelene	—	—	1–3.8
	HCl:HNO3	42°Be*	54	13–38	Polyethelene	—	—	1–3.8
Nimonic	$FeCl_3$ or	42°Be*	49	13–38	Polyethelene	1–3	51	1–3.8
	$FeCl_3$:HNO3: HCl	—	49	13–38	Polyethelene	1–3	51	1–3.8

Be*: Baumé specific gravity scale.
Adapted from Machinability Data Center [1].

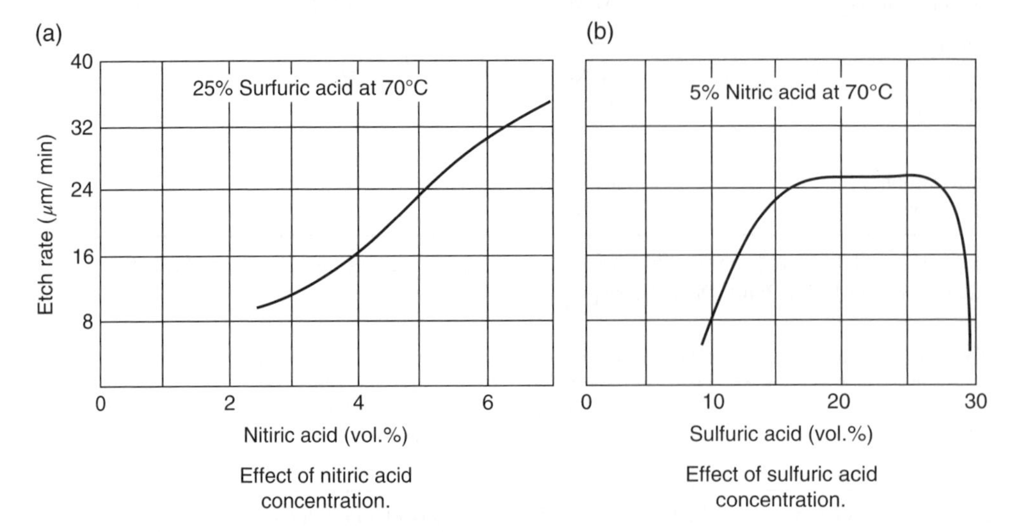

Figure 8.15 Effect of acid concentrations on etch rate in CH-milling: (a) effect of nitric acid concentration; (b) effect of sulfuric acid concentration. (Compiled from Metcut.)

depths. Deep cuts yield tolerances up to ±75 μm (Figure 8.16 and Figure 8.17). Minimum width of cut should be twice the depth plus the EF [1].

8.2.5.2 Surface Quality

The gentle chemical action of CH-Milling does not introduce stress into the workpiece, since it removes the material molecule-by-molecule, resulting in surfaces free from residual stresses. Surface roughness is influenced by the initial workpiece roughness condition. Too violent agitation can lead to uneven cut or grooving. Tables 8.11 and 8.12 show the surface roughness of stainless steels and super alloys attained by CH-Milling, respectively.

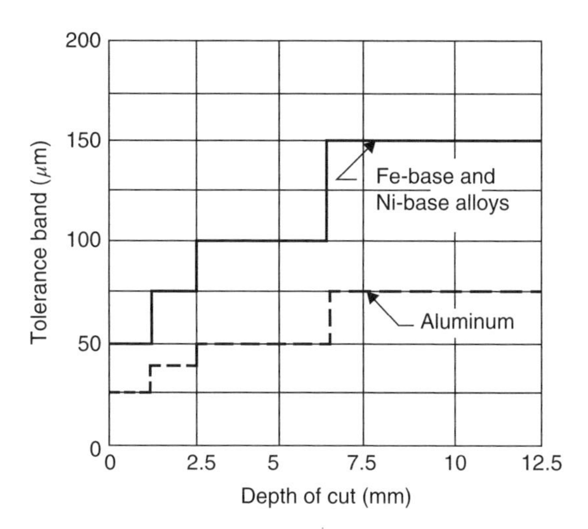

Figure 8.16 Achieved tolerances for different depths of cut in CH-milling of super alloys and aluminum – a comparison. (Adapted from Metcut.)

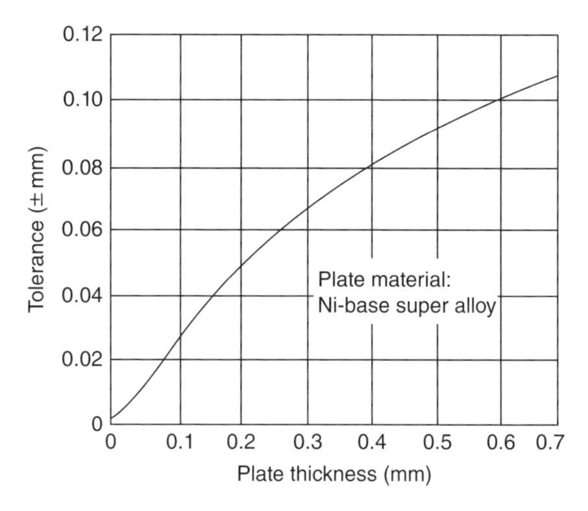

Figure 8.17 Achieved dimensional tolerances for thin plates of Ni-base super alloys, machined by CH-milling. (Adapted from Metcut.)

8.2.6 Photochemical Machining (Spray Etching)

Most of the commonly used industrial metals and alloys can be etched with the photochemical machining (PCM) process. More exotic alloys such as stainless steels and super alloys can be PCMed, but these require more sophisticated chemistry and operator knowledge. The etchability of a metal or alloy mainly depends on its chemical composition. Table 8.13 lists the etchability ratings of some stainless steels and super alloys, machined by PCM. Copper, brass, aluminum, and magnesium have good rating. Stainless steels and super alloys such as Inconels and Hastelloy B have good to fair rating. Udimet alloys have fair to poor rating. Hastelloy C and René 41 along with W, Ti, Nb, and Ta have poor etchability rating.

Table 8.13 Etchability ratings of some selected super alloys and stainless steels machined by PCM

Etchability rating	Super alloys and stainless steel grades (etchability in descending order)
Good	Copper, brass, aluminum, magnesium
Good to fair	AISI 215, 301, 302, 304, 305, 316, 321, 347
	PH 15-7, PH 17-7
	AISI 410, 420, 430
	Inconel alloys (e.g., Ni, 15% Cr, 7% Fe)
	Hastelloy B (Ni, 28% Mo, 5% Fe, 2.5% Co, 1% Cr, 0.5%V, 0.05%C)
Fair to poor	Udimet alloys (e.g., Ni, 42% Fe, 12.5% Cr, 2.7 % Ti)
Poor	Hastelloy C (Ni, 15% Mo, 14% Cr, 5% Fe, 3% W, 2.5% Co, 0,08%C)
	René 41 (Ni, 19% Cr, 11% Co, 10% Mo, 3%Ti, 1.5% Al)

Adapted from Machinability Data Center [1].

Table 8.14 Etchants for PCM of plates made from some selected super alloys and stainless steels

Plate material	Etchant formulation	Temperature (°C)
Super alloys		
HyMu 80, 800 (80% Ni, 4% Mo, Fe)	42 Bé, $FeCl_3$:HCl (9 : 1)	43–49
Inconels (Ni, Cr, Fe)	42 Bé, $FeCl_3$	54
Nimonics (about 80%Ni, 20% Cr)	42 Bé, $FeCl_3$:HNO_3:HCl	49
Stainless steels		
Mo-free stainless steel	35–48 Bé, $FeCl_3$	35–55
Mo-stainless steel	36–42 Bé, $FeCl_3$ with HNO_3-addition	35–55

Adapted from Metcut.

Etchants for PCM of stainless steels and super alloys are listed in Table 8.14, together with their operating characteristics. Ferric chloride ($FeCl_3$) solutions are used for PCM of a wide variety of metals and alloys and therefore it became the most widely used etchant in the PCM industry. Etchants are usually restricted to the less dangerous ones, namely ferric chloride, often modified with additives such as diluted mineral acids, and some alkaline etchants based on sodium hydroxide or ammonium salts. Sodium hydroxide is extensively used with Al and Al-alloys. Etchant compositions can be adjusted to meet the requirements of specific applications, and proprietary additives can be included to control foaming or wetting characteristics, increase or decrease etching rate, or make etching more uniform.

Etching machines are made from materials (such as polyvinyl chlorides and titanium) that can withstand corrosion from ferric chloride and other etchants. Etchant temperature must be maintained below 55 °C to avoid distortion of plastics used in machine construction.

Visser *et al.* [30] investigated the effect spray jet pressure on the etching speed and the etchant concentration on the surface roughness. Figure 8.18 illustrates the effect of jet pressure and temperature of sprayed etchant (acidic solution of concentration of 3.8 mol/l) on the etching speed, when machining martensitic stainless steel AISI-420, whereas, Figure 8.19 visualizes the effect of etchant concentration on the surface roughness of austenitic alloy AISI-304.

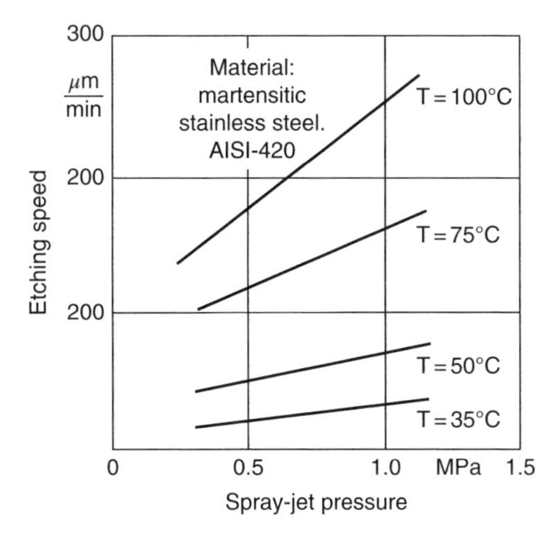

Figure 8.18 Effect of jet pressure and temperature in spray etching on etching rate of martensitic stainless AISI-420. (From Visser *et al.* [30], with permission.)

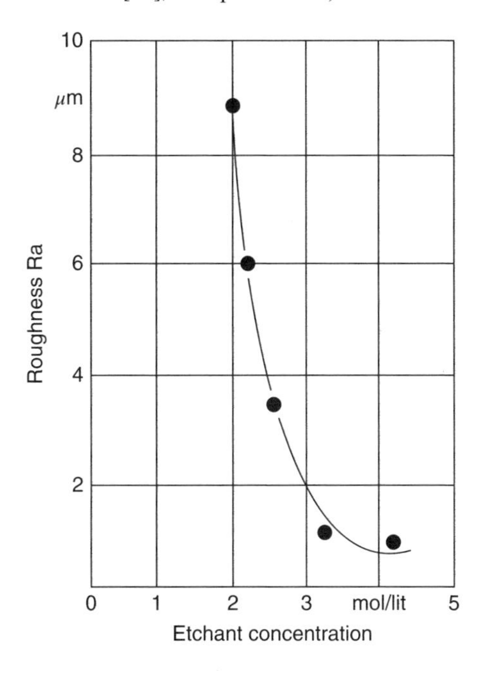

Figure 8.19 Effect of etchant concentration in spray etching on the surface roughness of austenitic stainless steel AISI-304. (From Visser *et al.* [30], with permission.)

8.3 Thermoelectric Machining Processes

8.3.1 Electric Discharge Machining (EDM)

Applications of EDM in turbo-machinery manufacture whether for aircraft jet engines or industrial land based gas turbines used for power generation have not changed significantly in the past 40 years. However, even today, despite the developments, that have taken place in generator technology, the thermal nature of the EDM process and the resulting workpiece damage, with consequent effects relating to fatigue life and performance, have stifled EDM expansion. Additionally, the development and take-up of laser systems for rapid hole drilling (initially partly dismissed due to shortcomings in accuracy and hole quality) have limited/restricted EDM utilization. Despite this, EDM provides better regulation of breakthrough detection/depth control and higher achievable aspect ratios compared to laser systems [31].

The reasons for choosing EDM over traditional processes are that productivity is not limited by the hardness or strength of the workpiece and complex features, or high aspect ratios of holes and cavities, can be readily machined. The turbo-machinery materials therefore specifically machined by EDM consist of the super-alloys Inconel 738, Inconel 939, CMSX4, MAR-M002, MAR-M247, Udimet 720, Nimonic 105, Nimonic 713, and so on. The main areas for application include the drilling of cooling holes and die sinking of slots, pockets, and grooves together with some currently limited wire cutting operations [10].

As stated before in CH_7, the two fundamental configurations that have essentially defined the development of EDM generator technology are relaxation and transistor-based pulse generators. Relaxation generators were the first introduced by Lazarenkos and remained popular due to their simplicity and ability to produce both high and very low pulse energies and discharge durations, making them suitable for roughing and finishing operations as well as accurate precision machining. Conversely, transistor-based systems offered the advantage of programmable pulse shape and greater flexibility in terms of peak current, pulse width, and current ignition slope settings, enabling substantial benefits in terms of increased MRR, reduced recast layer thickness/HAZ depth, and better workpiece surface quality [32].

Historically, transistor-based generators originated from the evolution of relaxation type systems towards controlled-pulse generators to produce high-frequency pulses and control the relaxation-discharge energy independently. For die-sinking EDM, discharge durations are longer than with wire electric discharge machining (WEDM) which can favor lower electrode wear, but the same principle can be applied to maintain a given current level by alternating current delivery from the power source at high frequency. For finishing, relaxation generators are commonly employed, as the goal is to minimize both the peak current and duration of machining sparks. Such high frequency, low energy discharges can also be produced, however, by controlled-pulse circuits using only transistors, the pulse energy being defined by the gap capacitance [10].

Modern transistor-based generators use high power transistors and ultra-fast recovery diodes for generating peak currents up to 1000 A with durations in the range of microseconds. In order to achieve such performance, state-of-the-art machines have very low line and machining zone inductances below 0.5 μH, allowing rising current slopes up to 600 A/μs. The transistor-based circuits can also be designed to produce trapezoidal pulses for increasing the pulse energy for roughing operations. However, triangular pulses with extremely short durations are often preferred in EDM of aerospace components as they minimize the heat transferred to the work-piece and produce high integrity surfaces devoid of cracks and with reduced tensile stress after finishing operations. For applications involving intermediate roughness (between 0.15 and 0.8 mm Ra), similar results can be achieved with modern capacitor-based generators by exploiting the line to the machining zone as a source of capacitance and as a means for minimizing the circuit inductance, thus achieving very high ratios of peak current to pulse width [33]. For polishing operations and micro-machining, very high speed and recovery diodes are used for achieving pulse-widths of the order of 30 ns at frequencies of the order of 10 MHz, with peak current values around 1 A. Under such conditions, surface roughness down to 0.03 µm Ra can be achieved in tungsten carbides and 0.08 µm Ra in steel, with almost zero white layer thicknesses [34].

Other key features of modern generators relate to their ability to tackle corrosion problems by introducing alternative ignition voltages such that the mean value during operation is nil (Figure 8.20). Such anti-electrolysis configurations are currently used for the machining of dedicated turbo-machinery alloys.

Commercial duplex/combination systems allowing for example the option for EDM drilling and laser ablation in a single machine already exist and are intended for applications such as HP turbine blade and vane machining where laser ablation can be used for removing any thermal barrier coating prior to ED (electric discharge)-drilling parent material [35].

In other commercial EDM drilling systems, the combination of an individual tube electrode with a holder and positioning guide in a single assembly enables electrode rotation up to 1000 rpm providing improved flushing and removal of debris [36].

Electrode materials include copper and graphite (≤10 µm grain size) with the dielectric fluid at present principally hydrocarbon oil (synthetic or mineral), although it is understood that the associated environmental issues are a concern.

As basic result it can be concluded that, in contrast to conventional cutting operations, Ni-based alloys generally possess higher maximum MRR during sinking electric discharge machining (SEDM) compared to Ti-based alloys.

Improved MRR has been reported with a number of hybrid EDM processes, the most significant employing continuous arcing or a combination of controlled arcing and discharges, as a consequence of the higher energy densities that are possible compared with spark discharges alone. Workpiece accuracy and roughness are less controllable than for EDM and machined surfaces are subject to greater thermal degradation which can encompass re-deposited material, cracking, and a recast layer (typically 50–100 µm but can be greater depending on arcing energy and electrolyte/fluid used), with consequent changes to workpiece microstructure and micro-hardness.

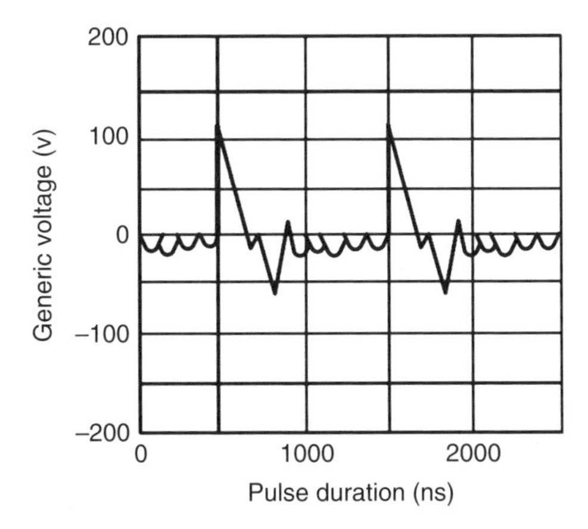

Figure 8.20 Principle of generic high frequency anti-electrolysis sub-microsecond pulse voltage profile. (Courtesy of GF Machine Solutions.)

Commercial machines utilizing arcing to rapidly section large forging blanks, castings, extrusion products, bar stock, and so on, in the range of difficult-to-cut materials including stainless steels, nickel, and titanium alloys, appeared during the 1960s both in the Soviet Union and USA and subsequently in Japan [37]. In contrast to EDM development and growth however, their use was not mainstream or widespread, the main focus of application during the late 1980s and 1990s appearing to be in the decommissioning of atomic reactor pressure vessels, where any workpiece damage (typically recast layer material and cracking) and relative inaccuracy resulting from the thermal erosion process, was of little concern [10].

When machining stainless steels with EDM, Cu-electrodes are generally used with reversed polarity, and kerosene as dielectric fluid. In the case of pulse generators being used, the recommended peak currents are ranging from 2 to 12A, pulse duration 50–200 μs, pulse-off 50–200 μs, and the duty factor 0.5. According to Rahman *et al.* [38] in their investigations have found that when EDM of stainless steel 304, the MRR, tool wear, and surface roughness increase with increasing pulse-on time, and peak current. The finest surface finish can be achieved when low peak current at long pulse-on time are secured. When machining stainless steel, minimum wear of electrode is realized if long pulse duration is provided. Data indicated that the fatigue life of SS-alloys such as 304 and 410, machined by EDM, can be significantly reduced compared to traditionally machined stainless steels [39].

With productivity as a key driver, the move to develop specialized EDM equipment in addition to adapting and optimizing standard systems for the manufacture of turbo-machinery components is likely to continue. There is little evidence at present of operating modes such as ED-milling, which is able to utilize simple electrodes, being used in practice [4].

8.3.2 Electrical Discharge Milling of SSs and SAs

The electric discharge sinking (SEDM), discussed in the previous section, requires a preliminary phase for producing specially shaped electrodes. These electrodes are very expensive, as they are difficult to design and manufacture, and therefore they add more than 50% to the total machining cost of the product. Recently, a revolutionary breakthrough in the EDM realm has been achieved through a new ED-Milling technology that makes use of simple and cheap standard rotating pipe electrodes. In this process, three-dimensional cavities are machined by successive sweeps of the electrode down to the desired depth, while the NC automatically compensates by means of powerful algorithms the electrode's front wear to ensure product accuracy along the three axes. Therefore there is no need to manufacture specially shaped electrodes as in the case of SEDM, which means saving of time and money. The theory of ED-Milling (also termed electric discharge scanning) is shown in Figure 8.21. The thickness of the layer removed per path ranges from 0.1 mm to several millimeters on rough paths and from 1 to 100 μm on finish paths.

8.3.2.1 Fields of Applications of ED-Milling

ED-Milling technology is particularly applicable for machining cavities with or without taper, including three-dimensional shapes. It is used notably for making molds of parts for electrical and electronic industries, household appliances, and the automotive and aeronautical components made of stainless steels and super alloys. Another technological

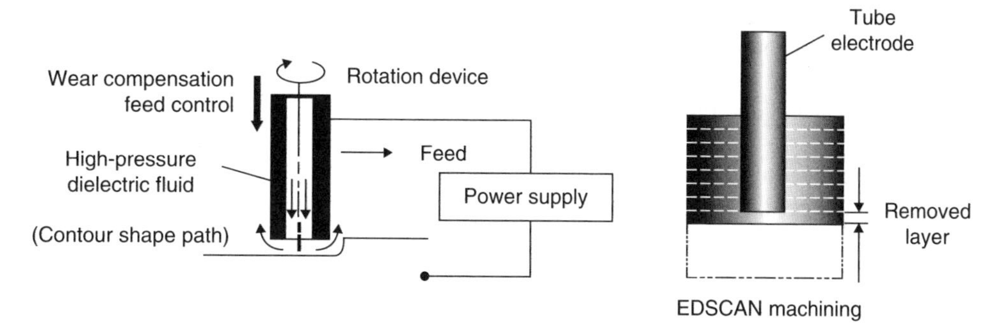

Figure 8.21 Theory of ED-milling. (Courtesy of Mitsubishi EDSCAN Technical Data, 1997.)

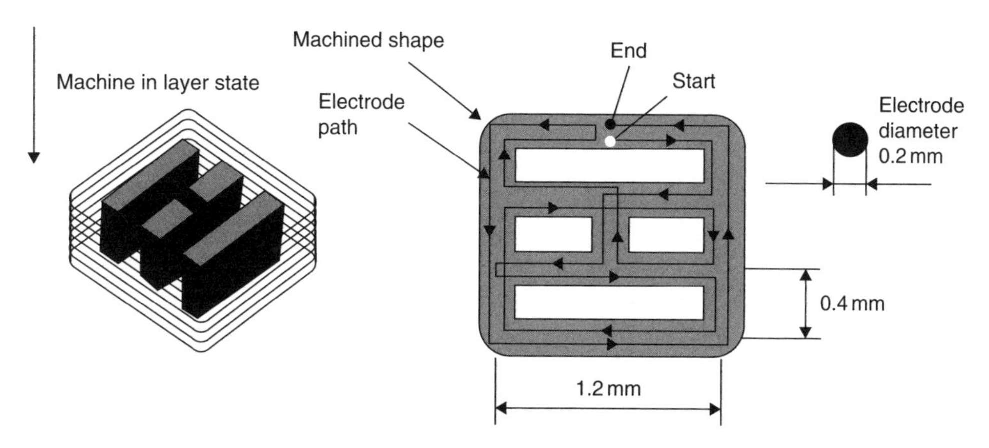

Figure 8.22 Micro-EDM-milling. (Courtesy of Mitsubishi EDSCAN Technical Data, 1997.)

breakthrough of ED-Milling is that the process has entered the domain micromachining, where it is possible to produce fine and intricate shapes with sharp corners. The path of the two-dimensional shape is created by the NC system built into the machine beforehand to allow the target shape to be stored. Layered machining is then carried out by executing the NC path program several times, until the required depth is achieved, as shown in Figure 8.22. ED-Milling technology was announced by Mitsubishi and Charmilles at Hannover Exhibition in October 1996. Since that time, this technology has gained attention in applications such as fabrication of micro-dies and aeronautical industries.

8.3.2.2 Advantages and Limitations of ED-Milling

The advantages include the following:

- Design and manufacture of electrodes is totally omitted.
- Fine shapes can be readily produced.
- Electrode wear does not need to be considered.

- NC data can be directly generated from the EDM part data.
- Sharp edges and corners can be readily produced due to the excessive frontal wear.

The disadvantages and limitations of ED-milling are

- The removal rate may be less than that achieved by SEDM.
- If there is a large side taper (10° or more), it is difficult to maintain side accuracy.

8.3.3 Electron Beam Machining

Operating parameters for drilling holes and cutting slots in different alloys of stainless steels, and super alloys are shown in Tables 8.15 and 8.16, respectively.

Cylindrical, conical, and barrel-shaped holes of various diameters can be drilled with consistent accuracy at rates of several thousand of holes per second. EB-drilled holes in super alloy turbine blade at angles of 60–90° to profile chord can be easily machined. Hole of inclination angle of 15° are possible. The largest diameter and depth of holes that can be accurately drilled by electron beam (EB) are 1.5 and 10 mm, respectively and the aspect ratio is typically 1:1 to 15:1 [40]. Rectangle slots of 0.2 mm×6 mm in 1.6 mm thick SSs plate are produced in 5 minutes using 140 kV, 120 µA, pulse duration of 80 µs, and frequency of 50 Hz. The traverse speed is inversely proportional to the work thickness (Table 8.16).

Figure 8.23 shows a proportional relationship between the volume of produced cavity in stainless 304 and the pulse charge Li.

8.3.4 Laser Beam Machining

Approximately 5% of all industrial laser material processing applications are laser drilling operations [42]. In this context, the generation of cooling holes in gas turbines for aircraft as well as for power plants is one of the most important, established drilling applications.

Table 8.15 Parameters for drilling holes in stainless steels and super alloys with EBM

WP-thickness (mm)	Hole diameter (mm)	Drilling time (s)	Acceleration voltage (kV)	Average beam current (µA)	Pulse width (µs)	Pulse frequency (Hz)
Ferritic and martensitic stainless steel alloys						
0.25	0.013	<1	130	60	4	3000
Other stainless alloys (austenitic)						
1.0	0.13	<1	140	100	80	50
2.0	0.13	10	140	100	80	50
2.5	0.13	10	140	100	80	50
6.4	0.5–1.0	180	145	4000	2100	12.5
Hastelloy (Ni-base alloy)						
10	2.5	70	130	5000	5300	100

Adapted from Metcut.

Table 8.16 Parameters for cutting slots in stainless steels with EBM

WP-thickness (mm)	Slot width (mm)	Rate of cut (mm/min)	Acceleration voltage (kV)	Average beam current (μA)	Pulse width (μs)	Pulse frequency (Hz)
0.05	0.05	100	130	20	4	50
0.18	0.10	50	130	50	80	50
1.57	0.2	1.25	140	120	80	50

Adapted from Metcut [1].

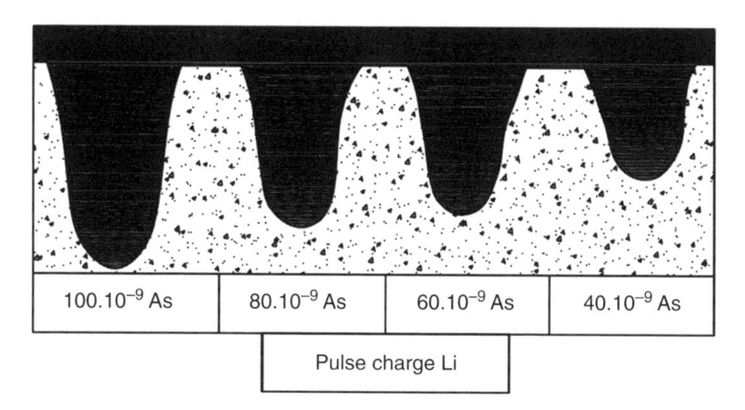

Figure 8.23 Proportional relationship between the volume of produced cavity in stainless 304 and the pulse charge Li [41].

The steady progression of laser-based system technology (e.g., laser sources, data preparation software, machine control, positioning system, sensor devices, etc.), and the development of novel laser drilling strategies, offering design freedom together with its cost effectiveness has increased significantly during the last decade. It is possible to drill hundreds/thousands of cooling holes with high precision and of variable diameter and shape in multi-material blades of complex geometry (Figure 8.24).

One of the main challenges is drilling through multilayer material systems composed of metal coatings with at least one ceramic wear protection layer without enhanced formation of microcracks and thermal induced damage causing a removal of the coating layers. Furthermore, to achieve significant cooling performance the cooling holes are generally tapered or shaped [43]. While EDM has comparatively lower machine tool costs, parallelized machining capability [27], and better process control in terms of reduced HAZs, it is limited to the machining of electrically conductive materials and therefore cannot be applied for ceramic-coated turbine blades.

Tables 8.17 and 8.18 provide the machining parameters and lasers for drilling and slotting operations performed in various stainless steels. It is predicted that nonsatisfactory results are achieved when cutting austenitic alloys with laser due to less fluidity of the molten metal as compared to other SS-alloys (Table 8.18). Also, it can be predicted that higher cutting rates are realized when O_2 is used as assisting gas (Tables 8.17 and 8.18). The traverse cutting speed v_t

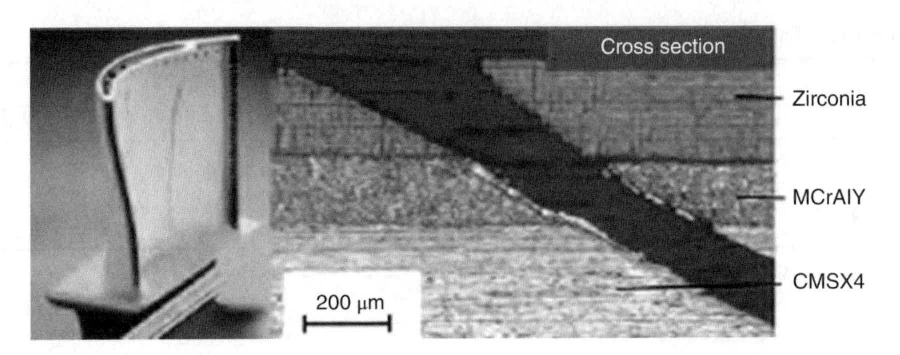

Figure 8.24 Laser drilling of cooling holes: nickel-based turbine blade coated with a ceramic wear resistance layer with cooling holes and cross-section of a cooling hole drilled in a CMSX-4 turbine blade coated with a MCrAlY and zirconia wear resistance layer. (Compiled from Beck [42] and Horn *et al.* [43].)

Table 8.17 Parameters for drilling 0.12 mm diameter holes in SS 304 using ND-YAG (1.06 μm) laser

WP-thickness (mm)	Assisting gas	Drilling time (s)	Lamp current (A)	Average power (W)	Maximum thickness (mm)
3.0	Oxygen	88	34	31	4.8
3.0	Argon	221	34	31	4.8

Source: Metcut [1].

Table 8.18 Parameters for gas-assisted CO_2 (10.6 μm) laser cutting various types of SS

Plate thickness (mm)	Assisting gas	Cutting rate (m/min)
250 W laser		
2.3	Oxygen	0.76
500 W laser		
0.3	Oxygen	3.71
1.0	Oxygen	1.65
3.2	Oxygen	0.89
1000 W laser		
0.5	Oxygen	19.00
0.8	Oxygen	16.50
1.6	Oxygen	11.40
3.2	Oxygen	5.08
1250 W laser		
1.0 (austenitic)	Oxygen	8.89
3.2 (austenitic)	Oxygen	3.05
3.2 (austenitic)	Air	1.52
5.2 (ferritic/martensitic)	Oxygen	1.78

Source: Metcut, Saunders [44], Compiled.

(up to 20 m/min for gas-assisted laser cutting of stainless steel) is a measure of machining efficiency. It depends on laser power P and plate thickness t (Table 8.18). These are correlated for a certain material by the following equation.

$$v_t \alpha \frac{P}{t}$$

In general pulsed laser systems are applied for laser drilling processes, where selection of the pulse duration depends on the hole characteristics and the material being processed. In the field of laser drilling of turbine components, pulse duration is normally of the order of nano- or milliseconds. The required pulse energy basically depends on the exact chemical composition of the material, the material thickness, and the desired hole diameter and shape.

Advanced designs of Nd:YAG lasers have cut Ni-base super alloys up to 50 mm thick at speeds greater than EDM. Table 8.19 illustrates the PD-time (s) when laser beam machining (LBM) of Inconel 718 plates of different thicknesses using Nd:YAG of pulse energy of 10 J/pulse at three different power levels. It is depicted from this table that the drilling time of thin sheets (up to 2.5 mm) does not depend on the power level. Therefore, small power levels are highly recommended for drilling thin plates, and that is the same reason why LBM is preferred for drilling and slotting of small sheets. Table 8.19 shows the PD-time (s) when LBM of Inconel 718 plates of different thicknesses using Nd:YAG of pulse energy of 10 J/pulse at three different power levels.

Different physical mechanisms take place inside the irradiated material volume depending on laser pulse duration and laser pulse peak intensity. The dominating effects causing material removal are melting and vaporization. Assuming a temporal and spatial Gaussian intensity distribution of the incoming pulsed laser beam, vaporization occurs in the hole-centre and melting in surrounding material sections. For a laser peak intensity of approximately 10^6 W/cm^2 and pulse durations of the order of several milliseconds, melting is the dominating effect being responsible for material removal and hole formation. With a further increase of laser peak intensity greater than 10^6 W/cm^2, sublimation drilling occurs. In this case, the dominating effect causing material removal is ablation by vaporizing plasma formation. In this context, the applied peak intensity has to exceed a material-dependent threshold value. To achieve such high intensities, shorter laser pulses are required. A system suitable for sublimation-drilling is a Q-switched Nd:YAG laser with pulse durations of the order of 10–100 ns. A detailed explanation of plasma formation and resulting material removal can be found in the literature [4, 45, 46].

Table 8.19 Percussion drilling time in seconds of Inconel 718 of different plate thicknesses using Nd:YAG laser of pulse energy of 10 J/pulse at various power levels

Nd:YAG laser of average power level (W)	Drilling time in seconds for plate thickness (s) in millimeters of Inconel 718					
	s = 2.5 mm	s = 5 mm	s = 10 mm	s = 15 mm	s = 20 mm	s = 25 mm
75	0.5	3	15	37	65	95
150	0.5	2	5	17	33	55
250	0.5	1	4	12	23	38

Adapted from Metcut [1].

Laser Drilling for Cooling Holes of Turbine Blades: Industrially established laser drilling operations include single-pulse drilling, percussion drilling (PD), trepanning, and helical drilling. For cooling holes in turbine blades the relevant drilling operations are percussion drilling and trepanning [42]. The principles of trepanning and percussion drilling are schematically shown in Figure 8.24. A detailed theoretical description of the drilling technologies is presented by Poprawe [46] and Majundar [45]. During the drilling process, an inert gas stream protects the focusing optics. The gas stream is also employed to assist material removal and to prevent oxidation and necking of the holes by melting deposits from ablated material.

During percussion drilling the laser spot is stationary at the same position on the workpiece in contrast to the trepanning operation. The diameters of the approximately cylindrical holes are commonly of the order 0.5–0.7 mm and the achievable aspect ratio is in the best case 1:20 [42]. An important point is the exact positioning of the focus plane relative to the workpiece surface, which is a function of workpiece material. The optimal position of the focus spot being located approximately 5–15% of the workpiece thickness under the workpiece surface [42]. In practice the best setting for a specific problem will be empirically identified by analyzing hole quality in terms of geometry, distribution of cracks and the amount of ablated material recombining at the hole edges and on the workpiece surface [47].

When trepanning, the laser beam is rotated relative to the workpiece, whereby the laser spot diameter is distinctly smaller than the diameter of the hole. A hole is generated by removing a cylindrical core during one circulation of the focused laser beam. The principle of a rotating laser spot provides the opportunity to generate holes with high reproducibility and high flexibility in terms of the hole design. Furthermore hole shape can be noncircular. In comparison to percussion drilling, trepanning is more time-consuming and the heat input into the turbine blade is larger.

As previously mentioned in Chapter 7, the carbon content of the work material affects the surface quality of deep holes produced by PD. It is enhanced with increased carbon contents for the previously given reasons. Therefore, a bad surface quality is expected when deep drilling stainless steels, and super alloys, since these alloys generally contain low carbon contents [4].

8.3.5 Plasma Arc Cutting

This process is characterized by its highest cutting rate and lowest specific cutting energy when cutting stainless and exotic materials as compared to other nontraditional machining processes (NTMPs). The gas used for the arc (primary gas) may be N_2, H_2, Ar, or various admixtures. Compressed air may also be used to increase cutting rates, depending on the thickness of SS plate, as shown in Table 8.20. However, using air as primary gas calls for oxidizing the cut surfaces.

For stainless steels, CO_2 may be used with N_2 as a primary gas (Tables 8.21 and 8.22). Water curtain may be used in place of the shielding gas or may be injected into the plasma stream to produce a cleaner cut with a reduced bevel, and narrow kerf, but without improvement of the cutting rate. However, nozzle life can be improved by cooling action of the water. Table 8.23 shows also the cutting speeds for Al, and carbon steel. It can be depicted from this table, that when using plasma arc cutting (PAC), Al has the best machinability rating, followed by stainless steels, then carbon steel

Table 8.20 Comparison of PAC cutting rates of SSs using Ar/H_2 or air as primary gases

Thickness (mm)	Cutting rate (m/min)		Remarks
	Ar/H_2	Air	
5	5.0	5.0	
10	2.6	3.4	
15	1.5	1.8	Increased cutting rates if air is used
20	1.1	1.2	
25	0.8	0.85	
30	0.65	0.6	
35	0.5	0.4	
40	0.4	0.3	
45	0.35	0.25	
50	0.3	0.2	
60	0.3	0.2	

Adapted from: Holden [48].

8.4 Economical Analysis of ECM and Thermo-electrical Processes of Turbo-machinery Components

The technical capabilities and areas of application of electro-chemical, electro-physical, and photonic processes have been have been previously analyzed showing the broad potential of ECM, EDM, and laser material removal for the manufacture of turbo-machinery components. Clear advantages have been identified for their use when machining advanced and difficult-to-cut materials, including high removal rates, superior geometrical precision, and acceptable surface integrity [10].

For efficient turbo-machinery component manufacture, an economic analysis of individual process technology alternatives has been investigated by Klocke *et al.* [50]. Machining processes with low MRRs but relatively of low machine running costs such as EDM must be evaluated in such a way in order to be competitive against other traditional or nontraditional manufacturing technologies with the objective to minimize the production costs. Alternatively, ECM and PECM (with high MRR) could be employed for the whole process, as they are capable of roughing, finishing, and polishing with the same base technology. While all other process technologies need separate machine tools, ECM and PECM can be realized in one set-up. Conversely, such machine tools together with the process and tool electrode design are complex, and therefore more expensive. In order to identify an economic alternative, an appropriate cost model considering all relevant boundary conditions has to be executed and constantly updated when key parameters are changing.

Such an economical comparison has been carried out for different roughing strategies for blisk gap slotting from solid via multi-axis milling, SEDM and ECM for titanium- and nickel-based blisks of a characteristic geometry and technological key parameters for a certain batch size. MRRs have the greatest influence one overall manufacturing costs, as they affect the direct process time and are thus responsible for machine hourly rates and wage costs. In contrast, tooling costs have limited influence although they have to be taken into account. To allow a comparison, other manufacturing parameters appropriate to blisk manufacture have to be kept constant [10].

Table 8.21 Cutting rate and current selection of PAC of SSs

Plate thickness (mm)	Cutting speed (m/min)	Power selection
		Amperage (A)
6	1.78	105
	2.54	140
13	0.51	135
	1.02	190
	2.54	270
	3.81	700
25	0.51	210
	0.76	270
	2.03	540
	2.79	1000
38	0.25	280
	0.51	420
	1.02	620
	1.78	1000
51	0.13	320
	0.25	610
	1.02	950
64	0.13	410
	0.25	550
	0.51	820
76	0.13	510
	0.25	675
	0.51	1020
89	0.25	730
	0.51	1110
102	0.13	675
	0.25	900
114	0.13	900
127	0.076	1100
140	0.076	1100

Adapted from Bagley [49], cited in *Machining Data Handbook*, Vol. 2, Metcut, 1980, p. 12.99–12.100.

For Inconel 718 the choice of the most economical roughing process is more difficult. With low machine tool investment costs and relatively high average MRRs, SEDM especially for batch sizes up to 400 blisks per year, is a viable alternative. In the case of larger batch sizes, ECM is the most cost-effective technology. Batch sizes of 100–300 require a more detailed analysis. For a batch size of 200, milling, SEDM and ECM with low tooling outlays reach the same cost level. With higher numbers of machine tools, the roughing costs per blisk decrease. This effect is due to the increase in capacity utilization for each machine tool so that single investment costs are normalized [50].

During application of ECM for example, no additional de-burring operation is required. On the other hand, additional washing operations may be necessary. Energy and recycling costs

Table 8.22 Cutting speeds and machining conditions of PAC for SSs

Machining conditions	Plate thickness (mm)	Cutting speed (m/min)
		Best/maximum
Amperage selection: 100 A	6	1.25/2.54
Primary gas N_2 (1.55 m^3/h, 2.07 bar)	13	0.51/0.76
Secondary gas CO_2 (5.8 m^3/h, 2.75 bar)	25	0.23/0.28
Amperage selection: 200 A	6	1.65/3.43
Primary gas N_2 (1.95 m^3/h, 2.07 bar)	13	1.27/1.78
Secondary gas CO_2 (5.8 m^3/h, 2.75 bar)	25	0.51/0.66
	38	0.30/0.40
Amperage selection: 400 A	13	1.91/3.05
Primary gas N_2 (1.4 m^3/h, 1.40 bar)	25	1.02/1.40
Secondary gas CO_2 (5.8 m^3/h, 2.75 bar)	38	0.64/0.97
	64	0.30/0.38
	76	0.20/0.25

Courtesy of Thermal-Dynamic Corp., cited in *Machining Data Handbook*, Vol. 2, Metcut, 1980, p. 12.102.

Table 8.23 Cutting speeds and machining conditions of water-injection PAC for SS, Al, and carbon steel

Machining conditions	Plate thickness (mm)	Cutting speed (m/min)		
		Carbon steel	SS	Al
		Best/maximum	Best/maximum	Best/maximum
Amperage selection: 300 A	6	1.5/2.8	1.9/3.3	2.2/3.7
Primary gas N_2 (2.1 m^3/h, 2.07 bar)	13	1.0/1.5	1.3/1.8	1.4/2.0
Water injection at	25	0.5/0.6	0.6/0.9	1.0/1.5
30–60 l/h	38	0.3/0.4	0.4/0.5	0.5/0.6

Courtesy of Thermal-Dynamic Corp., cited in *Machining Data Handbook*, Vol. 2, Metcut, 1980, p. 12.103 [1].

are similarly gaining importance and therefore have to be considered. Recycling of contaminated chips from traditional cutting operations will also need to be critically evaluated. The residual tensile stresses from EDM material removal could be neutralized by subsequent surface finishing operations such as etching, shot peening, or AFM. Finally, by using a nonmechanical material removal process, induced forces become less significant, allowing the machining and therefore the design of more filigree and complex geometries, which to date have not been possible by utilizing conventional means [10].

8.5 Nontraditional Micro-drilling of Deep Holes – a Comparison

The performance of modern air engines depends on a very large number of small holes for cooling hot components generally made of super alloys such as turbines and combustion chambers. A modern air turbine rotor and stator assembly may have more than 20 000 small cooling holes, of typical diameters and aspect ratios in the range of 1–4 and 20–200 mm, respectively. The rotor blades are subjected to high stress and vibration so that the form and surface finish may affect the fatigue strength. Turbine blades are subjected to very high temperatures, and they are made of difficult-to-machine super alloys. Consequently, deep cooling holes are usually drilled by NTM-techniques [51]. Table 8.24 compares the applicable EC and thermal nontraditional methods for producing micro deep cooling holes. It is depicted from this table that STEM appears to be the preferred technique in terms of:

1. required physical dimensions of the hole;
2. large drilling depth and aspect ratio;
3. surface quality and integrity;
4. relatively lower power consumption.

Figure 8.25 compares also the above competing methods for drilling holes, from which it is depicted that LBM and electron beam machining (EBM) are competing as regarding the machining speed, accuracy, and diameter ranges. Both LBM and EBM cannot be effectively used for thick materials or deep holes. However, it should be considered that laser beam (LB) does not necessitate a vacuum as EB [4]. EDM cannot produce cooling holes of high aspect ratios. Moreover it produces holes of poor surface finish and integrity.

ES has the disadvantage of producing holes of lower aspect ratio and smaller depths as compared to STEM. It has significantly higher operating voltage, that necessitates higher consumption. Furthermore, in ES breakage of capillaries during deep drilling creates an additional problem.

8.6 Thermally-Assisted Machining of Stainless Steels and Super Alloys

Hard-to-cut materials and super alloys are all candidates for TAM. Generally, work materials with hardness ranging from 40 to 70 RC should be considered for TAM. Poor thermal conductivity of the work material can reduce the heat loss between the time of heat application

Table 8.24 Characteristics of nontraditional micro drilling methods of deep holes – a comparison

Parameters	NTM-techniques				
	STEM	ES	EDM	LBM	EBM
Hole diameter (mm)	0.75–2.5	0.12–0.87	0.12–6.5	0.12–1.2	0.03–1
Hole depth (mm)	125	20	3	5	2.5
Aspect ratio	300 : 1	40 : 1	10 : 1	16 : 1	6 : 1
Cutting rate (μ/s)	25	25	12	<1000	250
Operating voltage (V)	5–25	150–850	30–100	—	150 kV
Surface roughness (Ra) (μm)	0.8–3	0.3–1.5	1.5–3	0.8–6	0.8–3
Surface integrity	No HAZ	No HAZ	HAZ	HAZ	HAZ

Adapted from Sharma *et al.* [51].

	Achieved accuracy	Machining speed
EDM	10 μm	
Mechanical drilling	> 20 μm	
STEM	0.1 d	
LBM	(0.05–0.2) d	
EBM	(0.05–0.1) d	

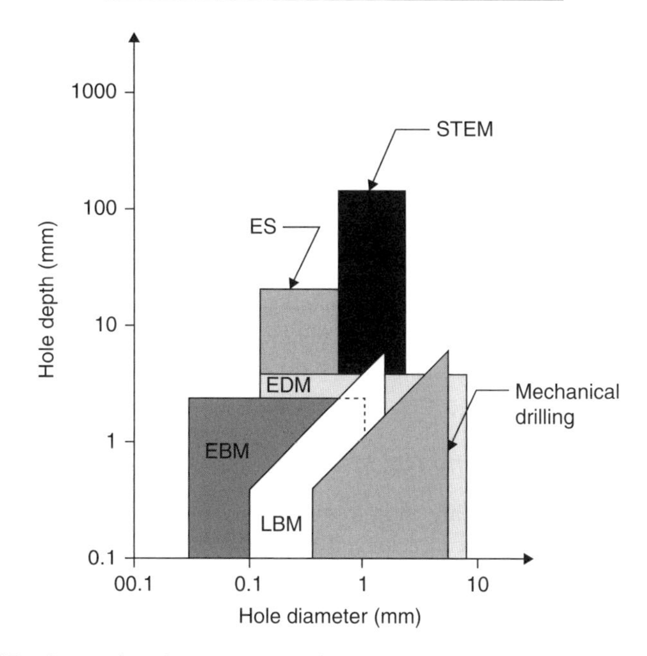

Figure 8.25 Comparison between nontraditional drilling operations of micro deep holes.

and the time of cutting. Al, Cu, and Ti with their relatively high thermal conductivities are not good candidates for TAM. Most applications of TAM have been for turning; however, experimental work has been carried out on milling and slotting. Cuts of long duration of difficult-to-cut materials such as stainless steels and super alloys have the greatest economical benefit when using TAM.

8.6.1 Surface Integrity and Removal Rates for TAM of Stainless Steels and Super Alloys

In TAM, the heat source parameters are added to conventional turning parameters. The stand-off-distance (SOD) and the angular position of the heat source should be considered. Moreover, a careful balance between heat input, depth of cut, and speed are needed for adequate, but not excessive temperatures are achieved at the necessary depth of cut [Metcut]. When the heat input and cutting rates are properly balanced, the work material below the cut is usually not heated enough to alter the metallurgical micro structure. Consequently, the

Table 8.25 Comparison of recommend speeds by conventional turning and TAM of super alloys and austenitic stainless steels

Work material	Cutting speed (m/min)		Speed ratio
	Conventional turning	TAM	$V_{TAM}:V_{Conv.}$
Nimonic 115	11	150	14
Stellite (soft)	6–15	120	20–8
Stellite (hard)	Grinding only	110	—
Inconel 718	30	120	4
Waspaloy	27	120	4.5
René 41	8	125	16
Austenitic stainless (Cr/Mn)	3	12	4

Adapted from *Metals Handbook* [2].

surface roughness, tolerances, plastic deformation, residual stresses, and so on, associated with conventional turning apply equally to TAM [Metcut]. However, the MRRs increase considerably because the workpiece is softened. The shear forces and the specific cutting energy are reduced. Table 8.25 lists the improved rates of TAM of super alloys and stainless steels as compared to conventional turning, which means that the machinability of these materials is considerably enhanced [1].

8.6.2 Laser Assisted Turning (LAM) of Inconel-718

During turning of Inconel-718 with ceramic inserts under laser assisted machining (LAM), better cutting performance has been achieved using the materials high absorptivity of CO_2 laser energy. The specific cutting energy is reduced to 25% by increasing the temperature from 30 °C (conventional machining) to 620 °C (LAM). The surface roughness decreased from 1.8 μm in conventional machining to 0.9 μm during LAM at 540 °C. Increasing the cutting speed from 60 to 180 m/min is beneficial during LAM, since the notch wear decreased by one-half. The average flank wear during LAM is significantly lower than conventional machining. A considerable advantage of LAM is that the cost of machining of inconel-718 with carbide tool decreases by 2/3 compared with conventional machining, and by almost by the half if ceramic tools are used compared with conventional machining at 180 m/min [8, CH4].

Also, plasma enhanced machining (PEM) has improved the machining performance of inconel-718 when turning with SiC-Whisker-reinforced aluminum oxide inserts. The effect of the temperature of the workpiece is found to be the most significant on the tool life. During experimental work with a fixed value of plasma and shield gases flow rates, the flank wear was measured [7, CH4]. The experiments considered the effects of plasma current I (A), Initial bulk temperature T_0 (K), workpiece diameter D (mm), cutting speed v (m/s), and feed rate f (mm/rev). The surface temperature T_s (K) is correspondingly determined based on the empirical equation:

$$T_s = 80.3 \frac{I^{0.6} T_0^{\,0.06}}{v^{2.2} D^{0.4} f^{0.2}}$$

8.6.3 *Plasma Assisted Turning (PAT) of Super Alloys and PH-Stainless Steel*

Table 8.26 lists the machining parameters, the MRR, and the specific removal rate of plasma assisted turning (PAT) in rough cutting and finishing of some types of Ni-base super alloys and precipitation hardenable-stainless steels (PH-SSs), which are difficult-to-machine traditionally. According to the machining duty, the rotational speed is selected in the range of 50–200 rpm to realize a cutting speed ranging from 10 to 100 m/min, whereas the cutting tool and the plasma torch are feeding in the rate of 1–5 mm/rev, depending on the type of cut required. The torch is located apart from the machining point by a suitable SOD. The depth of cut is determined by varying the inclination angle α (Figure 8.26).

Table 8.26 Machining parameters, the MRR, and the specific removal rate of PAT of some types of Ni-base super alloys and PH-SSs

Material	Type of cut	Current (A)	Voltage (V)	Speed (m/min)	Feed (mm/rev)	MRR (cm^3/min)	Specific RR (kW/cm^3/min)
Inconel 718	Rough	300	100	12	4.1	66	0.41
	Finish	170	90	23	1.3	25	0.69
René 41	Rough	325	95	16	4.1	82	0.37
	Finish	240	90	32	2.0	33	0.69
Hastelloy C	Rough	200	130	9	2.5	57	0.32
	Finish	140	100	18	1.3	25	0.69
PH stainless	Rough	300	92	15	4.1	74	0.37
steel	Finish	170	90	61	1.0	33	0.69

Adapted from Bagley [49], cited in *Machining Data Handbook*, Vol. 2, Metcut, 1980, p. 12–104.
PH = precepitation hardening.

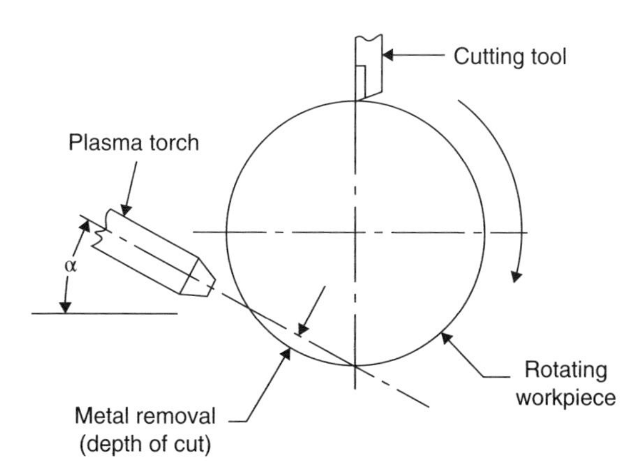

Figure 8.26 PAT – setup.

References

[1] Machinability Data Center (1980) *Machining Data Handbook* Metcut Research Associates, Inc, Vol. **2**, 3rd edn Cincinnati, OH.
[2] Editor Committee of ASM International (1989) *Metals Handbook: Machining* Vol. **16**, ASM International Materials Park, OH.
[3] Schwartz, B.L. (1985) Principles and applications of AWJ-Cutting. *High Productivity Machining Materials and Processes*. ASM, pp. 291–298.
[4] Youssef, H.A.(2005) *Non-Traditional Machining Processes-Theory and Practice,* 1st edn, El-Fath Press, Alexandria.
[5] Neppiras, E.A. and Foskett, R.D. (1957/1958) Ultrschall Material Bearbeitung – Prinzip und Apparatur, *Philips Tech. Rundschau* **19** (2): 98–110.
[6] Soo, S.L., Hood, R., Aspinwall D.K., *et al.* (2011) Machinability and surface integrity of RR1000 nickel based superalloy. *CIRP Ann. Manuf. Technol.* **60**:89–92.
[7] Klocke, F., König, W. (2007) *Fertigungsverfahren 3: Abtragen, Generieren und Lasermaterialbearbeitung*, Springer, Berlin.
[8] Reed, R.C. (2006) *The Superalloys*, Cambridge University Press, Cambridge.
[9] Burger, M., Koll, L., Werner, E.A., Platz, A. (2012) Electrochemical machining characteristics and resulting surface quality of the nickel-base single-crystalline material LEK94. *J. Manuf. Processes* **14**:62–70.
[10] Klocke, F., Klink, A., Veselovac, D., *et al.* (2014) Turbomachinery component manufacture by application of electrochemical, electro-physical and photonic processes, *CIRP Ann. Manuf. Technol.* **63:** 703–726.
[11] McGeough, J.A (1988) *Advanced Methods of Machining*, London, New York: Chapman & Hall.
[12] Bellows, G. (1967) *ECM Machinability Data and Ratings*, Technical Paper, SME, Metcut, Dearborn.
[13] Kalpakjian, S., Schmid, S.R. (2003) *Manufacturing Processes for Engineering Materials*, 4th edn. Prentice Hall, New York.
[14] Baumgärtner, M. (2013) Entwicklung des electrochemischen Senkens (ECM) und der mechanischen Bearbeitung von Titanaluminiden, Abschlussbericht BMBF-Forschungsvorhaben, http://refhub.elsevier.com/S00078506(14)001863/sbref0105 (accessed April 19, 2015).
[15] El-Hofy, H. (1995), Machinability indices for some non-conventional machining processes, *Aexandria Eng. J.*, **34**(3): 231–245.
[16] EMAG's ECM/PECM Machines (2013) http://www.online-amd.com/Article.aspx?article_id=132602 (accessed April 22, 2015).
[17] Platz, A., Feiling, N. (2013) *Precise Electrochemical Machining of Nickel Base Integrated Blade Compressor Rotors*, Precision and Microproduction Engineering Vol. **7**, Fraunhofer IWU, Chemnitz, pp. 23–32.
[18] ECM/PECM Technologie Polieren (2011) Entgraten, 3D-Konturen, EMAG ECM.
[19] MM (2010) Verfahren zur Reduzierung von Chrom(VI) bei der ECM-Bearbeitung, http://www.maschinenmarkt.vogel.de (accessed April 19, 2015).
[20] Steffens, K., Platz, A., Buckl, F. (2004) Feinbearbeitungsverfahren – Schlüsseltechnologien für moderne Luftfahrtverdichter. MTU Aero-Engines, www.mtu.de (accessed April 19, 2015).
[21] ECM (2014) Leistritz Turbo-maschinen Technik, www.leistritz.com (accessed April 19, 2015).
[22] Burger, M., Platz, A., Werner, E. (2007) Herstellung/Nachbearbeitung von Turbi- nenblisks durch Präzises Elektrochemisches Bearbeiten, TUM, http://www.wkm.mw.tum.de/forschung/postergalerie/ (accessed April 19, 2015).
[23] GKN Aerospace Capabilities (2010) www.gknaerospace.com (accessed April 19, 2015).
[24] Giese, C. (2005) Verfahrensvergleich EDM/ECM im industriellen Umfeld – Anwendungsgebiete dungsgebiete von ECM, Fachtagung Funkenerosion, RWTH Aachen.
[25] Dilba, D. (2012) Höchste Präzision. Report MTU Aero Engines, www.mtu.de (accessed April 19, 2015).
[26] Innovative Technologies for Future Alloys (2013) http://www.turbinentechnik.com/files/alloy_folder.pdf (accessed April 19, 2015).
[27] Thümmler, T. (2008) Herstellung von komplexen Kühlluftbohrungen in Hoch-druckturbinenschaufeln, MTU Aero Engines.
[28] Electrochemical Machining (2013) http://www.köppern-international.com (accessed April 22, 2015).
[29] Elektrochemisches Abtragen (2009) Verein Deutscher Ingenieure VDI-Richtlinie3401-1(Entwurf).
[30] Visser, A., Junker, M., Weissinger, D. (1994) *Sprühätzen mittallischer Werkstoffe*, 1st edn, Eugen G. Leuze Verlag, Bad Saulgau.

[31] Fleischer, J. (2011) Erodierbohren – Neue Wege und Anwendungsbeispiele. Fachtagung FunkenerosionWZL RWTH Aachen University, Aachen.

[32] Antar, M.T., Soo S.L., Aspinwall, D.K. *et al.* (2010) WEDM of aerospace alloys using 'clean cut' generator technology. *Proceedings of the16th International Symposium on Electromachining (ISEM XVI), Shanghai, China, April 19–23*, pp. 285–290.

[33] D'Amario, R. (2008) Method and apparatus for generating machining pulses for electrical discharge machining. European Patent EP 1719570.

[34] Han, F., Wachi, S., Kunieda, M. (2004) Improvement of machining characteristics of micro-EDM using transistor type isopulse generator and servo feed. *Precision Engineering* **28**: 378–385.

[35] Winbro Group Technologies (2014) Series 800 Laser & EDM Datasheet.

[36] NCMT (2013) Deep-Hole EDM Drilling of Turbine Components is Seven Times Faster, http://www.ncmt.co.uk (accessed April 19, 2015).

[37] Paul, M.A. and Aspinwall, D.K. (1998) Arc sawing performance evaluation and machine design. *Proceedings of the 12th International Symposium on Electro-machining (ISEM XII), Aachen, Germany, May 11–13*, pp. 407–416.

[38] Rahman, M.M., Khan, M.A.R., Kadirgam, K. *et al.* (2011) Experimental investigation into EDM of stainless steel 304. *Journal of Applied Sciences* **11**(3): 549–554.

[39] Field, M. (1966) The surface effects produced in nonconventional metal removal- comparison with conventional techniques, *Met. Eng. Q.* **6** p 32–45.

[40] El-Hofy, H.A. (2013) *Fundamentals of Machining Processes – Conventional and Nonconventional Processes*, 2nd edn, CRC Press.

[41] Visser, A. (1966) Werkstoffabtrag mittels Elektronenstrahl. Dr.-Ing. Dissertation. TH Braunschweig.

[42] Beck, T. (2011) Laser drilling in gas turbine blades: Shaping of holes in ceramic and metallic coatings. *Laser Tech. J.* **3**:40–43.

[43] Horn, A., Weichenhain, R., Albrecht, S. *et al.* (2000) Microholes in zirconia coated Ni-superalloys for transpiration cooling of turbine blades. *Proceedings of the SPIE 4065, High-Power Laser Ablation III*, p. 218.

[44] Saunders, R.J. (1984) *Laser Metalworking, Metal Progress*, p. 51.

[45] Majundar, J.D. (2012) *Laser Assisted-Fabrication of Materials*, Springer, Berlin.

[46] Poprawe, R. (2005) *Lasertechnik fu¨r die Fertigungsgrundlagen, Perspektiven und Beispiele fu¨r den innovativen Ingenieur*, Springer, Berlin.

[47] Leigh, S., Sezer, K., Li, L. *et al.* (2010) Recast and oxide formation in laser-drilled acute holes in CMSX-4 nickel single-crystal. *Proceedings of the Institution of Mechanical Engineers, Part B: Journal of Engineering Manufacture* **224**: 1005–1016.

[48] Holden, S. (1985) The plasma cutting of SS, *SS Industry* **13**(74): 13.

[49] Bagley, J.A. (1969) *Plasma Arc Cutting, Technical paper MR69-578*, Society of Manufacturing Engineers, Dearborn, MI, p. 23.

[50] Klocke F, Zeis M, Klink A, Veselovac D (2013) Technological and economical comparison of roughing strategies via milling, sinking-EDM, wire-EDM and ECM for titanium- and nickel-based blisks. *CIRP J. Manuf. Sci. Technol.* **6**(3):198–203.

[51] Sharma, S., Jain, V. K., Shekhar, R. (2002) Electrochemical drilling of inconel super alloy with acidified sodium chloride electrolyte *Int. J. Adv. Manuf. Technol.* **19**:492–500.

9

Current and Recent Developments Regarding Machining of Stainless Steels and Super Alloys

9.1 General Considerations

Many research works have been adopted in the last two decades to highlight the associated problems regarding the machinability of difficult-to-cut materials. These have been focused on the traditional machining of stainless steels and super alloys. The main objective of these research works was the enhancement of the productivity by increasing the machinability through many strategies, previously discussed in Chapter 4, which include:

- Adoption of free-machining and enhanced versions of stainless steels, which were previously discussed in Chapter 2.
- Implementation of thermally assisted machining (TAM), or hot-machining which is based on that difficult-to-cut materials can be machined more easily at elevated temperatures thus lowering cutting forces and increasing tool life. Most applications of TAM are found in turning and milling. Propane torches, oxyacetylene torches, plasma beams, and induction coils have been tried with various degrees of success, most limited by heat control. The main disadvantage of this development is that a uniform temperature distribution may be difficult to maintain and control; consequently, the original microstructure of the work may be adversely affected leading to heat affected zone (HAZ).In the late 1970s, lasers emerged as a viable heat source capable of providing intense heat, concentrated in a very precise region. In 1978, Bass *et al.* [1] showed the feasibility of hot spot laser-assisted machining (LAM) using a 1.4 kW-CO_2 laser to assist machining of SS and Udimet 700. No modeling of the process was performed at that time, providing little insight into temperatures achieved during the process. It was observed that coordination between cutting speed and laser heating must be optimized. Moreover, there were some problems in heating the metal as a result of its reflectivity.

Machining of Stainless Steels and Super Alloys: Traditional and Nontraditional Techniques,
First Edition. Helmi A. Youssef.

- Adopting high speed machining (HSM) technology, which necessitates the application of advanced tool materials, and powerful machine tools operating at high speeds. This promotes productivity when machining difficult-to-cut materials including stainless steels and super alloys. The research emphasized on the variables that directly affect the machinability of stainless and super alloys, such as tool life, tool material, cutting fluids, and process parameters during HSM.
- Application of advanced cooling techniques, which include cryogenic cooling, minimum quantity lubrication (MQL), and high pressure cooling (HPC) to improve the performance of machining operations.
- Application of ultrasonic-assisted machining (UAM) to significantly improve the machinability.

9.2 Recent Research Work Related to Traditional Machining of Stainless Steels

Current research works [1–16] focused on exploring the machinability of some types and categories of SSs, that have predominant industrial application, such as austenitic types P550, AISI 303, AISI 304, AISI 316, and the martensitic TETHETE, and so on.

Chandrasekaran and Johnson [2] investigated the machinability of four high strength austenitic stainless (AISI 316 LN and similar alloys) using carbide tools (P30), operating at cutting speeds below 75 m/min, and feed rates between 0.08 and 0.3 mm/rev. The tools showed rapid notch wear failure often with built-up edge (BUE). The critical initiation of the notch seems to be related to some factors such as transverse stress, temperature distribution, and chemical interactions with the work material. The high cutting forces, and significant strain hardening reported during machining of these N-strengthened austenitics make LAM an alternative candidate for machining such alloys.

High-N, Ni-free, manganese austenitic steel P550 has much higher yield strength than typical stainless steels and their high corrosion resistance and high strength make them especially attractive in medical, aerospace, and oil drilling industries. The high strength also makes these steels difficult-to-cut owing to the high cutting forces, and the relatively small tool durability. Anderson and Shin [3] investigated LAM as an economic alternative for traditional machining of these stainless steels. The major elements of P550 stainless steels are given in Table 9.1. Nitrogen, and manganese stabilize the austenitic phase. While Mn increases the solubility of N in iron, Cr promotes ferrite formation; moreover, it is a critical element for corrosion and oxidation resistance.

Turning tests have been performed on a 45 kW turret lathe. For LAM, two laser beams, translating jointly with the cutting tool. A 0.5 kW-Nd:YAG laser irradiates the machined chamfer 10–12° circumferentially ahead of the tool. Owing to the temperature limitation achievable for the Nd:YAG laser, a second laser was used to provide additional heating using 1.5 kW-CO_2 laser, positioned 55° ahead of the cutting tool and irradiating the unmachined work surface (Figure 9.1) [3]. Force data were collected using Kistler dynamometer and an

Table 9.1 Chemical composition (wt%) of the special high-nitrogen Mn-Cr austenitic stainless steel, P550

Ni	C	Mn	Cr	Mo	N	Fe
max 1.5	max 0.06	19–20.5	17.5–19	max 0.45	0.5–0.6	Balance

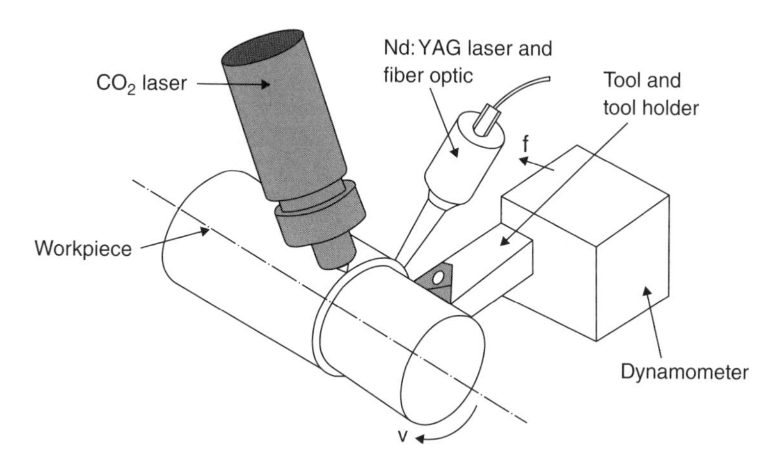

Figure 9.1 Experimental set-up for LAM using two lasers, CO_2 and Nd:YAG. (From Anderson and Shin [3].)

Table 9.2 Tool materials and tool geometry for machining Mn-Cr austenitic SS P550 [3]

Tool type	Rake angle (°)	Side cutting angle (°)	Inclination angle (°)	Nose radius (mm)	Clearance angle (°)
Coated carbide	5	15	0	0.8	7
Cermet	5	15	0	0.8	7
Uncoated carbide	5	15	0	0.8	11
Ceramic	5	15	0	0.8	11

amplifier, and a labView program was used to process and record force signals. Temperature measurements were performed using a FLIR SC 3000 IR-camera.

Four different types of tool inserts (Table 9.2) were investigated to determine the optimum tool material for machining of austenitic P550. The geometries are very similar, except that the clearance of the uncoated carbide and ceramic is 11°, compared with 7° for the coated carbide and cermet. Ceramic inserts are reinforced with SiC-whiskers, [3].

The feed rate and depth of cut were kept constant all over the test at 0.1 mm/rev and 0.76 mm, respectively, while the cutting speed was varied from 0.5 to 4 m/s for traditional machining (0.5–2 m/s for coated and uncoated carbides, and 2–4 m/s for ceramic tools). Figure 9.2 presents the specific cutting energy, and the average flank wear in case of traditional machining.

In case of LAM, SiC-whisker-reinforced ceramics are only used for a speed range of 2–4 m/s. In traditional machining, tool inserts with a larger clearance, show less than 50% of the flank wear of other tools. Even at higher cutting speeds (LAM, 2–4 m/s), ceramic inserts show the lowest flank wear due to the higher hardness of the tool material at elevated temperatures. In LAM, where ceramic is used, the specific cutting energy is considerably lower than that with traditional machining using carbides and cermet inserts, owing to the higher cutting speed used in LAM, and the lower coefficient of friction of ceramics [3].

Ceramic tools have often been considered unsuitable for machining austenitic stainless steels, since its ductility and toughness create a gummy nature during machining. It has been

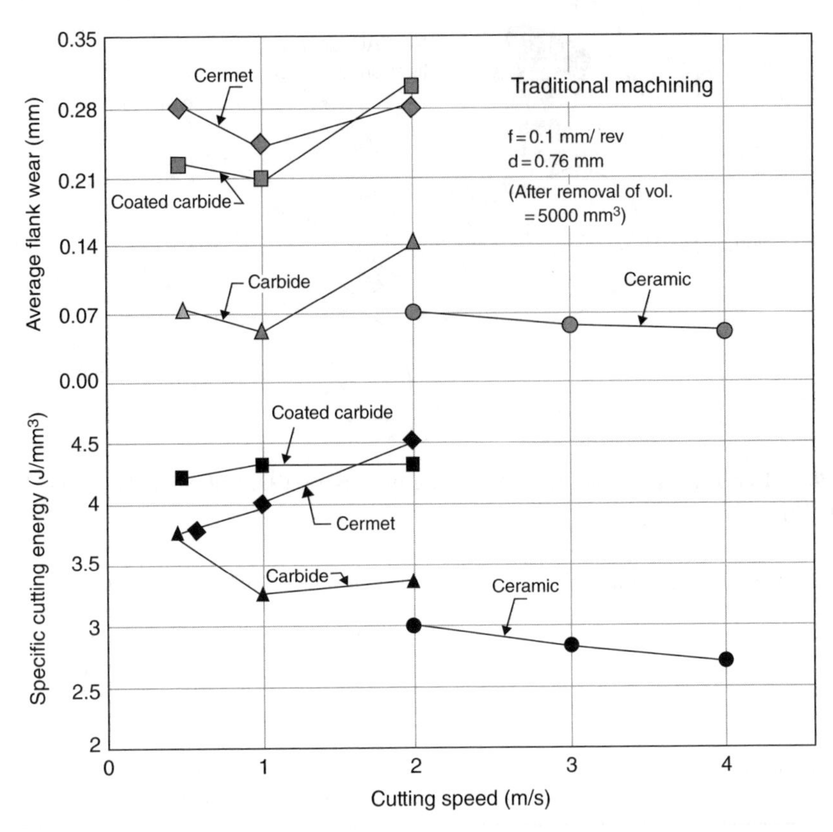

Figure 9.2 Specific energy and average flank wear for traditional machining of austenitic SS P550 using different tool materials versus cutting speed. (Adapted from Anderson and Shin [3].)

recommended that a highly positive carbide tool is to be employed for machining high strength austenitic stainless steels to increase shearing and reduce tearing of the material. However, there is a little difference in surface roughness of the workpiece machined by various cutting tools. The combination of low tool wear and lower specific cutting energy of ceramics (Figure 9.2) indicates that tearing of the material is not an issue and whisker-reinforced ceramics should be recommended for HSM of high strength austenitic stainless steels such as P550. However, tests have been performed, using ceramic inserts at speeds ranging from 2 to 4 m/s to determine the appropriate cutting speed for LAM [3].

During traditional HSM of P550 using ceramic inserts, the roughness (Ra-value) remained below 1 μm throughout the tests, even after 17 000 mm^3 of material removal. The roughness remained more or less constant all over the machining time. The specific cutting energy when using ceramics decreased slightly as the speed increased (Figure 9.2). According to their observations [3], the major tool wear concentrated on the primary flank, and on the rake face of the tool. The average primary flank wear steadily increased with the material removal from 0.6 mm for volume of material removed = 5000 mm^3 to 0.9 mm for volume of material removed = 20 000 mm^3. The high cutting speeds allowed with ceramics possibly eliminated the BUE that was responsible for the tool failure and the deterioration of the surface quality of the workpiece.

During LAM, it was essential to achieve the optimal temperature of the workpiece. Too high temperature may prematurely degrade the cutting tool, and cause a substrate damage of the workpiece, while too low temperature would not realize the maximum benefit of the LAM. A two-laser transient thermal model has been suggested and used to predict temperatures during LAM, and consequently the laser power would be controlled to realize the optimum machining conditions. For LAM, a cutting speed of 3 m/s was recommended [3]. Figure 9.3 illustrates the specific cutting energy at different average bulk material temperatures as determined at material removal of 5000 mm$^{3.}$ The specific energy decreases gradually with increasing bulk temperature, attaining a value of 3 J/mm^3 at room temperature (traditional machining) and 2.4 J/mm^3 at a bulk temperature of 425 °C during LAM. Therefore, the machinabilily of P550 (as based on specific cutting energy) using LAM is improved by more than 20%, if compared to traditional machining.

Throughout all traditional and LAM machining tests, using whisker reinforced ceramic tooling, the surface roughness (Ra-values) remained below 0.75 μm. Detrimental precipitates are not observed on the surface during LAM owing to the short interval of time of high peak temperatures. Accordingly, hardness values of workpieces that have undergone traditional machining and LAM exhibited the same trend. The average hardness of the base material was 43 HRC. A slight reduction of hardness was exhibited from the surface to a depth of 75–100 μm. Continued improvements of lasers will provide the potential for future benefits for LAM of stainless steels [3].

The modes of application of cutting fluids, when turning weldable austenitic stainless steel AISI 316L, have been investigated by Leppert [5]. In this work, tests were planned to be performed dry (D), with MQL, and with emulsion (E). MQL (aerosol on basis of Accu-Lube LB8000 oil, of kinematic viscosity = 37 mm^2/s at 40 °C) was generated by a Minibooster II applicator (Accu-Lube Manufacturing GmbH). It is then transferred to the cutting edge by two

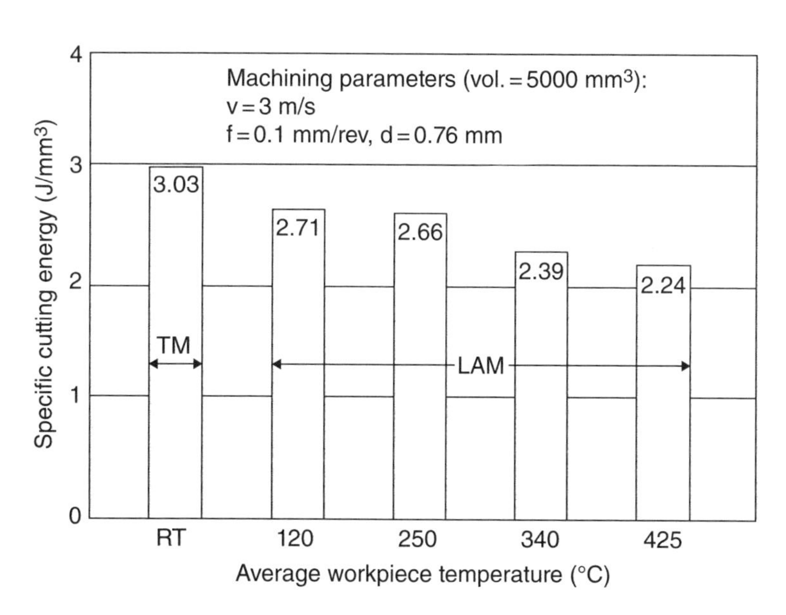

Figure 9.3 Specific energy at various machining temperatures using ceramic tools to machine austenitic SS P550. (Adapted from Anderson and Shin [3].)

Table 9.3 Test conditions according to Leppert [5]

Workpiece	Material: AISI 316L (austenitic stainless steel) diameter: 60 mm
Tool	Tool holder: MSS 2525-12-EB (Mircona AB) Carbide insert: SNMG 120408-TF, grade IC 907 PVD coating: TiAlN (Iscar Ltd)
Tool geometry	Rake angle = 5°, clearance angle = 10° Cutting edge angle = 45°, cutting inclination angle = 0° Nose radius = 0.8 mm
Cutting parameters	Cutting speed: 82, 164, 255 m/min Feed rate: 0.08, 0.27, 0.47 mm/rev Depth of cut: 0.5, 1, 2 mm

through holes built inside the tool holder, and two nozzles of 0.8 mm diameter, directed to the rake face and principal and auxiliary flanks at velocity of 30 m/s (near to the sonic speed). Three oil rates of consumptions were tested, namely, 10, 20, and 50 ml/h. A 6% (by vol.) emulsion was made using emulsifying oil ARTEsol Super EPy, which was supplied to the cutting zone at a flow rate of 4 l/min.

The workpiece (60 mm diameter, 300 mm long) was segmented by grooves in 15 mm length for each cutting test. The bars were pre-machined with a 1 mm depth to ensure similar surface properties. The tool was a physical vapor deposition (PVD)-TiAlN coated inserts. Each set of tests was conducted using a new insert edge to minimize the effect of tool wear. The test conditions are given in Table 9.3. These tests are considered HSM, since they are performed at speeds up to 255 m/min using high efficient coated carbide inserts.

In these tests, cutting force measurements were performed using a Kistler 9247 B piezo-dynamometer. Its signal was sent to computer equipped with Dyno-software and Kistler 5017 B amplifier. The surface parameters were measured on a Hommel-tester T2000, while the surface topography was measured on laser scanning electron microscope (LSEM).

Leppert [5] reported that both feed rate and depth of cut exerted the greatest effect on the cutting force. However the cutting speed experienced a limited effect on the cutting force, which may be attributed to the increased strain hardening effect when machining austenitic stainless steels [6, 7]. It was also depicted that the feed rate significantly affected the surface roughness which increased from 2 to 8 μm (Ra-value), if the feed increased from 0.08 to 0.47 mm/rev, which is in full agreement with the expected results in machining references. Similarly, an increase in depth of cut caused an increase of the surface roughness. This can be related to an increase of chip cross section and consequently the cutting force, which could affect the dynamic stability of machine-tool-workpiece system. On the other hand, the cutting speed showed no significant effect on the surface roughness in the investigated range [5].

Leppert [5] finally investigated the effect of modes of cutting fluid applications on the cutting force and the surface roughness. It was found that, no significant changes of the values of cutting force and surface roughness as a function of lubricant modes (E, D, and MQL) was observed. Figure 9.4 shows the surface roughness (Ra-value) as ranging from 1.5 to 2.2 μm for the different modes of lubrication. Increasing the oil supply rate from 10 to 50 ml/h in MQL did not significantly affect the surface roughness. His results, concerning the effect of lubrication modes on the surface roughness, and the cutting force are, however, unpredictable, and totally unexpected.

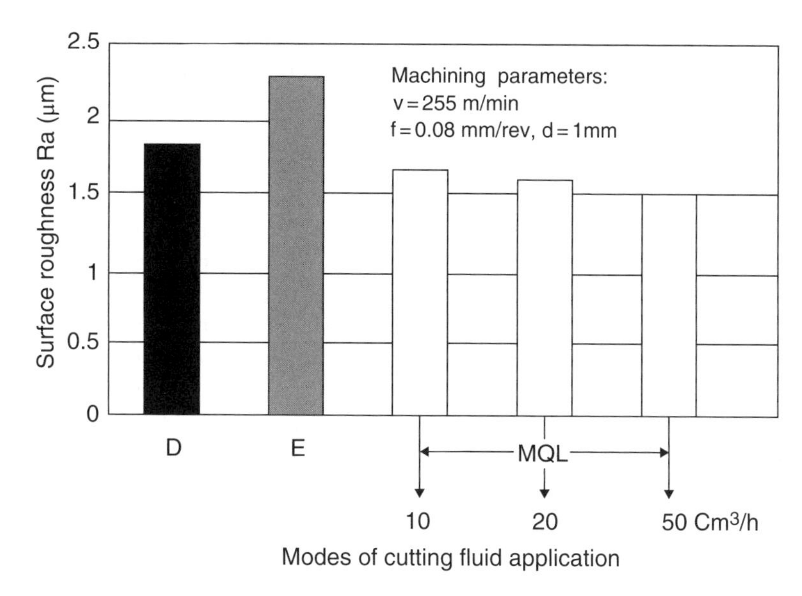

Figure 9.4 Influence of modes of application of cutting fluids on the surface roughness of AISI 316L. (From Leppert [5].)

Tekiner and Yesilyurt [8] investigated the optimum working conditions during high speed turning of austenitic stainless steel AISI 304 using cemented carbide tips ISO P10. The machining was performed on a CNC-turning center, and the work specimens were bars of 30 mm diameter, and 200 mm length. Five cutting speeds, namely, 120, 135, 150, 165, and 180 m/min were used. For each three different feed rates of 0.20, 0.25, and 0.30 mm/rev were selected. The machining was performed dry, while a new cutting edge was used for each test. A constant depth of cut of 2.5 mm was used for all tests.

A sound analysis of the machining process was recorded by a setup composed of a sensitive microphone, and amplifier switched between the sound source at the cutting edge and a personal computer PC. If the machine was running idle (no cutting), the total sound pressure level attained a lowest value. During cutting, the sound levels changed depending upon the cutting parameters. It was found that the sound pressure level decreased as the cutting speed up to a value of 165 m/min, where it attained a minimum value, after which it increased again when the cutting speed increased to 180 m/min. The measured surface roughness of the specimen showed the same trend. The feed rates affected the sound signal, such that while it was more or less constant at feed rates of 0.20, 0.25 mm/rev, it significantly increased at a feed rate of 0.3 mm/rev. The outcome of these investigations was that the optimum working conditions were realized at a cutting speed of 165 m/min and a feed rate of 0.25 mm/rev.

Finally, it was depicted that chip curl radii decreased and the chip thickness increased at lower cutting speeds and higher feed rates. Moreover, the liability of built-up formation decreased with increasing cutting speed, and increased with increasing feed rate.

A tangential cutting force in excess of 3500 N has been observed by Mesquita and Marques [9] when machining martensitic stainless steel of 200 HV with coated carbide tools at a cutting speed of 130 m/mm. Such a high cutting force caused chipping/fracture of the cutting tool and excessive frictional heat during cutting. It was claimed that such condition can lead to surface deterioration, which reduces the fatigue strength of machined components by up to 30% [10], with consequent damage in landing gear applications accounting for approximately 37% of all craft accidents [11]. These statistics limit HSM of martensitic SSs, where a rapid rate of nose wear is a major problem [12].

The machining of martensitic SS AISI 414 used in the JETHETE engine using coated carbide tools has been investigated by Jawaid *et al.* [13]. Three commercially available trigon-shaped inserts (one PVD- and two CVD (chemical vapor deposition)-coated) (Table 9.4) were used by Jawaid *et al.* in turning the martensitic SS AISI 414. They evaluated the effect of machining parameters on the tool life and failure modes of the tested inserts. The workpiece bar was 200 mm diameter, and 400 mm length. Up to 6 mm of the material was removed prior to turning tests to minimize any effect of material inhomogeneity and associated experimental scatter. The bars were originally chamfered prior tests to avoid insert damage on entry as it touches the work. Turning was carried out without coolant on a CNC-lathe, under the following machining conditions.

- *Cutting speed*: 10 (low) to 150 (mild) m/min to 200 (high) to 250 (severe) m/min
- *Feed rate*: 0.2–0.4 mm/rev,
- *Depth of cut*: 2 mm.

The statistical regression technique experienced by Jawaid *et al.* [13], revealed that the cutting speed and feed rate had significant effects on tool performance, with the cutting speed showing the largest effect. Very low rates of wear occurred when machining martensitic JETHETE using the three grade of coated carbide inserts, T1, T2, and T3 (Table 9.4) at low speed of 100 m/min, while significant nose wear was the dominant failure mode when

Table 9.4 Cutting tool inserts used by Jawaid *et al.* [13]

Tool code	Substrate	Others
T1 (CVD)	Grain size: 1–2 μm Hardness: 92 RA Grade: P05/K05-P15/K15	Coating: Ti(C,N) (5 μm), Al_2O_3 (8 μm) Binder content: 5.9% Cubic carbides: 8.4% Thermal conductivity: 90 W/m k Coating technique: CVD Chip geometry: medium finishing
T2 (CVD)	Grain size: 1–8 μm Hardness: 90 RA Grade: P20/M20-M40/M40	Coating: Ti(C,N) + Al_2O_3 + TiN (10 μm) Binder content: 18.0% Cubic carbides: 8.4% Thermal conductivity: 95 W/m k Coating technique: CVD Chip geometry: roughing
T3 (PVD)	Grain size: 1–3 μm Hardness: 93 RA Grade: P05-K15, M05-M20	Coating: TiN (2 μm) Binder content: 6.0% Cubic carbides: Traces Coating technique: PVD Chip geometry: medium finishing

Cutting geometry: back rake angle –6°, cutting rake angle –6°, and approach angle 95°.

machining at mild, high, and severe speed conditions of 150, 200, and 250 m/min (Figure 9.5). Plastic deformation and chipping fracture of the cutting edge were additional failure modes observed on the CVD (T2), and PVD (T3) coated inserts. The triple layer coated CVD (T2) inserts exhibited the poorest tool performance owing to the severe effects of thermo-mechanical loads and coarse grain size and higher Co-content (18% Co) of carbide substrate. A better performance was recorded when machining with the dual-coated CVD (T1), and the single coated PVD (T3) inserts, because of the substrate quality and the thermal stability of the alumina coating on T1-insert and the improved micro-hardness of the PVD-TiN coating of the T3-insert.

Finally, Jawaid *et al.* [13] reported that attrition wear was the principal mechanism at lower cutting speed conditions, while abrasion and diffusion wear mechanisms jointly controlled the failure modes at higher speed conditions.

Lin [14] examined the possibility of high speed fine turning of three types of austenitic steels (free-machining AISI 303, AISI 303 Cu, and nonfree-machining AISI 304), using cermet inserts under dry cutting conditions. The machining has been performed at high speeds ranging from 250 to 450 m/min, and feed rates from 0.02 to 0.1 mm/rev, and a depth of cut 0.1 mm. The minimum roughness (Ra-value) that could be achieved was 0.4–0.6 μm. However, at critical feed rate of 0.02 mm/rev, and less, the surface quality has been suddenly deteriorated due to occurrence of chatter. At higher speeds and feeds (v = 450 m/min, and feeds 0.06–0.1 mm/rev), the surface quality also deteriorated, probably due to thermal tool failure.

Fernandez-Abia *et al.* [15] investigated the behavior of the free-machining austenitic steel AISI 303, which widely used for automotive parts, during high speed turning under dry condition. The influence of the cutting speed, particularly, on the tool wear, surface roughness, cutting forces, and chip morphology has been examined.

Turning tests were performed at a wide range of cutting speeds of 37, 75, 150, 300, 450, 600, 750, 845, and 870 m/min. Feed rate and depth of cut were kept constant at 0.2 mm/rev, and 1 mm, respectively. The tool was PVD multilayer-coated (TiCN, Al_2O_3, and TiN) on cemented

Figure 9.5 Coating delamination and nose fracture of a T1-grade insert after machining AISI 414 martensitic SS at high speed. (From Jawaid *et al.* [13].)

carbide substrate. The workpiece was 60 mm diameter, and 180 mm length. A three component Kistler 9121 piezoelectric turning dynamometer, adapted with Kistler amplifier 5070 A, and a Dap Board/2000 PCI data acquisition board have been used to measure the cutting force components.

According to their observations [15], a critical cutting speed of 450 m/min was identified, above which the machinability improved due to the lower value of the main cutting force, leading to less power consumption, and less deformation and stress of the cutting tool. Additionally, the surface quality at such high speeds has been improved. Also, it was found that the chip thickness was significantly decreased, which means higher shear angle, and consequently improved machinability. However, side material flow was depicted when machining above this critical speed. Moreover, the depth of microstructure affected zone increased in this high speed range.

Karenk *et al.* [16] investigated burr formation in drilling of the weldable austenitic stainless steel AISI 316L, of yield strength 300–330 MPa, and a hardness of 170 BHN, using Taguchi model and Artificial Neural Network (ANN). Burr formation, especially in drilling has been the most troublesome problem that affect the assembly tasks and consequently the automation and productivity of manufacturing processes. The main objective of this research work was to determine the best combination of the feed rate and point angle for a specified drill diameter that minimize the burr size, so as to reduce de-burring time and cost. High speed steel (HSS) drills were used at low speed (8–12 m/min) and low feed rates (0.04–0.12 mm/rev). In order to minimize the burr size, a smaller point angle (118°) was found to be suitable for the twist drill diameter in the range of 8–17 mm, while a larger point angle (134°) is necessary for drill diameters beyond 17 mm.

During turning of stainless steel with coated carbide insert K20, the tool life with cutting speeds of 100–300 m/min was investigated by Khan and Ahmed [17]. It has been found that with conventional coolant, the tool life was 13.5 min at a cutting speed of 100 m/min and 1 mm depth of cut, whereas under the same cutting conditions in cryogenic cooling, the tool life was 32 min. The reduction in tool wear was probably due to the reduction of diffusion and adhesion wear, affected by the liquid nitrogen jet. Flaking of the rake surface of coated carbide inserts just at the end of the crater wear region was observed, especially under the cryogenic machining condition. This is attributed to higher thermal gradient at the end of the crater contact.

The cutting force required in cryogenic cooling is less than that required for dry cutting. This is because application of cryogenic fluid reduces the coefficient of friction at the interface of the tool–chip over the rake face. At higher feed rate, chip thickness is higher; plastic deformation at the shear zone takes place at a faster rate, generating more heat. Therefore, cryogenic cooling is more effective at higher feed rate [18].

9.3 Recent Research Works Related to Traditional Machining of Super Alloys

The relevant researches [19–35] have focused on super alloys, especially Ni-base alloys, such as RR1000, Haynes, Inconel, and so on. Recent alloy development for gas turbine components has produced materials able to maintain strength and integrity at operating temperatures up to 1050 °C.

It is believed that next-generation RR1000 Ni-based super alloy reflect this philosophy, although at expense of machinability [19]. RR1000 is a new aerospace super alloy processed via powder metallurgy route. Its properties can be enhanced by heat treatment. It has good

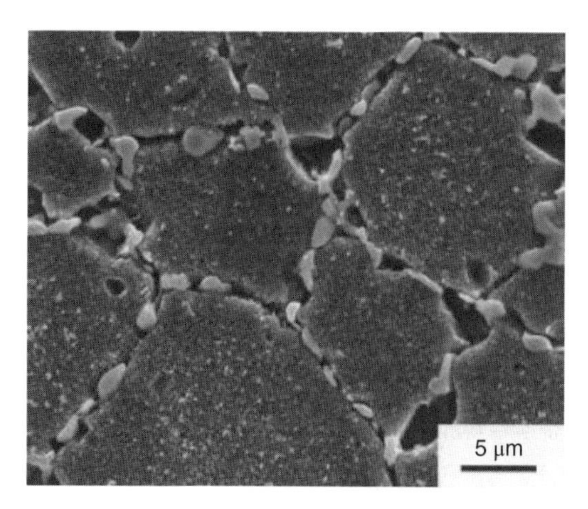

Figure 9.6 Scanning electron micrograph of the super alloy RR1000. (Courtesy of Rob Mitchell.)

strength and creep resistance at elevated temperatures as well as good toughness at all temperatures. This alloy features good oxidation and corrosion at high temperatures. Figure 9.6 illustrates a scanning electron micrograph of this alloy showing extensive precipitation of σ phase at γ-grain boundaries, after a heat treatment of 5000 h at 750 °C. RR100 is used in disc applications as Inconel, Waspaloy, and Udimet.

RR1000 is difficult to machine due to its strength, low thermal conductivity, abrasiveness, and especially, its tendency to work harden. Work hardening occurs whenever a tool makes a cut. Sharp, more positive cutting angles, often PVD coated tools, are preferred since they reduce this phenomenon. Compared to other Ni-based alloys (such as IN 718, and Udimet 728), RR1000 shows higher sensitivity to cutting tool geometry, coating, and machining parameter to achieve acceptable surface conditions.

Axinte and Andrews [20] performed hole making process on the Ni-based alloy RR1000 of chemical composition: 52.5% Ni, 15% Cr, 18.5% Co, 5% Mo, 3.5 Ti, 3% Al, 2% Ta, 0.5% Hf, 0.03% C.

In this research, the hole making process comprised two main operations:

1. Roughing D, using uncoated carbide drills (140° point angle, 30° helix).
2. Finishing, with three alternatives.
   (a) Reaming R1 with a carbide reamer (14° relief and 4° helix).
   (b) Reaming R2 with a special carbide reamer (1° relief and 1° helix).
   (c) Milling M with solid carbide plunge-milling cutter.

Summary of integrated hole making, along with machining conditions are given in Table 9.5. Figure 9.7 presents the tool flank wear versus the number of machined holes while drilling and reaming lead to surface overheating (white layers) and material dragging, the change of edge preparation (double relief angle on the special reamers, or the use of alternative cutting strategies on (plunge milling) can generate finished holes within the acceptable surface integrity. Table 9.6 illustrates the surface roughness Ra as obtained after the different hole-making processes.

Table 9.5 Parameters of integrated hole making according to Axinte and Andrews [20]

Tooling	n (rev/min)	v (m/min)	f (mm/min)
Roughing			
D-drilling	686	13	50
Finishing			
R1-normal reaming	250	5	38
R2-special reaming	298	6	60
M-milling	713	15	43

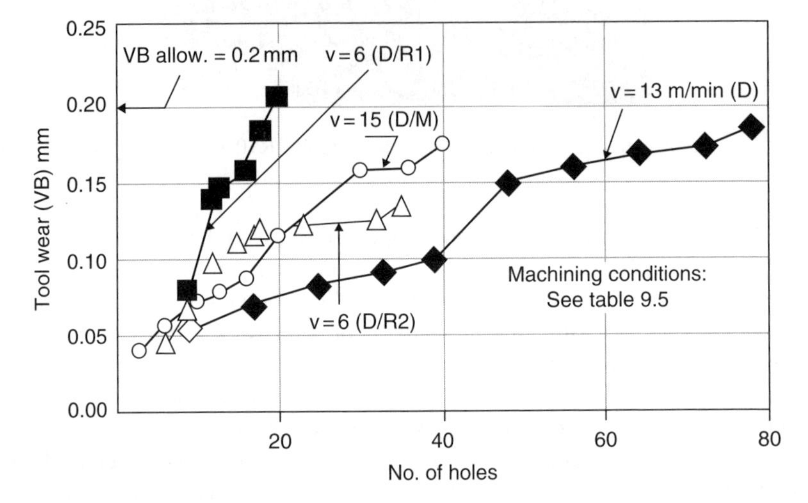

Figure 9.7 Tool flank wear versus No. of holes when drilling RR1000Ni-base super alloy. (Adapted from Axinte and Andrews [20].)

Table 9.6 Surface roughness Ra as obtained from different hole making processes in RR 1000 according to Axinte and Andrews [20].

Hole-making process	Achieved surface roughness, Ra (μm)	Achieved number of holes
D-drilling	1.1	80
D/R1-Normal reaming	0.6	22
D/R2-Special reaming	0.2	35
D/M-Milling	0.2	40

Referring to Figure 9.7, although drilling D achieved a good tool life (80 holes), it failed to fulfill the surface quality criterion (Ra > 1 μm) (Table 9.6). This is, however, expected because it is well known that drilling is a roughing operation, and it is not surprisingly, white layer appeared on the machined surface proving the sensitivity of RR1000 to overheating phenomenon. Normal reamers (R1) showed an acceptable tool life (22 holes) (Figure 9.7) and surface roughness of Ra = 0.6 μm (Table 9.6). In addition, white layers, and significant surface drags have been observed. This method (R1) is also not adequate for hole finishing as it fails to fulfill surface quality criterion. Both special reamer (R2) and plunge-milling (M), clearly demonstrated the capability of fulfilling the tool life (35 and 40 holes), as well the quality requirements (Ra = 0.2 μm) (Table 9.6).

Finally, Axinte and Andrews [20] claimed that a good indication of the reduced machinability of RR1000 relative to the more conventional Ni-based alloys such as Inconel 718 was recognized by comparing the thrust force and torque when drilling RR1000, and Inconel 718. Under the same working conditions, they registered when drilling RR1000 an increase of 18% for thrust, and 10% for torque as compared to their relevant values when drilling Inconel.

Experimental data [21] when drilling the same Ni-based alloy RR1000 as in the research [20], however at a cutting speed of 45 m/min, showed flank wear to be less than 100 μm for a distance cut of 1800 mm (150 holes). Thrust force was 1700 N. It has been reported that the roughness of end-milled specimens of RR1000, was less than 0.8 μm Ra, with minimal damage if new tools are used; however, when using worn tools, significant burning/increased micro-hardness (about 150 HK 0.05), and white layer occurred [21].

Among the commercially available super alloys, Inconel 718 stands out as the most dominant alloy in production. It accounts for 45% of wrought Ni-based and 25% of cast Ni-based production. Its primary uses are in the aircraft gas turbines, and many other applications such as nuclear power plants, and medical applications [22]. However, it is very difficult to machine. This difficulty resolves into short tool life and poor surface integrity. Dudzinski *et al.* [23] reported that abrasion is the wear mechanism for all tested tools. BUE formed was repeatedly removed leading to severe notching. Machining induced plastic deformation, and heat generation, causing metallurgical transformations and residual stresses in the machined surface layer. The residual stress distribution exhibits a maximum tensile stress near the machined surface, then a compression stress. The depth of affected layer and the tensile and compressive stresses increase with increasing cutting speed.

At low cutting speeds of 20–30 m/min, K20 grade appeared to be the best for cutting Inconel 718. Higher cutting speeds up to 100 m/min, the PVD (TiAlN) coated carbides, under dry condition was found to be most suitable, because they displayed high oxidation resistance, high temperature chemical stability, high hot hardness, and low thermal conductivity. The use of coolant is undesirable for environment and human health; furthermore, it induces high additional costs. PVD-coatings of carbides have reduced friction at elevated temperatures with high wear resistance. They [23] showed excellent performance in drilling high strength materials.

Much higher cutting speeds (from 200 to 700 m/min) are attained with ceramic tools. The Al_2O_3-TiC is the most chemical stable to Inconel718, since it has the most thermal resistance to withstand HSM. Ceramics are poor conductors and sensitive to thermal cracks; therefore, dry machining is recommended for them [23].

Choudhury and El-Baradie [22] investigated the machinability of Inconel 718 when turning using Sandvik coated GC3015 and uncoated H13A inserts under dry cutting conditions. The workpiece was a fully heat-treated (solution treated and age hardened) IN 718 of 55 mm diameter. The tool geometry: angle of approach 95°, rake –6°, inclination angle –6, major flank clearance 0°. The coated insert: GC3015 two layers, TiC and Al_2O_3 on the top. The total thickness of the coating = 10 μm. The inserts were in the form of 80° rhomboid shapes without any chip-breaker. The 10 HP Colchester M1600 lathe had max spindle speed of 1600 rpm, and a feed range of 0.06–1 mm/rev. The forces were measured using a three-component dynamometer (Kistler, 92625 A1); the force components are recorded by a UV recorder. The tool wear was measured by Mitutoyo TM300 Toolmakers Microscope.

Choudhury and El-Baradie [22] studied the effect of cutting parameters (speed, feed, and depth of cut) on cutting forces and tool life. These parameters have been optimized, where the flank wear was considered as the criterion for the tool life. A comparison between uncoated and coated tools has been made using generalized Taylor tool life equation. Generally, no significant difference in tool life values was observed for coated and uncoated tools. The use of coated tools is only justified when depths of cut exceed 1.0 mm. It was depicted, that recommended cutting speed when using uncoated carbides should be within 20–25 m/min, the feed should be 0.15–0.200 mm/rev, whereas the depth of cut should be higher than 1.0 mm.

Bushlyaa *et al.* [24] examined the machinability of aged Inconel 718 during high speed turning using coated and uncoated poly-cubic boron nitride (PCBN) tools. The machinability was evaluated in terms of cutting forces, tool life, tool wear, and generated surface integrity. Their results indicated that PCBN coated tools were highly sensitive to cutting speed, where the tool life decreased by 250% with an increase in speed from 250 to 350 m/min. Therefore, it is not recommended to machine with coated PCBN tools over cutting speeds greater than 250–300 m/min. The application of coated PCBN tools instead of uncoated tools, leads to the formation of tensile stress surface layer instead of compressive stress layer.

Attia *et al.* [25] investigated the laser-assisted high-speed finish turning of Inconel 718 under dry conditions, with the objectives (i) to optimize the LAM process in terms of tool life, surface integrity, and productivity and (ii) to assess the use of SiN ceramic tools (SiAlON) in dry HSM to minimize environmental impact and to reduce cost.

The finish turning tests were performed on Böhringer NG200, CNC-turning center of 36 kW main spindle, and 4000 rpm maximum speed. A high power, Nd:YAG laser 1006-D 4 kW Trumpf was generated. The temperature field near the cutting zone was measured using IR camera suitable for temperature measurements up to 900 °C. A three-component Kistler dynamometer, type 9121, was used to measure the force components. Tool wear was measured using stereoscopic microscope. Surface roughness was measured after each pass, using a portable Taylor Hobson Surtronic 3+. The workpiece (IN 718 of 28 HRC) was solution heat-treated and aged. The tool was a silicon nitride/aluminum oxide/aluminum nitride (SiAlON) round ceramic insert with $\gamma = -5°$, and $r = 6.35$ mm (Kennametal, KY 1540), which is insensitive to notch wear.

The analysis of chip morphology showed continuous ribbon-like chips (Figure 9.8). It was indicated that chips produced by LAM exhibited more tendency to shear localization and larger strain in the primary shear zone due to the thermal softening effect of LAM [23]. Increasing the cutting speed resulted, however, in thinner chips (lower strain). During LAM, as the cutting speed increase, the cutting temperature increases but the absorption of the laser heat was reduced such that the net effect was a reduction of the surface temperature Ts (Figure 9.9). At speeds above 300 m/min, the temperature level required to cause significant softening effect (650–700 °C) was not reached. The formation

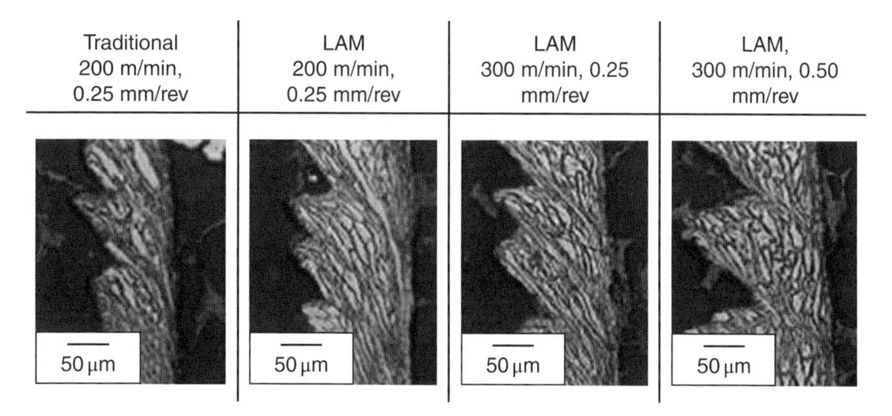

Figure 9.8 Effect of cutting on chip morphology of Inconel 718 under dry condition. (From Attia *et al.* [25].)

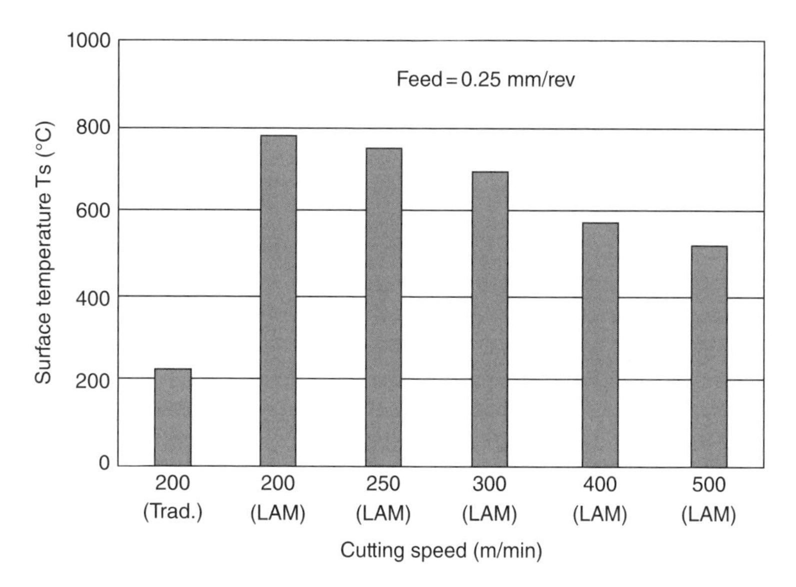

Figure 9.9 IR-surface temperature measurements at different cutting speeds for traditional and LAM of Inconel 718. (From Attia *et al.* [25].)

of shear localized chips (saw-tooth type) (Figure 9.8) was attributed to the favorable cutting conditions of large undeformed chip thickness, negative rake, and high cutting speeds that promote the material fracture [25].

The effect of feed on LAM was found to be significant; At v = 300 m/min, an increase in the feed from 0.25 to 0.5 mm/rev resulted in an increase of hardness in HRC from 39.1 to 48.9. The investigations have also revealed that a significant drop in the main force component in case of LAM is achieved as compared to traditional machining. The radial and feed components were not significantly affected during LAM by the cutting speed (v = 200–500 m/min); however, they showed a significant drop as compared to traditional machining where the cutting speed v = 200 m/min. When machining under LAM condition at optimum cutting speed (300 m/min), Attia *et al.* [25] depicted that the force components increased with increasing feed, as thicker chips were produced. A more interesting result was that the tool flank wear and the surface roughness during LAM decreased with increasing the feed from 0.25 to 0.4 mm/rev, which may be considered as a main advantage of LAM (Figure 9.10). This was attributed to thicker chips that creating more contact pressure distribution between the chip and the cutting edge. Above a feed rate of 0.4 mm/rev, however, an increase of tool wear and surface roughness were observed, which could be attributed to lower radiation heat absorption at higher feeds. In LAM, the roughness was reduced by 25% if the feed increased from 0.25 to 0.4 mm/rev (Figure 9.10). Attia *et al.* [25] have estimated that the material removal rate (MRR) increased by 800% in LAM, as compared to traditional machining.

In terms of surface integrity, provided optimum conditions, LAM did not introduce phase change due to overheating, or micro-defects that makes LAM is a promising technology for machining super alloys, stainless steels, and other difficult-to-cut materials. Anderson *et al.* [26] also investigated the LAM of Inconel 718 under varying conditions. The tool wear, forces, surface roughness, and specific cutting energy were examined. The benefit of LAM was demonstrated by a 25% reduction in specific cutting energy, a two- to threefold decrease in surface roughness, and 200–300% increase in ceramic tool life over traditional machining. An economical analysis depicted significant benefits of LAM Inconel 718 over traditional machining with carbide and ceramic insets [26].

Germain *et al.* [36] also investigated the LAM of Inconel 718 (NiCr19FeNb at 46 HRC), using carbide and ceramic inserts. They found that the machinability improved by locally

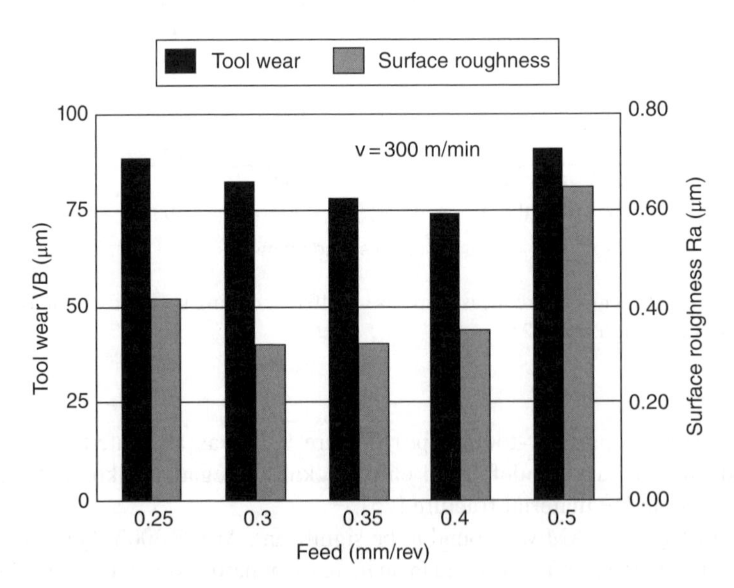

Figure 9.10 Effect of feed on average flank wear and surface roughness Ra for LAM optimum cutting speed (300 m/min) of Inconel 718. (From Attia *et al.* [25].)

heating the material prior to its removal. LAM reduces significantly the cutting forces. Tests have shown that, no matter which insert is used, LAM reduces the cutting force with up to 40%. The integrity of the machined surface in terms of surface roughness is not improved with the use of ceramic inserts in LAM compared to traditional machining. However, this gain is not visible with carbide inserts. Unlike carbide inserts, ceramic inserts allow a very good performance during LAM. The tool life of carbide inserts in LAM is considerably lower than that in traditional machining, whereas the life of ceramic inserts increases by about 25% as compared to traditional machining.

Most published works on machining Inconel 718 (and generally super alloys (SAs)) have been concerned mainly with single-point tools (turning), while milling has received little attention due to the process complexity [22]. Alauddin *et al.* [27] presented a study into the machinability of In 718 (hot forged and annealed-260 BHN) during end-milling when using uncoated carbide inserts under dry conditions. Their main objective was to optimize the machining conditions regarding tool life and surface quality. In this research, the tool insert was uncoated carbide, ISO-K20 or Sandvik Grade H13A (94% WC (tungsten carbide), 6% Co). The end-mill diameter was 25 mm, with two inserts, with inclination angle 5° and nose radius of 0.8 mm. A Cincinnati Universal Milling Machine of 8 kW rated power was used. Though the relative machinability of IN 718 is poor, it could be milled satisfactorily within a cutting speed range of 15–30 m/min, feed range of 0.04–0.1 mm/tooth and an axial depth of cut up to 2.0 mm [27]. In slot-milling, a tool life in the range of 5–9.5 min, and a surface roughness in the range of (0.35–1.2 μm) has been achieved at a cutting speed range of 19–29 mm/min, feed rate of 0.09 mm/tooth, and axial depth cut of 1.0 mm. As expected, the surface finish increased with the speed and decreased with feed rate [27].

HSM using ball nose end mills to machine aero-foils made of Inconel 718 was investigated by Ng *et al.* [28]. The best tool life was realized when cutting was performed using high pressure cutting fluid (70 bar, 26 l/min). The tool life was twice that realized under dry condition. All tools investigated were solid micro-grain WC, 8 mm diameter, 30° helix end mills. They are PVD-coated as shown in Table 9.6 [28].

During dry machining, TiAlN coating performed better than CrN coating, since the former had higher oxidation resistance, higher hardness, and lower coefficient of friction (Table 9.7). When using high pressure cutting fluid, the TiAlCrN coated tool performed the best (larger length of cut) (Figure 9.11). Moreover, its tool life improved 450% when compared to dry machining with TiAlN monolayer coated tool, at a cutting speed of 90 m/min. This demonstrates the advantage of using high pressure cutting fluids when using both coated and uncoated carbide tools [28].

Table 9.7 Properties of the PVD-coatings used in high speed end-milling of Inconel 718 according to Ng *et al.* [28]

PVD coatings	CrN (mono-layer)	TiAlN/CrN (multi-layer)	TiAlN (mono-layer)	TiAlCrN (mono-layer)
Coating thickness (μm)	3–4	2–5	1–3	3
Coefficient of friction	0.5	0.4	0.4	0.7
Micro-hardness (HV 0.05)	1750	3000	3500	3000
Max working temperature (°C)	700	800	800	700

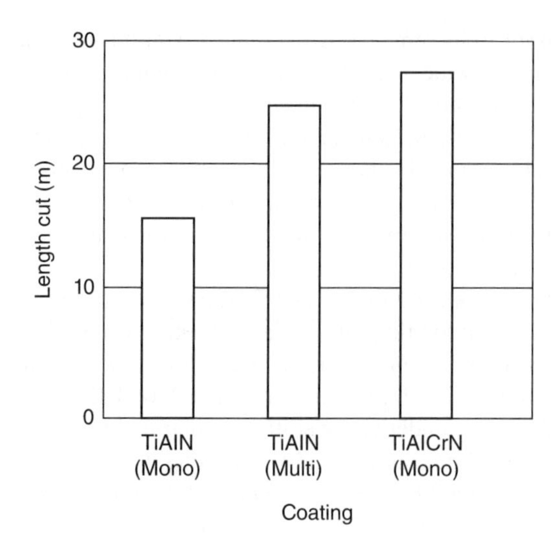

Figure 9.11 Effect of tool coating on tool life, expressed in length of cut (m), in high speed end milling of Inconel 718. (From Ng *et al.* [28].)

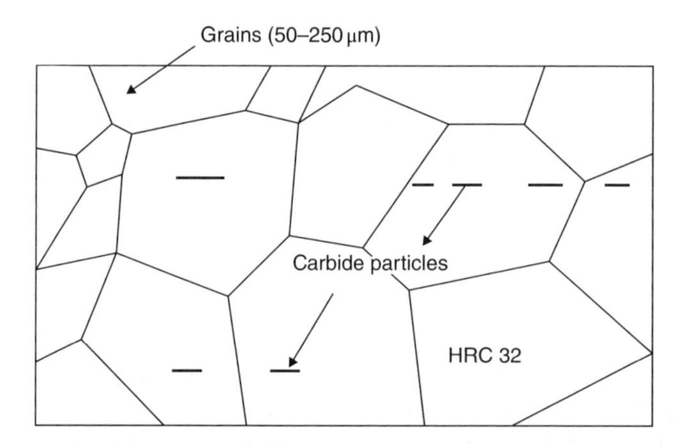

Figure 9.12 Illustrative sketch of the micro-structure of Ni-base super alloy Hynes 282 [29].

Hood *et al.* [29] detailed the effects of operating variables on tool life and surface integrity when end-milling of Ni-based Haynes 282 (Ni – 20 Cr – 10 Co – 8.5 Mo – 2.1 Ti – 1.5 Al – 1.5 Fe – 0.3 Mn – 0.15 Si – 0.06 C – 0.005 B), precipitation heat treated to provide a bulk hardness of about 32 HRC, and was supplied in plates of a thickness 6.5 mm. Grain size typically ranged from about 50–250 μm with carbide particles of up to 12 μm in length interspersed within the microstructure (Figure 9.12). The cutter was a 20 mm diameter provided with three round, TiAlN-coated carbide inserts. Cutting forces were measured using a Kistler 8654A three-component piezo-electric dynamometer. Tool wear was measured using a toolmakers digital microscope of resolution 1 μm, and the surface roughness Ra was measured using Talysurf Hobson 120 L. The machine tool used was a 15 kW rated power CNC – Matsuura vertical milling center having a maximum spindle speed of 20 000 rpm.

According to their plan [29], only four machining levels were chosen based on suggested operating parameters (two speeds and two feeds) specified by tooling manufacturers. The highest operating parameters were high cutting speed and a high feed rate of 0.1 mm/tooth, whereas the lowest parameters were low cutting speed and a small feed rate of 0.05 mm/tooth. Unfortunately, commercial restrictions preclude the reporting of the exact cutting speeds used, although these may be within the range of 15–75 m/min [29]. Accordingly, at the highest operating parameters, a flank wear of 213 μm after 45 min of machining was measured. Reduction of cutting speed or feed rate, typically, caused a pro rata increase in tool life. Milling with the lowest operating parameters, typically resulted in surface roughness levels of 0.15–0.3 μm Ra, which increased to twofold, as average flank wear progressed from 50 to 180 μm. It was also depicted that the three force components increased with increasing flank wear, and the machinability performance of Haynes 282 during milling was found to be similar to IN 718, but proved to be lower as compared to the powder processed Ni-based RR1000 super alloy.

Hood *et al.* [30] performed a research similar to their previous one [29], however, it was performed using drilling instead of milling of the same Ni-base alloy, Haynes 282,which is applicable as aero-engine casing material. Coated carbide twist drills were used, and drilling is performed under high pressure (50 bar) cutting fluid. Testing was performed using the same machine and experimental equipment as in reference [29]. It also involved variation in cutting speed and feed during drilling. Accordingly, it was depicted that at lower operating parameters as previously discussed in [29], the tool flank wear was generally uniform. However, extensive wear/fracture of the tool corner chamfer was evident on the majority of tests. Increasing flank wear from 30 to 100 μm caused curiously, a reduction in roughness Ra by 33%. Changes in operating parameters approved to have a limited effect on the surface roughness. Burrs were up to 250 μm in height, and generated on hole entry and exit on the majority of tests. Micro-hardness results showed an increase in hardness up to 50 HK0.05 above the bulk within the first 50 μm. Surface/subsurface micro-structural damage was detected up to a depth of 15 μm of the deformed grain boundaries with a discontinuous white layer of up to 6 μm from the surface [30].

Imran *et al.* [31] investigated deep-hole micro-drilling of a single-crystal Ni-based alloy CMSX4 (US-Patent 6996695). Deep-hole micro-drilling is very challenging for difficult to cut materials, especially for deep holes of aspect up to 10, because the strength of such materials makes tool breakage a major issue.

The nominal composition and the mean micro-hardness of the alloy CMSX4 are listed in Table 9.8. Its history is reported in reference [32]. This alloy is currently one of the strongest Ni-base single crystal casting super alloys.

Table 9.8 Composition (wt%) and micro-hardness of Ni-base single crystal casting super alloy CMSX4 (US-Patent: 696695)

Cr	Co	Mo	W	Ta	Al	Ti	Hf	Re	Ni	HV (mean)
7	9	0.6	6	7	5.6	1	0.1	3	Balance	375

Average grain size: 1.5 μm, density: 9.0 g/cm^3, Co should be limited to 0.1–9.5%.

Although EDM (electric discharge machining), LBM (laser beam machining), and EBM (electron beam machining) can sometimes replace micro-drilling, there is concern about surface integrity and sputter when using these nontraditional thermal processes. Moreover, mechanical micro-drilling has the potential to achieve competitive roundness, smoother surface, and better lead times. There is not general consensus on what can be classified as micromachining. However, Masuzawa [33] broadly defined it as machining of features in the range of 1–999 μm, which is still quite acceptable.

Owing to the poor machinability of the Ni-based alloy CMSX4, coolant was used to reduce cutting temperature; moreover a customized pecking cycle was employed. The drilling was done on a Micron HSM-400 Machining Centre. A warm-up program was used at the start to stabilize the temperature of the machine and tooling.

Figure 9.13 shows the cutting strategy. A pilot hole of the same diameter was first made up to 0.13 mm deep. A center drill with point angle of 120° and helix of 30° was selected. The main drill with point angle at 150°, and helix 30° was used. Its chisel edge does not touch the bottom of the pilot hole, instead the main cutting lips are engaged first. That helps, gradually load the drill, stabilizing the axis of rotation on entry without promoting drill wander and breakage.

The drill was made of ultrafine WC of grain size 0.6 μm, with 8% Co, hardness of 93 HRA, coated with TiAlN, which is generally recommended for machining of difficult-to-cut materials. The drill diameter was 0.5 mm and of a shank of 3 mm. the hole depth was 5 mm (L/D = 10) [31].

The effect of process parameters were evaluated, and the tool wear mechanism was also investigated. The number of drilled holes before drill breakage was taken as a measure of cutting performance [32]. It is found that the micro-drilling was sensitive to feed, spindle speed, and peck depth. Figures 9.14 and 9.15 show the effect of speed (expressed in spindle rev/min) and drill feed. It appears that the optimum conditions realized at rotational speed n = 3000 rpm (corresponding to cutting speed $\nu \approx 5$ m/min), and a feed rate of 5 μm/rev). The dimensional

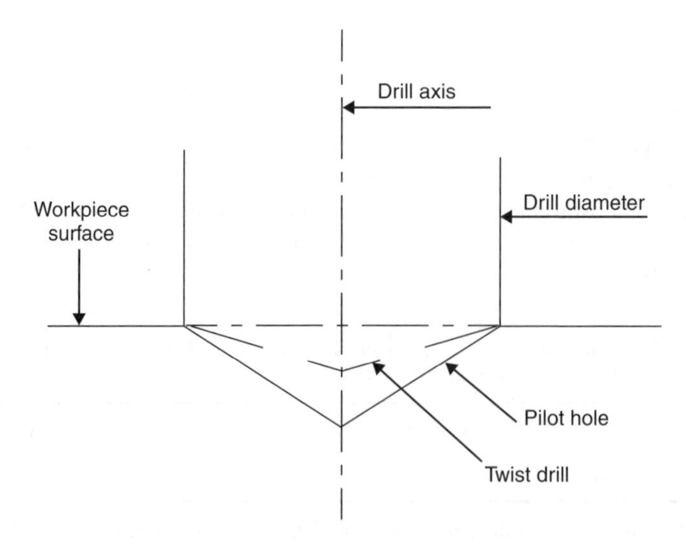

Figure 9.13 Tool strategies for micro-drilling difficult-to-cut materials.

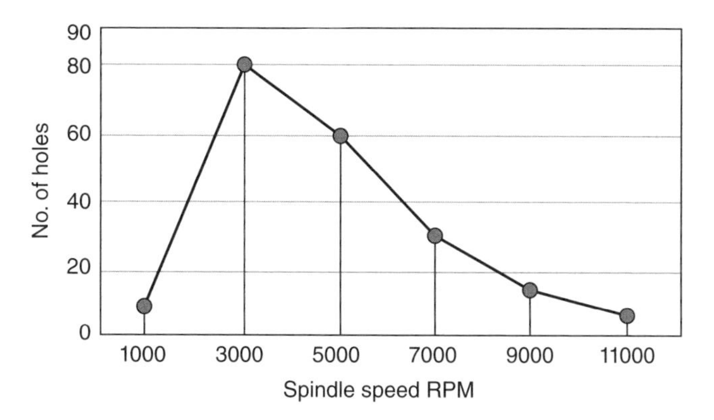

Figure 9.14 Effect of cutting speed on drill life (no. of holes), when micro-drilling of Ni-base CMSX4 super alloy. (From Imran *et al.* [31].)

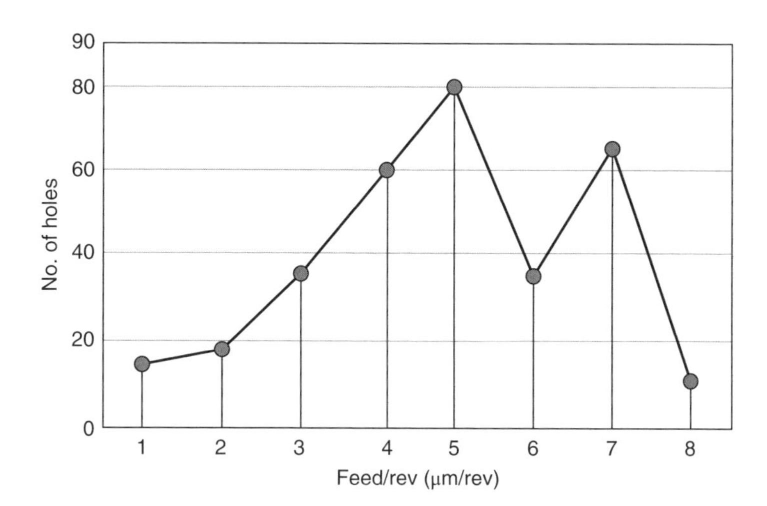

Figure 9.15 Effect of feed rate on drill life (no. of holes), when micro-drilling of Ni-base CMSX4 super alloy. (From Imran *et al.* [31].)

values of holes drilled at 3000 rpm and a feed of 5 μm/rev showed a gradual decrease with the number of holes due to the gradual side wear of the micro-drill [31].

Imran *et al.* [31] also tested the feasibility of micro-drilling of Ni-base alloy CMSX4, which has opened opportunities for future developments regarding issues like tool wear, surface integrity, and optimizing the process performance super alloys.

UAM is a technique involving oscillating the cutting tool ultrasonically to improve the machining performance. US-oscillations, produced by a HF-transducer (frequency $f = 20$ kHz and amplitude $\xi = 10$ μm), are directly superimposed in the direction of feed and speed movements. In case of turning, the application of US-oscillations along the feed direction enables the machining parameters to be adjusted independent of the workpiece diameter. A substantial decrease in cutting forces and improvement in surface finish up to 50% when turning Ni-base super alloy Inconel 718 was observed by Babitskya *et al.* [34], when applying UAM in the feed direction as compared to traditional turning. This was attributed to that traditional turning was changed over to a high frequency impact process of higher dynamic stiffness leading to improved machinability.

A significant improvement of machinability has also been reported when applying UAM in the tangential direction. An improvement in the surface finish and a reduction of tool flank to one-fifth of its value by traditional turning, have been realized. These improvements are attributed that BUE rarely occurs with UAM in the tangential direction [34].

Ezugwu *et al.* [35] compared the HPC delivery with conventional coolant supplies on the tool life while machining Inconel-718 with a whisker-reinforced ceramic tool. It has been found that the tool life improved as much as 740%, while machining at 200 bar coolant pressure at a speed of 50 m/min. Chip segmentation depends on the cutting conditions employed and further, to a greater extent, on the coolant pressure. Lower cutting forces were generated while machining owing to improved cooling and lubrication at the cutting interface. Once a critical value of pressure has been reached, any further increase in coolant pressure only resulted in marginal increase in tool life. When turning Inconel-718 and other difficult-to-cut alloys using HPC, the surface finish and accuracy were acceptable [35].

9.4 Recent Research Work Related to Nontraditional Machining of Stainless Steels and Super Alloys

In the past two decades, the EDM of difficult-to-cut materials, especially super alloys has been considerably developed. Ni-base super alloys such as Inconel 718, Hastelloy, IN-100, and so on, have been successfully EDMed [37].

Jeelani and colons [38] have reported that the fatigue life of EDMed Inconel 718 decreased slightly as compared with that of the virgin material. The micro-hardness and roughness of this Ni-base alloy increased marginally producing a hard recast layer (RL). The response parameters such as MRR, electrode wear rate (EWR), surface roughness (SR), and RL of super alloys machined by EDM have been examined.

Hastelloy-X was investigated under various EDM working conditions and analyzed in terms of surface integrity (SI) by Kang und Kim [39]. The main parameters in their investigations were the pulse on-time. They have reported that MRR and EWR behaved nonlinearly with respect to the pulse duration, whereas the morphological and metallurgical features showed rather a constant trend with pulse duration. The pulse on-time, however, affected the number of micro-cracks and the thickness of HAZ.

When investigating Inconel 718, Wang *et al.* [40] explored the feasibility of removing the RL using etching or grinding. Their results proved that positive polarity could create a thicker RL than negative polarity. It has been found that corrosion affected by phosphoric or hydrochloric acids could significantly enhance the RL removal rate of Inconel 718. The micro-hardness test revealed that such corrosion could only damage the structure of the RL without effecting the base material.

Rajesh *et al.* [41] investigated SI after EDM of Inconel 718. The morphology of the machined surface (RL) was characterized by enormous amount of heat discharge during sparking that causes melting and vaporization of the material, followed by swift cooling. Surface quality was deteriorated at high pulse current and pulse on-time. In the same research,

Rajesh *et al.* investigated also the effect of the machining parameters on MRR when EDM the same Ni-base alloy using a tubular pure Cu-electrode, and commercial grade kerosene as a dielectric fluid. The investigations were carried out on a CNC-EDM machine and the process parameters were pulse current, pulse on-time, gap control, and flushing pressure. It was concluded that the pulse current is the most significant parameter, followed by pulse on-time.

Bharti *et al.* [42] investigated the machinability of Inconel 718 during ED-sinking using Cu-electrode. Discharge current and pulse on-time were identified as common influencing parameters for MRR, SR, and EWR. Duty cycle and tool electrode withdrawal time were found to be the least effective parameters.

Kristen [43] in his study on Ni-base alloy IN-100 used two different graphite electrodes Poco EDM3, and Poco AF5, and tested the effect of pulse duration. The physical properties of these electrodes used by Kristen are listed in Table 9.9.

Graphite electrodes are widely used in EDM due to their high electrical and thermal conductivities, along with their good machinability. They are graded based on their grain size, densities, and mechanical and electrical properties. The results demonstrated that the finer grade Poco AF5 of smaller grains and less porosity performed very well giving significantly higher MRR with acceptable relative electrode wear, while the other coarser-grained grade Poco EDM3 gave significantly lower electrode wear and MRR. Two levels of discharge current (20 and 15 A) and pulse duration (20 and 30 μs) were tested. The OCV was 240 V, with positive polarity. As a conclusion if the electrode wear is an important issue, optimum conditions can be realized by some increase of pulse duration from 20 to 30 μs and with sacrificing only 6% of MRR.

Klocke *et al.* [44] investigated the influence of powder suspended dielectrics on the RL of Inconel 718 EDMed surface. They reported that the physical property of the powder additives plays an important role in changing the RL composition and morphology. Kumar *et al.* [45] realized the potential of graphite powder as additive in enhancing machining capabilities of powder suspended EDM on Inconel 718, and found that the addition of graphite powder enhanced the machining rate was improved 27% with 12 g/l of fine graphite at the best parametric settings. Prabhu and Vinayagam [46] investigated the EDM process on Inconel 825 using carbon nano tube (CNT) mixed with dielectric and correspondingly analyzed the surface characteristics. Atomic force microscope analyses revealed that CNT improved the surface characteristics (Surface morphology), SR, and microcracks from micro level to nano level.

Kuppan *et al.* [47] investigated small deep hole drilling of Inconel 718 using EDM process, and revealed that the MRR is more influenced by peak current, duty cycle, and electrode rotational speed, whereas the SR (Ra-value) is strongly influenced by the peak current and

Table 9.9 Physical properties of graphite electrodes Poco EDM3, and Poco AF5, according to Kristen [42]

Physical properties	Electrode	
	Poco EDM3	Poco AF5
Particle size (μm)	<5	<1
Hardness (Shore)	76	87
Apparent density (g/cm^3)	1.8	1.8
Compression strength (MPa)	148	186

pulse on-time. An increase of electrode rotational speed leads to an increase of MRR, whereas Ra attains a minimum value at a rotational speed of 200–300 rpm then increases with increasing rotational speed. Better SI was achieved by moderate current values.

Bozdana *et al.* [48] presented a comparative study on machining surface characteristics of through and blind holes produced on aerospace alloys Ti-6Al-4 V and Inconel 718 by fast rotating EDM using tubular hollow Cu-and brass-electrodes. It was revealed that the achievement of desirable MR (machinability rate) and EWR values and acceptable topography of EDMed surfaces depends upon the appropriate selection of electrode material and the choice of making through or blind hole. The brass electrode has provided the best results regarding MRR for through and blind holes in both materials as compared to Cu-electrode.

Arun Muthu *et al.* [49] investigated the EDM of Inconel 800 using Al-electrode in conventional as well as magnetic force assisted EDM. The response characteristics including SR, EWR, and MRR were evaluated and compared. It was concluded that the response parameters are significantly improved using the magnetic force assisted EDM rather than conventional EDM.

It is difficult to produce large deep holes by gas assisted LBM, because it is difficult to eject the molten and recast metal from the hole bottom by the gas, consequently, the machinability and accuracy are reduced. Ultrasonic assisted laser beam machining (USALBM) may be the solution of that problem, and may be preferred than gas-assisted LBM. Lau *et al.* [50] investigated USALBM of difficult-to-cut materials, and it was depicted that an increase of the drilling speed and the accuracy of the produced deep holes with less taper and increased aspect ratios were obtained.

Research involving ultrasonic machining assisted electric discharge machining (USM/EDM) and latterly that relating to lower frequency vibration assisted EDM, has shown there to be significant benefits in drilling/die-sinking operations in terms of increased MRR/reduced machining time, increased penetration depth, a reduction in the incidence of arcing together with improvement in sparking efficiency, thinner RL, and HAZ and in some cases reduced tool electrode wear, principally as a result of improved flushing and clearance of debris from the spark gap. In the early 1980s, Kremer *et al.* [51] investigated some of the limiting issues of EDM such as poor dielectric circulation and the evacuation of debris and gasses especially with intricate electrodes. It was found that using an electrode vibrating at ultrasonic frequency (~20 kHz) allowed deeper penetrations and higher feed rates in die sinking operations. Feed Penetration rates when using a graphite electrode increased 30% when roughing and 300% when finishing. Subsequent publications [52, 53] all involving vibration at ultrasonic frequencies (20–23 kHz) with amplitudes ranging from 3 to 30 μm, both with and without abrasive particles dispersed in the dielectric fluid, detailed positive benefits for the USM/EDM approach. The research of Lin *et al.* [52] is particularly relevant. The MRR when USM/EDM Ti–6Al–4 V using 3 μm SiC in distilled water dielectric (90 g/l) being twice that of conventional EDM. Here the concentration of abrasive was reported as critical; if too much, it produces unstable discharges.

Results for lower frequency longitudinal tool vibration reported by Prihandana *et al.* [54] and Uhlmann and Domingos [55] are similarly positive, with data for the former when operating at 600 Hz and 0.75 μm amplitude with a 12.5 mm Cu tool electrode, suggesting a 23% increases in MRR and lower workpiece roughness and tool wear rate when machining SS304 stainless steel. In the latter paper, twin piezo-actuators are detailed for the machining of high aspect ratio seal slots in MAR–M247 using graphite electrodes vibrating at up to 1000 Hz with amplitudes from 2 to 16 μm achieving increased MRR and reduced wear.

Salt electrolytes such as $NaClO_3$, NaCl, and $NaNO_3$ commonly used in electrochemical machining (ECM) should not be used in STEM (shaped tube electrolytic machining) and ES (electro-stream) because they produce a large amount of sludge which clogged the flow of electrolyte and limits the minimum diameter to be drilled. Consequently, acids are used for drilling deep holes in STEM and ES.

Dilute hydrochloric acid has been used for drilling IN-100 and Ti-alloys, whereas dilute sulfuric acid is preferred for Co-base super alloys, and stainless steels 304, 316, and 321. However, a problem with acid electrolytes is the work corrosion, and poor surface finish, especially with HCl. To overcome this problem Sharma *et al.* [56] used an acidified NaCl electrolyte (10% NaCl + 1% HCl) to minimize sludge formation in the inter-electrode gap, pumped at 3 atm. The flow rate through the system depends primarily upon the magnitude of the inter-electrode gap.

Relevant investigations show that good holes can be obtained by a combination of low OCV (15–30) and comparatively high feed rates (0.65–1 mm/min), resulting of low values of equilibrium gap. The cathode tool is a hollow Cu-tube (1.75 mm/0.68 mm). To avoid any stray machining, the tool should be coated with an insulating layer of Perspex. The cathode is tipped with a tool bit, serving as the de facto cathode, of land 1 mm, and a diameter 3 mm. The other end of the cathode tube is soldered to the head of a syringe, which is connected to the electrolyte supply tube. A good, uniform hole with an aspect ratio of 11 was obtained in Inconel 718 at OCV of 17 V, and tool feed rate of 1 mm [56]. Under these machining conditions, however, sparking was observed at feed rates exceeding1.25 mm/min. The workpiece Composed of four Inconel strips, each 30 × 7 × 7 mm. To achieve good contact, each strip was finished by grinding both sides, then soldered at the ends. The composite workpiece facilitated measurement of hole diameters, oversize, and out-of-roundness as a function of the hole depth.

References

[1] Bass, M. Beck, D., Copley, S.M. (1978) Laser-assisted machining SPIE. *4th European Electro-Optics Conference*, Vol. **164**, pp. 233–240.

[2] Chandrasekaran, H., Johnson, J.O. (1994) Chip flow and notch wear mechanism during the machining of high austenitic stainless steels. *Ann. CIRP* **43**(1), 101–104.

[3] Anderson, M.C., Shin, Y.C. (2006) Laser-assisted machining of an austenitic stainless steel: P550. *Proc. Inst. Mech. Eng.* **220** (Part B), 2055–2067.

[4] Saller, G., Aigner, H. (2004) High nitrogen alloyed steels for nonmagnetic drill collars. Standard Steel Grades and latest developments. *Mater. Manuf. Processes* **19**(1), 41–49.

[5] Leppert, T. (2011) Surface layer properties of AISI 316L steel when turning under dry and with MQL conditions. *Proc. Inst. Mech. Eng.* **226** (Part B), 617–631.

[6] Ciftci, I. (2006) Machining of austenitic stainless steels using CVD multi-layer coated cemented carbide tools. *Tribol. Int.* **39** (6), 565–569.

[7] Bruni, C., Forcellese, A., Gabrielli, F., Simoncini, M. (2006) Effect of the lubrication-cooling technique, insert technology and machine bed material on the workpart surface finish and tool wear in finish turning of AISI 420B. *Int. J. Mach. Tools Manuf.* **46** (12–13), 1547–1554.

[8] Tekiner, Z., Yesilyurt, S. (2004) Investigation of the cutting parameters depending on process sound during turning of AISI 304 austenitic stainless steel, *Mater. Des.* **25,** 507–513

[9] Mesquita, R.M.D., Marques, B. (1992) Effect of chip-breaker geometries on cutting forces. *J. Mater. Process. Technol.* **31**, 317–325.

[10] Flower, H.M. (1995) *High Performance in Aerospace Materials* Chapman & Hall, London.

[11] Campbell, G.S., Lahey, R.T.C. (1984) Survey of serious aircraft accidents involving fatigue fracture. *Int. J. Fatigue*, **6** (1), 25–30.

[12] Olajire, K.A. (1999) Machining of aerospace steel alloy with coated carbides. PhD thesis. Coventry University, August 1999.
[13] Jawaid, A., Olajire, K.A., Ezugwu, E.O. (2001) Machining of martensitic stainless steels (JETHETE) with coated carbides. *Proc. Inst. Mech. Eng.* **215** (Part B), 769–779.
[14] Lin, W.S. (2008) The study of high speed fine turning of austenitic stainless steels. *J. Achiev. Mater. Manuf. Eng.* **27**(2), 191–194.
[15] Fernandez-Abia, A.I., Barreiro, J., Lopez de Lacalle, L.N., Martinez, S. (2011) Effect of very high cutting speeds on shearing, cutting forces, and roughness in dry turning of austenitic stainless steels. *Int. J. Adv. Manuf. Technol.* **57**, 61–71.
[16] Karenk, S.R., Gaitonde, V., Davim, J.P. (2007) Integrating Taguchi principle with genetic algorithm to minimize burr size in drilling of AISL 316L stainless steel using an artificial neural network mode l. *Proc. Inst. Mech. Eng.* **221** (Part B), 1695–1704.
[17] Khan, A.A., Ahmed, M.I. (2007) Improving tool life using cryogenic cooling. *J. Mater. Process. Technol.* **196**, 149–154.
[18] Venugopal, K.A., Paul, S., Chattopadhyay, A.B. (2007) Growth of tool wear in turning of Ti-6Al-4V alloy under cryogenic cooling. *Wear* **262**, 1071–1078.
[19] RR1000-SecoTools (2013) https://www.secotools.com (accessed June 15, 2015).
[20] Axinte, D.A., Andrews, P. (2007) Some considerations on tool wear and workpiece surface quality of holes finished by reaming or milling in Ni-base super alloys, *Proc. Inst. Mech. Eng.* **221** (Part B), 591–603.
[21] Soo, S.L., Hood, R., Aspinwall, D.K., *et al.* (2013) Machinability and surface integrity of RR1000 Ni-base super alloy. *CIRP-Ann.* **60**(1), 89–92.
[22] Choudhury, I.A., El-Baradie, M.A. (1998) Machining nickel-base super alloys: IN718, *Proc. Inst. Mech. Eng.* **212** (Part B), 195–206.
[23] Dudzinski, D., Devillez, A., Moufki, A., Larrouquère, D., Zerrouki, V., Vigneau, J., A review of developments towards dry and HSM of Inconel 718, *Int. J. Mach. Tools Manuf.* **44**, (2004), 439–456.
[24] Bushlyaa, V., Zhoua, J., Stahla, J.E. (2012) Effect of cutting conditions on machinability of Inconel 718 during high speed turning with coated and uncoated PCBN tools, *CIRP-Ann.* **3,** 370–375.
[25] Attia, H., Tavakoli, S., Vargas, R., Thomson, V. (2010) Laser-assisted high-speed-finish turning of super alloy Inconel 718 under dry conditions, *CIRP-Ann. Manuf. Technol.* **59**, 83–88.
[26] Anderson, M., Patwa, R., Shin, Y.C. (2006) Laser-assisted machining of Inconel 718 with an economic analysis, Center for Laser-based Manufacturing, School of Mechanical Engineering, Purdue University, USA, November 2005, *Int. J. Mach. Tools Manuf.* **46**, 1879–1891.
[27] Alauddin, M., El-Baradie, M.A., Hashmi, M.S.J. (1996) End-milling machinability of Inconel 718, *Proc. Inst. Mech. Eng.* **210** (Part B), 11–23.
[28] Ng, E.G., Lee, D.W., Charman, A.R.C. *et al.* (2000) High speed nose end milling of Inconel 718, *CIRP-Ann.* **49**, 41–50.
[29] Hood, R., Soo, S.L., Aspinwall, D.K. *et al.* (2012) Radius end milling of Haynes 282 Ni-base alloy. *Proc. Inst. Mech. Eng.* **49**, 226, 1745–1753.
[30] Hood, R., Soo, S.L., Aspinwall, D.K. *et al.* (2011) Twist drilling of Haynes 282 super alloy, *Proc. Inst. Mech. Eng.* **19**, 150–155.
[31] Imran, M., Mativenga, R.T., Kannan, S., Movovic, D. (2008) An experimental investigation of deep-hole micro-drilling capability for a Ni-base Super Alloy CMSX4. *Proc. Inst. Mech. Eng.* **222** (Part B), 1589–1596.
[32] Bhadeshia, H.K.D.H. (2003) 1st, 2nd, and 3rd Generation Single Crystal Ni-based Super Alloys, http:www.msm.cam.ac.uk/phase-trans/2003/super alloys/SX/SX.html (accessed 17 April 2015).
[33] Masuzawa, T. (2000 State of the art of micro-machining., *CIRP-Ann. Manuf. Technol.* **49**(2), 473–488.
[34] Babitskya, V.I., Kalashnikov, A.N., Meadows, A., Wijesundara, A.A. (2003) Ultrasonically assisted turning of aviation materials. *J. Mater. Process. Technol.* **132**, 157–167.
[35] Ezugwu, E. O., Bonney, J., Fadare, D.A., Sales, W.F. (2005 Machining of nickel base Inconel-718 alloy with ceramic tools under finishing conditions with various coolant supply pressure. *J. Mater. Process. Technol.* **162–163**, 609–614.
[36] Germain, G., Lebrun, J.L., Braham-Bouchnak, T, *et al.* (2008) Laser-assisted machining of Inconel 718 with carbide and ceramic inserts *International Journal of Material Forming* **1**: 523–526.
[37] AK. Singh, S. Kumar, V.P. Singh Electrical discharge machining of superalloys: a review, *Int. J. Res. Mech. Eng. Technol.*
[38] Jeelani S., Collins M.R. (1988) Effect of electric discharge machining on the fatigue life of Inconel 718, *Int. J. Fatigue* **10**(2), 121–126.

[39] Kang, S.H., Kim, D.E. (2003) Investigation of EDM characteristics of nickel-based heat resistant alloy, *The Korean Society of Mechanical Engineers KSME International Journal* **17**(10): 1475–84.
[40] Wang C.C., Chow H.M., Yang L.D., Lu C.T. (2009) Recast layer removal after electrical discharge machining via Taguchi analysis: a feasibility study, *J. Mater. Process. Technol.* **209**, 4134–4140.
[41] Rajesh, S., Patnaik, P.K., Sharma, A.K., Kumar, P. (2010) Surface integrity evaluation of electro discharge machined Inconel 718. *Proceedings of the 3rd International 24th AIMTDR Conference*, pp. 259–264.
[42] Bharti P.S., Maheshwari S., Sharma C. (2010) Experimental investigation of Inconel 718 during die-sinking electric discharge machining, *Int. J. Eng. Sci. Technol.* **2**(11), 6464–6473.
[43] Kristen L.A. (2004) Performance of two graphite electrode qualitied in EDM of seal slots in a jet engine turbine vane, *J. Mater. Process. Technol.* **149**, 152–156.
[44] Klocke F., Lung D., Antonoglou G., Thomaidis D. (2004) The effect of powder suspended dielectrics on the thermal influenced zone by electro discharge machining with small discharge energies, *J. Mater. Process. Technol.* **149**, 191–197.
[45] Kumar A., Maheshwari S., Sharma C., Beri N. (2010) Realizing potential of graphite powder in enhancing machining in AEDM of nickel based super alloy 718, *Proc. Int. Conf. Adv. Mech. Eng.* 50–53.
[46] Prabhu S., Vinayagam B.K. AFM surface investigation of Inconel 825 with multi-wall carbon nano tube in electric discharge machining process using Taguchi analysis, *Arch. Civil Mech. Eng.*, **XI** (1), 149–170.
[47] Kuppan P., Rajadurai A., Narayanan, S. (2011) Influence of EDM process parameters in deep hole drilling of Inconel 718, *Int. J. Adv. Manuf. Technol.* **38**, 74–84.
[48] Bozdana, A.T., Yilmaz, O., Okka, M.A., Filiz, I.H. (2009) A comparative experimental study on fast hole EDM of Inconel 718 and Ti-6Al-4V. *5th International Conference and Exhibition on Design and Production of Machines and Dies/Moulds*, pp. 18–21.
[49] Arun Muthu, B., Karthik, K.M., Soundararajan, R., Palanisamy, A. (2009) Characteristics of magnetic force-assisted electric discharge machining on Inconel 800. *Sixth International Conference on "Precision, Meso, Micro and Nano Engineering", COPEN* **6**, pp. G7–G11.
[50] Lau, W.S., Yue, T.M., Wang, M. (1994) US-aided laser drilling of Al-based metal matrix composites. *CIRP Ann.* **43/**1, 177–180.
[51] Kremer, D., Basine, G., Moisan, A. *et al.* (1983) Ultrasonic machining improves EDM technology. *Proceedings of the 7th International Symposium on Electromachining (ISEM VII),* Birmingham, UK, *2–14 April*, pp. 67–76.
[52] Lin YC, Yan BH, Chang YS (2000) Machining characteristics of titanium alloy (Ti–6Al–4V) using a combination process of EDM with USM. *J. Mater. Process. Technol.* **104**:171–177.
[53] Thoe TB, Aspinwall DK, Killey N (1999) Combined ultrasonic and electrical discharge machining of ceramic coated nickel alloy. *J. Mater. Process. Technol.* **92–93**:323–328.
[54] Prihandana GS, Mahardika M, Hamdi M, Mitsui K (2011) Effect of low-frequency vibration on workpiece in EDM processes. *J. Mech. Sci. Technol.* **25**(5):1231–1234.
[55] Uhlmann E, Domingos DC (2013) Investigations on vibration-assisted EDM – Machining of seal slots in high-temperature resistant materials for turbine components. *Procedia CIRP* **6**:71–76.
[56] Sharma, S., Jain, V.K., Shekhar, R. (2002) Electrochemical drilling of Inconel super alloy with acidified sodium chloride electrolyte. *Int. J. Adv. Manuf. Technol.* **19**:492–500.

Appendix

Table A.1 Symbols and characteristics of selected elements according to their atomic number and atomic mass unit in ascending order

Element	Symbol	Atomic number	Atomic mass unit (amu)	Melting point (°C)	Density (Mg/m^3) (=g/cm^3)	Crystal,* at 20 °C	Valence (most common)
Hydrogen	H	1	1.0078	−259.14	—	—	1+
Helium	He	2	4.003	−272.2	—	—	Inert
Lithium	Li	3	6.94	180	0.534	bcc	1+
Beryllium	Be	4	9.01	1289	1.85	hcp	2+
Boron	B	5	10.81	2103	2.34	—	3+
Carbon	C	6	12.011	>3500	2.25	hex	—
Nitrogen	N	7	1.007	−210	—	—	3−
Oxygen	O	8	15.999	−218.4	—	—	2−
Fluorine	F	9	19.00	−220	—	—	1−
Neon	Ne	10	20.18	−248.7	—	—	Inert
Sodium	Na	11	22.99	97.8	0.97	bcc	1+
Magnesium	Mg	12	24.31	649	1.74	hcp	2+
Aluminum	Al	13	26.98	660.4	2.70	fcc	3+
Silicon	Si	14	28.09	1414	2.33	Dc	4+
Phosphorus	P	15	30.97	44	1.8	—	5+
Sulfur	S	16	32.06	112.8	2.07	—	2−
Chlorine	Cl	17	35.45	−101	—	—	1−
Argon	Ar	18	39.95	−189.2	—	fcc	Inert
Potassium	K	19	39.10	63	0.86	bcc	1+

(*continued overleaf*)

Machining of Stainless Steels and Super Alloys: Traditional and Nontraditional Techniques,
First Edition. Helmi A. Youssef.

Table A.1 (*Continued*)

Element	Symbol	Atomic number	Atomic mass unit (amu)	Melting point (°C)	Density (Mg/m³) (=g/cm³)	Crystal,* at 20 °C	Valence (most common)
Calcium	Ca	20	40.08	840	1.54	fcc	2+
Titanium	Ti	22	74.90	1668	4.51	hcp	4+
Chromium	Cr	24	52.00	1875	7.20	bcc	3+
Manganese	Mn	25	54.94	1246	7.20	—	2+
Iron	Fe	26	55.85	1538	7.88	bcc	2+
						fcc	3+
Cobalt	Co	27	58.93	1494	8.90	hcp	2+
Nickel	Ni	28	58.71	1455	8.90	fcc	2+
Copper	Cu	29	63.54	1084.5	8.92	fcc	1+
Zinc	Zn	30	65.37	419.6	7.14	hcp	2+
Germanium	Ge	32	72.59	937	5.35	Dc	4+
Arsenic	As	33	74.92	~809	5.73	—	3+
Krypton	Kr	36	83.80	−157	—	fcc	Inert
Silver	Ag	47	107.87	961.9	10.5	fcc	1+
Tin	Sn	50	118.69	232	7.3	bct	4+
Antimony	Sb	51	121.75	630.7	6.7	—	5+
Iodine	I	53	126.9	114	4.93	ortho	1+
Xenon	Xe	54	131.3	−112	2.7	fcc	Inert
Cesium	Cs	55	132.9	28.4	1.9	bcc	1+
Tungsten	W	74	183.9	3410	19.4	bcc	4+
Gold	Au	79	197.0	1064.4	19.32	fcc	1+
Mercury	Hg	80	200.6	−38.86	—	—	2+
Lead	Pb	82	207.2	327.5	11.34	fcc	2+
Uranium	U	92	238.0	1133	19	—	4+

*bcc = body-centered cubic; bct body-centered tetragonal; Dc = diamond cubic; fcc = face-centered cubic; hcp = hexagonal close-packed; hex = hexagonal.

Table A.2 Properties of selected metals and alloys (20 °C)

Some selected metals and alloys	Density (Mg/m³) (=g/cm³)	Thermal conductivity, (W/mm³)/(K/mm)	Linear expansion (°C⁻¹ × 10⁻⁶)	Electrical resistivity, ρ (Ω·m × 10⁻⁹)	Average modulus of elasticity, Eav. (MPa)
Aluminum (99.9+)	2.7	0.22	22.5	29	70.000
Aluminum alloys	~2.7	0.16	22	~45	70.000
Brass (70Cu-30 Zn)	8.5	0.12	20	62	110.000
Bronze (95 Cu-5 Sn)	8.8	0.08	18	~100	110.000
Cast iron (gray)	7.15	—	10	—	140.000
Cast iron (white)	7.7	—	9	660	205.000
Copper (99.9+)	8.9	0.40	17	17	110.000
Iron (99.9+)	7.88	0.072	11.7	98	205.000
Lead (99+)	11.34	0.033	29	206	14.000

(*continued overleaf*)

Table A.2 (*Continued*)

Some selected metals and alloys	Density (Mg/m³) (=g/cm³)	Thermal conductivity, (W/mm³)/(K/mm)	Linear expansion (°C⁻¹ × 10⁻⁶)	Electrical resistivity, ρ (Ω·m × 10⁻⁹)	Average modulus of elasticity, Eav. (MPa)
Magnesium (99+)	1.74	0.16	25	45	45.000
Monel (70 Ni-30 Cu)	8.8	0.025	15	482	180.000
Silver (sterling)	10.4	0.41	18	18	75.000
Steel (1020)	7.86	0.050	11.7	169	205.000
Steel (1040)	7.85	0.048	11.3	171	205.000
Steel (1080)	7.84	0.046	10.8	180	205.000
Stainless steel 18/8	7.93	0.015	16	700	205.000
Super alloy	—	—	—	—	—

Table A.3 Si units for important quantities

Quantity	SI-unit	Quantity	SI-unit
Acceleration	m/s^2	Mass	kg
Angle	rad	Power	W
Area	m^2	Pressure, stress	Pa
Density	kg/m^3	Thermal conductivity	W/m·K
Energy	J	Torque	N·m
Force	N	Velocity	m/s
Length	m	Volume	m^3

Table A.4 Standard multiples and submultiples

Prefix	Symbol	Multiplication factor	Prefix	Symbol	Multiplication factor
tera	T	10^{12}	milli	m	10^{-3}
giga	G	10^{9}	micro	μ	10^{-6}
mega	M	10^{6}	nano	n	10^{-9}
kilo	k	10^{3}	pico	p	10^{-12}
deci	d	10^{-1}	femto	f	10^{-15}
centi	c	10^{-2}	atto	a	10^{-18}

Review Questions

Solved Questions

Q.1 What makes stainless steel stainless?
Ans. Stainless steel must contain at least 10.5% Cr. Cr is the element that reacts with O_2 in the air to form a complex chrome-oxide surface layer that is invisible but strong enough to prevent further O_2 from staining (rusting) the surface. Higher levels of Cr and the addition of other allowing elements such as Ni and Mo enhance the surface layer and improve the corrosion resistance of stainless steel.

Q.2 What is the difference between 18/8, 18/10, and 18/0 SS?
Ans. The first number is the amount of chromium that is contained in the stainless steel, that is, 18 is 18% chromium. The second number is the amount of nickel, that is, 8 stands for 8% nickel. So 18/8 means that this stainless steel contains 18 chromium and 8% nickel. SS 18/10 is 18% chromium and 10% nickel. The higher the numbers, the more corrosion resistant the material. Both 18/8 and 18/10 contain nickel and are part of the grade family "300 series" stainless. SS 18/0 means that there is 18% chromium but zero nickel. When there is no nickel the stainless grade family is the "400 series." 400 series are not as corrosion resistant as the 300 series and are magnetic where the 300 series are nonmagnetic.

Q.3 Can stainless steel rust? Why?
Ans. Stainless steel does not rust as you think of regular steel rusting with a red oxide on the surface that flakes off. If you see red rust it is probably due to some iron particles that

Machining of Stainless Steels and Super Alloys: Traditional and Nontraditional Techniques,
First Edition. Helmi A. Youssef.

have contaminated the surface of the stainless steel and it is these iron particles that are rusting. Look at the source of the rusting and see if you can remove it from the surface. If the iron is embedded in the surface, you can try a solution of 10% nitric and 2% hydrofluoric acid at room temperature or slightly heated. Wash area well with lots and lots of water after use. Commercially available "pickling paste" can also be used.

Q.4 What is the difference between 304 and 316 stainless steel?
Ans. 304 SS contains 18% Cr and 8% Ni. 316 contains 16% Cr, 10% Ni, and 2% Mo. Molybdenum is added to help resist corrosion to chlorides (like sea water and de-icing salts).

Q.5 Is stainless steel magnetic?
Ans. There are several types of stainless steels. The 300 series (containing Ni) is NOT magnetic. The 400 series (which just contains Cr, and no or small amounts of Ni) are magnetic.

Q.6 What is passivation?
Ans. When the amount of chromium (in an iron matrix) exceeds 10.5%, a complex chrome oxide forms instantaneously that prevents the further diffusion of oxygen into the surface and results in the "passive" nature of stainless steel and its resistance to oxidation (or corrosion), A chemical "dip" into 10% nitric acid plus 2% hydrofluoric acid bath will enhance the development of this "passive" oxide.

Q.7 Can stainless steel be welded?
Ans. Yes, Stainless steel is easily welded, especially if carbon content is limited by 0.03% C. However, the welding procedure is different from that used with carbon steel. The filler rod (electrode) must also be stainless steel.

Q.8 What does the "L" designation mean?
Ans. The use of the letter L after the grade number, that is, 304 L, means that the carbon content is restricted to a MAXIMUM of 0.03% (normal levels are 0.08% max. and in some grades can be as high as 0.15% max.). This lower level of carbon is usually used where "welding" will be performed. The lower level of carbon helps to prevent the chromium from being depleted (by forming chrome carbides at the weld site), and therefore allow it to remain over 10.5% so it can form the "passive" oxide layer that gives stainless its corrosion resistance.

Q.9 Can stainless steel be hardened ?
Ans. Yes, The 300 series stainless steel can be "hardened" but only by "work hardening." That is by cold working the material, either by cold rolling down to lighter and lighter gauges, or by "drawing" through a die or other size altering operation. "Annealing" stainless steel will remove the work hardening effect. The 400 series have two different stainless steel structures. One is called "ferritic" (409, 430, 434, 439), which cannot be hardened by heat treatment. The other is called "martensitic" (403, 410, 416, 420, and 440 A,B,C), which can be hardened by heat treatment.

Q.10 What is the recycle content of SS?

Ans. Stainless steel can be recycled 100%. That is, all stainless steel can be remelted to make new stainless steel. The typical amount of recycled stainless steel "scrap" that is used to make new stainless steel is between 65% and 80%. Many recycling companies will want the various SS-grades to be kept separate (all 300 series together, and so on).

Q.11 What is the "annealed" condition of SS?

Ans. Stainless steel is usually sold in the "annealed" condition. It just means that the material is in the "soft" or annealed condition. The 300 series of stainless steels can not be hardened by heat treatment (like carbon steels) but can be hardened by cold working. This cold work can be eliminated by a heating treatment (annealing) that will restore the original soft condition.

Q.12 What does the term "CRES" mean?

Ans. CRES is something used to designate stainless steel. It stands for Corrosion RESistant steel. It does not necessarily mean that the steel in fact is stainless steel, as there are other materials that are corrosion resistant but not stainless steel.

Q.13 Can SS be used at very low and very high temperatures?

Ans. Yes, stainless steel has excellent properties at both extremes of the temperature scale. Stainless steel can be used down to liquid nitrogen temperatures and for some types up to about 1000 °C.

Q.14 What are AISI specifications of stainless steel?

Ans. The AISI (American Iron and Steel Institute) was the originator of the 300 and 400 series numbering system (i.e., type 304 stainless steel). They also published a stainless steel products manual that listed these designations and the chemical analysis as well as most mechanical and physical properties of each individual grade.

Q.15 Can stainless steel be machined?

Ans. Yes. However the standard grades of stainless steel are usually "gummy" and will not produce a clean chip when machined or turned. To solve this problem, many companies produce "free-machining" grades of stainless where they add a "chip-breaker" to the matrix. Grade 303 is the free-machining equivalent to grade 304.

Q.16 Define and describe the basic (standard) alloys of SSs?

Ans. These are classified as ferritic, martensitic, and austenitic stainless steels. Ferritic contains low percentage of carbon (0.08%C) and as such these can't be hardened by heat-treatment and can be hardened only by working. To make them corrosion resistant, 14% Cr and 0.45% Mn are added. These are used for spoons, forks, and so on. Martensitic stainless steels contain more carbon content (>0.4% C, 13% Cr and 0.5% Mn). Due to high carbon content, these can be hardened by heat treatment, and are used for cutlery and similarly edged tools. Austenitic stainless steels contain 0.1% C, 18% Cr, and 8% Ni. They cannot be hardened by heat treatment and respond only to work hardening. They are nonmagnetic.

Q.17 Give reasons why free-machining alloys are not currently available in the duplex-or PH-SSs?

Ans. That is because duplex alloys are noted for excellent corrosion resistance, but have somewhat less hot-workability. The addition of free-machining elements, which would likely degrade both properties, would be undesirable. Similarly, PH-alloys are noted for good toughness at high strength levels, making it undesirable to add free-machining element, which would degrade toughness and strength.

Q.18 What is flash attack in passivation of stainless steel, and why does it occur?

Ans. Flash attack is heavily etched or darkened surface. It occurs due to the contamination of the passivating bath resulting from nonthoroughly cleaning of parts before passivating.

Q.19 What properties of SS are degraded by addition of free-machining elements?

Ans. The benefit of improved machinability due to the addition of these elements is not of course obtained without degrading other important properties of stainless steels such as:

- Corrosion resistance
- Strength, ductility, and toughness
- Hot workability
- Cold formability
- Weldability
- The improvement in machinability must be balanced against the possible reduction of the degraded properties, especially corrosion resistance.

Q.20 Differentiate between S, and Se as free-machining additives in SS.

Ans. Se is another commonly used free-machining agent after S in stainless steels, forming inclusions of MnSe analogous to sulfides. Sulfur is used in Europe, whereas Se is frequently used in USA. Se is less effective than an equivalent weight percent of S in improving the overall machining characteristics of stainless steel. However, Se-bearing alloys provide a better machined surface finish than S-bearing alloys. In addition, Se-bearing stainless steels may offer improved cold formability and somewhat improved corrosion resistance compared to the corresponding S-bearing alloys. Se additions have also been used in S-bearing alloys to promote sulfides that are larger and more globular, that are more beneficial to machinability.

Q.21 Define super alloys.

Ans. Super alloys, are relatively new class of materials that exhibit high mechanical strength, ductility, creep strength at high operating temperatures, high fatigue strength, and typically superior resistance to corrosion and oxidation even at elevated temperatures. These features make super alloys ideal for applications in aircraft, submarines, nuclear reactors, dies for hot working of metals, and petrochemical equipment. It will be noted that not all applications require elevated temperature strength capacity. Their high strength coupled with corrosion resistance has made certain super alloy standard materials applicable for biomedical joint implants, and cryogenic applications.

Q.22 Classify super alloys in their basic groups.
Ans. These alloys are usually classified into three main groups which are Fe-based, Ni-based, and Co-based alloys. The physical, mechanical, and machining behavior of each group varies considerably due to the chemical compositions of the alloy and the metallurgical processing it receives during manufacturing. Super alloys are generally identified by trade names or by special numbering systems.

Q.23 What is the duty of the following alloying elements in super alloys: Ni, Co, Mo, Cr, Al, Si, and C?
Ans. Super alloys generally contain Ni, Cr, Co, Mo, and Fe as major alloying elements. Others are Al, W, Ti, and so on. The role of these alloying elements is to enhance the characteristics of super alloys in the following manner:

- Ni stabilizes alloy structure and properties at high temperatures.
- Co, Mo, and W increase strength at elevated temperature.
- Cr, Al, Si enhance resistance to oxidation and provide high temperature corrosion.
- C increases creep strength.

Q.24 What is the difference between cutting stainless steel with a laser or a waterjet?
Ans. The most basic differences can be seen in (i) heat affected zones, which laser cutting has and waterjet does not have. (ii) laser cutting typically cuts up to 15 mm thick stainless steel. Anything thicker than that and you may need to use waterjet cutting. (iii) Laser cutting typically has tighter cutting tolerances, so if your part includes intricate designs and is less than 15 mm thick, then laser cutting is the ideal choice.

Q.25 Does abrasive waterjet cutting cost the same as laser cutting?
Ans. No, waterjet cutting is actually the more expensive process. This is due to the slower processing rates, coupled with the large amount of consumable material involved. Although multiple cutting heads, tight nesting capabilities and the ability to layer some materials do help to offset these additional costs, generally speaking, abrasive waterjet cutting is a more costly process than laser cutting.

Q.26 Can you cut titanium?
Ans. Yes, abrasive waterjet is an ideal process for cutting titanium.

Q.27 How thick of material does abrasive waterjet cut?
Ans. Abrasive waterjet can cut materials from 0.015 mm to 300 mm

Q.28 Does abrasive waterjet cut hardened metals?
Ans. Yes, It can be used to process hardened metals ranging from 20 to 70 Rockwell C.

Q.29 What are the limitations of waterjet cutting?
Ans. Four materials that we do not use waterjet cutting for are; tungsten, tungsten carbide, lead, and beryllium copper. Tungsten is an extremely hard material, and beryllium copper releases poisonous beryllium into the air. Because of this, these are materials that we do not process with waterjets.

Q.30 What kind of abrasive is used in abrasive waterjet cutting?

Ans. Garnet is the most commonly used abrasive for waterjet cutting. Since abrasive is one of the most expensive things in abrasive waterjet cutting, garnet is a nice choice as it is one of the least expensive options.

Q.31 What types of materials can you laser cut?

Ans. We can use laser cutting to process many different types of materials such as stainless steel, Hastelloy, aluminum, and carbon steel.

Q.32 Can you describe abrasive waterjet cutting?

Ans. As we know, waterjet cutting has been around for years. The process is a means of cutting countless types of material, much faster and with much higher quality than traditional methods of tooling. Waterjet has the capability to easily process materials like stainless steel, titanium, alloy steels, and many more. The best part about waterjet cutting is the fact that it has no heat-affected zones. This means that the metals do not get hardened or damaged in any way during the process of cutting, leaving no further work to be done with the parts. With that being said, it is important to point out that there are two types of waterjet cutting; abrasive waterjet cutting, and water-only cutting (also known as pure waterjet). In water-only cutting, the cutting stream erodes the material. However, in abrasive waterjet cutting, the abrasive particles shear the material after being accelerated through the waterjet stream. This process is much more powerful than water-only cutting. The abrasive is often made up of tiny garnet particles. The particles start as a pile of abrasive, and are delivered through tubing to arrive in the mixing chamber. From there, the abrasive is mixed in with the water and the two are rapidly projected down through the nozzle. The nozzle creates a perfect cutting stream to produce exact cuts with exceptionally smooth edges.

Q.33 What are the main functions of cutting fluids?

Ans. Cutting fluids are commonly applied to machining operations, chiefly to:

- cool the cutting zone to increase tool life, and to improve the accuracy and surface finish;
- lubricate the area of contact on tool flank and face to reduce friction, and accordingly tool wear, cutting forces, and consumed power;
- flush away chips and swarf;
- protect machine from corrosion;
- control BUE formation on the tool.

The action of cutting fluid, being a lubricant or coolant, depends mainly on temperatures encountered, and consequently depends on the type of cutting operation and cutting speed.

Q.34 What are the characteristics of a proper cutting fluid?

Ans. A proper cutting fluid should be characterized by:

- High heat absorption capacity
- Good lubricating quality
- High flash point.
- High chemical stability

- No fumes emitted while in contact which hot surfaces
- Of less biological and environmental hazards
- Of no corrosive effects on the workpiece and on the machine guide ways
- Available and nonexpensive.

Unsolved Questions

1. What are some of the various alloy additions that are used to improve the machinability of stainless steels?

2. Differentiate between S and Se as free-machining additives for stainless steels, regarding the following aspects:Machinability, surface finish, cold formability, weldability, and corrosion resistance.

3. What do you understand by a free-machining stainless steel? What elements are usually added to make SS, and explain how they make the steel free-cutting?

4. Why ferritic SSs and austenitic SSs do not respond to quench-hardening?

5. Define and describe the derived alloys of stainless steels.

6. Which SS would you recommend for:
   - Very high strength
   - Hardness and strength
   - Good weldability
   - Corrosive environment
   - Good machinability

7. What are the special properties of SSs that make them difficult to machine?

8. What are general rules to be considered when machining SSs conventionally?

9. When do you recommend nontraditional machining conditions for machining SSs instead of traditional machining?

10. Why should ferritic-SSs be given the first consideration when selecting a stainless steel?

11. Why might a martensitic-SS rust when exposed to a hostile environment?

12. What is duplex stainless steel?

13. What is sensitization of a SS, and how can it be prevented?

14. What are tool materials, that commonly used when machining stainless steels?

15. What are the two most commonly elements that are used to enhance machinability of stainless steels?

16. Differentiate between ferritic-, martensitic-, and austenitic- stainless steels.

17. What are the Mn-austenitic SSs and what special virtue do they possess?

18. Describe the three classes of PH-SSs.

19. Holes of 6 mm diameter and 25 mm depth are to be drilled into AISI-304. Estimate the recommended cutting speed (m/min), and the recommended feed (mm/rev.). Calculate, the spindle speed (rpm), feed (mm/min), drilling time (min), and consumed power if the specific cutting energy of the SS = 3700 N/mm^2. Check whether the recommended values are feasible on commercially available equipment.

20. Describe how the machinability be rapidly assessed using a simple drill-penetration test.

21. Describe how you can assess the machinability of a material through measurement of the cutting power.

22. What are disadvantages of free-machining grades of SSs?

23. Mark [Y] for and [X] for false statements
[ ] Austenitic SSs are nonmagnetic and hardenable by heat treatment.
[ ] PH-SSs are hardenable by heat treatment.
[ ] Martensitic alloy is one class of PH-SSs.
[ ] Ferritic alloy is one type of PH-SS.
[ ] Ferritic grades of SSs are only hardenable by cold working.
[ ] HSS and carbide tools are most widely used in machining SSs.
[ ] HSS tools are seldom used to machine SAs. They may be only used for interrupted cuts.
[ ] HSS broaches are used for broaching SAs and SSs.
[ ] HSS drills and milling cutters can not be used for machining SAs.
[ ] CBN is better than diamond for machining steels.
[ ] Grinding can be followed by milling operation.

24. Explain the corrosion types that can occur in stainless steels.

25. Justify the need of nontraditional machining in today's industries.

26. What are the basic limitations of traditional machining processes?

27. What are the basic factors upon which the nontraditional machining processes are classified? Explain.

28. List five traditional machining processes commonly used in industry.

29. Nontraditional machining processes yield low rates of material removal compared to traditional machining processes even though they have gained wide popularity. Discuss why.

30. ECM, EDM, USM, etc. are commonly referred to as nontraditional machining processes. What is nontraditional in these processes? Explain

31. Describe the basic mechanism of material removal in EBM and LBM.

32. Identify major components of EBM and LBM equipment.

33. State the working principle of EBM and LBM equipment.

34. Draw schematically the EBM and LBM equipment.

35. Identify the process parameters of EBM and LBM.

36. List three applications of EBM and LBM.

37. List three limitations of EBM and LBM

38. Choose the correct answer:
   (a) Mechanism of material removal in EBM is due to:
      (i) Mechanical erosion due to impact of high of energy electrons.
      (ii) Chemical etching by the high energy electron.
      (iii) Sputtering due to high energy electrons.
      (iv) Melting and vaporization due to thermal effect of impingement of high energy electron.
   (b) Mechanism of material removal in LBM is due to:
      (i) Mechanical erosion due to impact of high of energy photons.
      (ii) Electro-chemical etching.
      (iii) Melting and vaporization due to thermal effect of impingement of high energy laser beam.
      (iv) Fatigue failure.
   (c) Generally Electron Beam Gun is operated at:
      (i) Atmospheric pressure.
      (ii) At 1.2 bar pressure above atmosphere.
      (iii) At 10–100 mTorr pressure.
      (iv) At 0.01–0.001 mTorr pressure.

39. What are the free-machining additives of stainless steels?

40. What are the main factors that generally affect the machining characteristics of super alloys?

41. What are the practical guidelines to be observed when machining super alloys?

42. What are considerations to be observed when broaching super alloys?

43. Differentiate between super alloys and refractory metals. Why is the latter not widely used in aircraft applications?

44. Numerate five different types of each basic group of super alloys.

45. List the important alloying elements used in HSS. Explain why they are used.

46. What precautions would be considered when machining with ceramics?

47. Classify carbide tools according to ISO-Standard.

48. Discuss the suitability of ceramics to be used as a cutting tool for machining the following materials:
- Stainless steels
- Super alloys
- Aluminum
- Titanium.

49. Why are ceramics not suitable for interrupted cuts?

50. Define the machinability of a material. What are the assessments used for its evaluation?

51. What are the important techniques that enhance the machinability of DTC materials?

52. Explain what is meant by HSM. What is the main purpose of it?

53. Discuss the main types of chips associated with HSM.

54. What are the main recommendations of cutting tools used in HSM? Select suitable tool materials for HSM of super alloys and titanium alloys.

55. What are the advantages and limitations of HSM?

56. The tool life can be almost infinite at low cutting speeds. Would you then recommend all machining be done at low speeds? Explain.

57. For HSM, discuss the effect of cutting speed on:
- Cutting force
- Cutting temperature
- Surface finish.

58. Provide your justification for why the tool wear during UAM is lower as compared to traditional machining.

59. Explain the importance of advanced cooling techniques in machining DTC materials such as stainless steels and super alloys rather than low strength steels.

60. Define cryogenic cooling.

61. Define MQL technique. Give your expectation regarding the flow rate in this case.

62. Define cryogenic treatment of cutting tool materials. What is the advantage of cryogenic treated HSS, and carbide tools as compared to conventional ones.

63. What are the main characteristics of an ideal cutting tool material?

64. What is cermet?

65. What is CBN?

66. Discuss the role of Mo as an important replace to W in HSS?

67. Discuss the role of WC, TiC, TaC, and Co in cutting carbides and their impact on the properties of carbide tips.

68. What are advantages and limitations of PM-HSS?

69. Differentiate between Stellites and UCON as cutting tool materials.

Index

Machining of Stainless Steels and Super Alloys: Traditional and Nontraditional Techniques, First Edition. Helmi A. Youssef.